W9-BQK-977

Extended Interactions between Metal Ions

in Transition Metal Complexes

Leonard V. Interrante, *Editor*

A symposium sponsored by
the Division of Inorganic
Chemistry at the 167th
Meeting of the American
Chemical Society, Los Angeles,
Calif., April 3–5, 1974.

ACS SYMPOSIUM SERIES **5**

AMERICAN CHEMICAL SOCIETY

WASHINGTON, D. C. 1974

Library of Congress CIP Data

Extended interactions between metal ions in transition metal complexes.
(ACS symposium series; 5)

Includes bibliographical references and index.

1. Transition metal compounds—Congresses. 2. Complex compounds—Congresses. 3. Solid state chemistry—Congresses.

I. Interrante, Leonard V., 1939- ed. II. American Chemical Society. Division of Inorganic Chemistry. III. Series: American Chemical Society. ACS symposium series; 5.

QD172.T6E49 546'.6 74-30455
ISBN 0-8412-0224-9 ACSMC8 5 1-408 (1974)

PRINTED IN THE UNITED STATES OF AMERICA

ACS Symposium Series

Robert F. Gould, *Series Editor*

FOREWORD

The ACS SYMPOSIUM SERIES was founded in 1974 to provide a medium for publishing symposia quickly in book form. The format of the SERIES parallels that of its predecessor, ADVANCES IN CHEMISTRY SERIES, except that in order to save time the papers are not typeset but are reproduced as they are submitted by the authors in camera-ready form. As a further means of saving time, the papers are not edited or reviewed except by the symposium chairman, who becomes editor of the book. Papers published in the ACS SYMPOSIUM SERIES are original contributions not published elsewhere in whole or major part and include reports of research as well as reviews since symposia may embrace both types of presentation.

CONTENTS

Preface ix

1. **Extended Interactions in Transition Metal Oxides and Chalcogenides** 1
   Aaron Wold

2. **High Temperature Crystal Chemistry of V_2O_3 through the Metal-Insulator Transition and of $(Cr_{0.01}V_{0.99})_2O_3$, an Insulator** 16
   William R. Robinson

3. **Synthesis and Properties of Some New Organic Intercalation Complexes of Tantalum Sulfide** 23
   R. L. Hartless and A. M. Trozzolo

4. **Intermolecular Magnetic Exchange Interactions in Molecular Crystals** 34
   J. M. Grow and H. H. Wickman

5. **Computation of Field-Swept EPR Spectra for Systems with Large Interelectronic Interactions** 40
   R. L. Belford, P. H. Davis, G. G. Belford, and T. M. Lenhardt

6. **Multisite Magnetic Interactions in Hexaurea-Metal(III) Halide Lattices** 51
   Phillip H. Davis and R. L. Belford

7. **Polymeric, Mixed-Valence Transition Metal Compounds** 66
   Gilbert M. Brown, Robert W. Callahan, Eugene C. Johnson, Thomas J. Meyer, and Tom Ray Weaver

8. **Exchange Interactions in Nickel(II), Copper(II), and Cobalt(II) Dimers Bridged by Small Anions** 76
   David N. Hendrickson and D. Michael Duggan

9. **Structural and Magnetic Properties of Chromium(III) Dimers** ... 94
   Derek J. Hodgson

10. **Superexchange Interactions in Copper(II) Complexes** 108
    William E. Hatfield

11. **Electronic and Magnetic Properties of Linear Chain Complexes Derived from Biscyclopentadienyl Titanium(III) and of the Infinite RMX_3 Linear Chain Complexes** 142
    D. Sekutowski, R. Jungst, and G. D. Stucky

12. **Optical Properties of Linear Chains** 164
    S. L. Holt

13. **Spectroscopic and Magnetic Properties of $CsMI_3$ Type Transition Metal Iodides** 182
    G. L. McPherson and L. J. Sindel

14. **Magnetic and Thermal Properties of the Linear Chain Series $[(CH_3)_3NH]MX_3 \cdot 2H_2O$** 194
J. N. McElearney, G. E. Shankle, D. B. Losee, S. Merchant, and R. L. Carlin

15. **The Ferromagnetic Ordering of Ferrous Chloride Polymers of Diimine Ligands** 205
W. M. Reiff, B. Dockum, C. Torardi, S. Foner, R. B. Frankel, and M. A. Weber

16. **The Unusual Magnetic Properties of an Imidazolate Bridged Polymer of an Iron(III) Hemin** 221
Irwin A. Cohen and David Ostfeld

17. **Local and Collective States in Single and Mixed Valency Chain Compounds** 234
P. Day

18. **Evidence for Extended Interactions between Metal Atoms from Electronic Spectra of Crystals with Square Complexes** 254
Don S. Martin, Jr.

19. **The Nature of the Lowest-Energy Allowed Electronic Transition in Crystals of Certain d^8 Transition Metal Complexes that Possess Extended Metal Chains** 276
Basil G. Anex

20. **The Single Crystal Polarized Reflectance Spectra and Electronic Structures of $Rh(CO)_2$acac and $Ir(CO)_2$acac** 301
Thomas A. Dessent, Richard A. Palmer, and Sally M. Horner

21. **Directed Synthesis of Linear Chain Metal Complexes with Well-Defined Cooperative Properties** 314
R. Aderjan, D. Baumann, H. Breer, H. Endres, W. Gitzel, H. J. Keller, R. Lorentz, W. Moroni, M. Megnamisi-Bélombé, D. Nöthe, and H. H. Rupp

22. **Magnus Green Salt Solid Solutions Containing Mixed-Valence Platinum Chains: An Approach to 1-D Metals** 331
B. A. Scott, R. Mehran, B. D. Silverman, and M. A. Ratner

23. **Studies on Some 1-D Metal Group VIII Complexes** 350
K. Krogmann and H. P. Geserich

24. **Some Comments on the Electronic Structure of Krogmann Salts and the Stability of Pt 2.3+** 356
Aaron N. Bloch and R. Bruce Weisman

25. **The Peierls Transition in One-Dimensional Solids** 372
H. R. Zeller and P. Brüesch

26. **The Preparation of and the Anisotropic Dielectric Properties of $K_2Pt(CN)_4Br_{0.30} \cdot 3H_2O$** 376
R. B. Saillant and R. C. Jaklevic

27. **A SCF-Xα-SW Investigation of Solid State Interactions in $Pt(CN)_4^{n-}$ Complexes** 382
L. V. Interrante and R. P. Messmer

28. **One-Dimensional and Pseudo-One-Dimensional Molecular Crystals** **392**
Ulrich T. Mueller-Westerhoff and Friedrich Heinrich

Index **403**

PREFACE

The papers that comprise this volume span two major areas of research —solid state chemistry and physics, and the study of transition metal complexes. The former has been concerned largely with relatively simple ionically or covalently bound inorganic solids such as the metal oxides, halides, and sulfides, where extended interactions are the rule rather than the exception. On the other hand, until quite recently, the study of transition metal complexes has been almost entirely a study of molecular systems, where the interactions in the solid state are generally of the weak van der Waals type and where the solid state properties reflect closely those of the constituent molecular units in solution.

In the last few years it has become clear that there are a number of transition metal complexes which do not fit this description very well. In particular, these materials usually exhibit some of the characteristics typical of molecular systems but display certain optical, magnetic, or electrical properties which evidence "extended interactions" of appreciable magnitude in the solid state. The study of such complexes constitutes the main subject of this volume.

Many of these systems show quite anisotropic and even pseudo-"one-dimensional" solid state behavior, reflecting strong intermolecular interactions of a highly directional character in the crystal. Such one-dimensional systems, based on both metal complexes and organic charge-transfer compounds, are currently of considerable interest within the solid state physics community and are under active study in laboratories here and abroad (*1, 2*). The intense interest in these materials can be attributed, in part, to suggestions regarding the possibility of superconductivity in such systems (*3, 4*). The actual likelihood of superconductivity here is a matter of some controversy; however, the wealth of new concepts and the physical understanding that has already resulted from the study of these systems, not to mention the high conductivities and highly anisotropic optical, magnetic, and electronic properties that many of them exhibit, promise continued growth of interest.

One important conclusion from the work on these and other types of transition metal complex solids is that interactions between the metal ions in the crystal, propagated either directly or through a bridging group, are often of major importance in determining the solid state properties. The nature of these metal ion interactions and the manner in which they relate to the observed properties is the principal focus of the papers in this volume.

The first three chapters, which are concerned with transition metal oxides and sulfides, introduce the topic of extended interactions in transition metal compounds from a solid state viewpoint and provide an account of current work on the simple inorganic, non-molecular systems.

The discussion of transition metal complexes begins with Chapter 4, which deals with exchange interactions in, and pathways for exchange in some weakly associated molecular systems. The next two papers describe the application and interpretation of electron spin resonance measurements for studying exchange interactions in such systems.

Chapters 7 through 11 deal with systems in which the exchange interactions are localized largely within dimeric or higher polymeric units and provide some fundamental information regarding the influence of the metal ion, the bridging group, and the coordination geometry on the type and magnitude of the exchange process. Chapters 10 and 11 also introduce the topic of extended exchange interactions in one-dimensional, ligand-bridged systems—a topic which is developed further in Chapters 12–17.

Beginning in Chapter 17, which provides a general survey of current work on one-dimensional metal complex systems, the discussion shifts to direct metal–metal interactions in solids containing stacked planar metal complexes. The remaining chapters continue this discussion, concluding with some recent work on the mixed valence platinum salts, which have lately been of much interest for their one-dimensional metallic characteristics.

The authors include both academic and industrial research scientists from the United States and Europe and represent virtually every phase of the current work on transition metal complex solids.

The symposium on which this volume is based was supported, in part, by grants and travel assistance provided by the United States Air Force, through its European Office for Aerospace Research and Development, General Electric Corporate Research and Development, and the Inorganic Division of the American Chemical Society. This support made it possible for several European authors to travel to the symposium and contributed greatly to the international flavor and overall success of the program.

Literature Cited

1. Zeller, H. R., *Advan. Solid State Phys.* (1973) **13**, 31.
2. Garito, A. F., Heeger, A. J., *Acc. Chem. Res.* (1974) **7**, 232.

3. Little, W. A., Ed., *Proc. Intern. Symp. Phys. Chem. Problems Possible Org. Superconductors, J. Polymer Sci. C* (1969), No. 29.
4. Coleman, L. B., Cohen, M. J., Sandman, D. J., Yamagishi, F. G., Garito, A. F., Heeger, A. J., *Solid State Comm.* (1973) **12**, 1125.

L. V. INTERRANTE

Schenectady, N. Y.
November 6, 1974

1

Extended Interactions in Transition Metal Oxides and Chalcogenides

AARON WOLD

Division of Engineering, Brown University, Providence, R.I. 02912

As early as 1937 de Boer and Verwey (1) had indicated that in transition metal oxides, in which the d-bands of the transition metal ions were partially filled, the potential energy barrier between atoms was high enough to reduce the conductivity by an enormous amount. This was probably the first indication that the Block-Wilson band theory of solids (1,2) could not describe in a realistic way the transport properties of transition metal compounds. Indeed, the classical band theory had predicted that these compounds with partially filled d-bands would show high electrical conductivity ($\sigma = n e \mu$); the number of carriers n would show temperature independence and the mobility μ would decrease as the temperature T increased. The electrical conductivity observed for these compounds would be derived from these latter temperature dependencies.

In 1957 (3) Morin reported on the electrical properties of several vanadium and titanium oxides, namely VO, V_2O_3, VO_2 and Ti_2O_3. These oxides contain 3,2,1 and 1-d electrons per transition metal cation respectively. These compounds show metallic behavior at high temperatures and then become semiconducting when cooled through a critical temperature T_c, (114°K, 153°K, 340°K and 450°K). For most of these compounds, the electrical resistivity was observed to drop by several orders of magnitude over a small temperature range. The oxides of vanadium show a reduction in symmetry associated with the resistivity changes. The lattice parameters of Ti_2O_3 vary quite rapidly in the vicinity of the transition temperature (T_t) but there is no change in symmetry (4). Figure 1 summarizes these findings. It should also be indicated that for pure stoichiometric single crystals, the temperature range over which the transition occurs is greatly reduced and the magnitude of the discontinuity in electrical conductivity is increased by a factor of 10^3.

The 3d transition metal oxides may therefore be either metallic at all temperatures, semiconducting at all temperatures or undergo a semiconductor $\rightleftarrows$ metal phase transition on heating through a critical temperature T_t. The list of compounds belonging to these three classes may be enlarged to include 4d and 5d-transition metal oxides as well as 3d-transition metal sulfides. This is shown in Table 1 (5).

It is not the purpose of this presentation to discuss the various theoretical models used to correlate the electronic and structural properties of these compounds. One or two of the models, useful to the chemist, will be referred to later in the discussion of specific compounds. Neither is it possible to treat, in the available space, more than several compositions that have been studied at Brown University. However, it would be appropriate to mention before proceeding, an important property of many transition metal compounds i.e., that for both the semi-conducting and metallic phases, the electrons responsible for the observed electronic behavior of these compounds move in narrow d-bands. As a result, the usual formulas attributed to band theory break down.

From an experimental, or specifically synthetic, point of view the crux of the problem in studying these compounds is to prepare pure stoichiometric single crystals. This has involved the development of new techniques both for their preparation and characterization. An understanding of the electronic properties of these materials has had to depend upon the availability of such crystals.

A number of typical transition metal compounds selected from each of the classes previously described will now be discussed in some detail. They will include examples of binary transition metal oxides and oxyfluorides, perovskite "bronzes" and transition metal chalcogenides. No attempt will be made to discuss compounds other than those which have been prepared at Brown University. A relatively simple one-electron model proposed by Goodenough (6,7) will be used to correlate the various structural and crystallographic properties observed for these materials.

I. Tungsten (VI) Oxide and the Cubic Tungsten Bronzes. The ReO_3 type structure consists of ReO_6 regular octahedra joined together by sharing corners to form a three-dimensional lattice shown in Figure II. WO_3 does not crystallize according to this scheme. Braekken (8) has indicated that this compound has a structure of low symmetry, consisting of deformed WO_6 octahedra joined in the same way as the regular octahedra of the ReO_3 structure. The cell is monoclinic (pseudo-orthorhombic) with a_o = 7.285Å, b = 7.517 Å, c = 3.835Å, $\alpha = \gamma = 90^\circ$, $\beta = 90.90^\circ$.

The cubic tungsten bronzes crystallize in the perovskite structure, ABO_3. This structure, like ReO_3, is formed by the corner sharing of BO_6 octahedra. However, in the case of the perovskite structure all of the large A sites are occupied. For the cubic bronzes, M_xWO_3, the alkali metal cations are statistically distributed over the A-sites.

The electrical properties of ReO_3, the structurally related

Table I. Classification of Transition Metal Oxides according to Their Electrical Behavior

-CLASS I -

Metallic Compounds

3d-compounds:	TiO, CrO_2, TiS, CoS_2, CuS_2.
4d-compounds:	NbO, RuO_2.
5d-compounds:	ReO_3.

-CLASS II -

Semiconducting Compounds

3d-compounds:	NiO, CoO, MnO, FeO, Fe_2O_3, Cr_2O_3, MnS, MnS_2, FeS_2.

-CLASS III -

Transitional Compounds

3d-compounds:	Vo, V_2O_3, VO_2, V_3O_5, V_4O_7, V_6O_{13}, Ti_2O_3, Ti_3O_5, Fe_2O_4, NiS, CrS, FeS.
4d-compounds:	NbO_2.

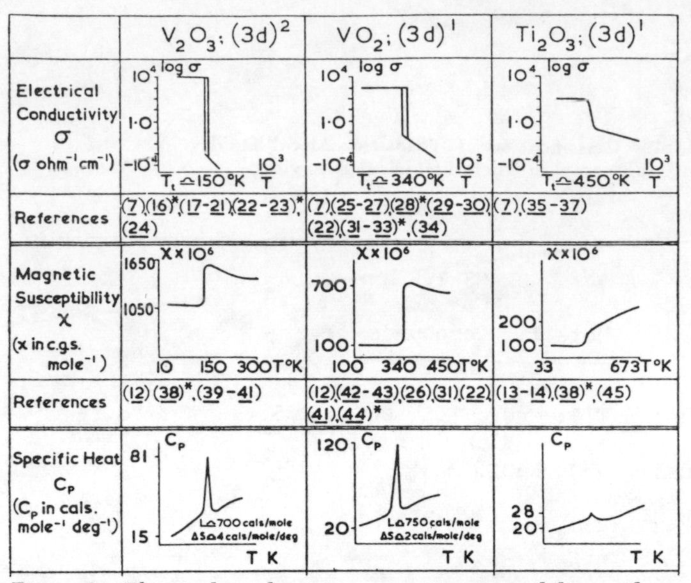

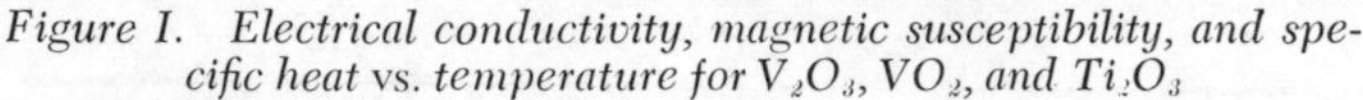
Figure I. Electrical conductivity, magnetic susceptibility, and specific heat vs. temperature for V_2O_3, VO_2, and Ti_2O_3

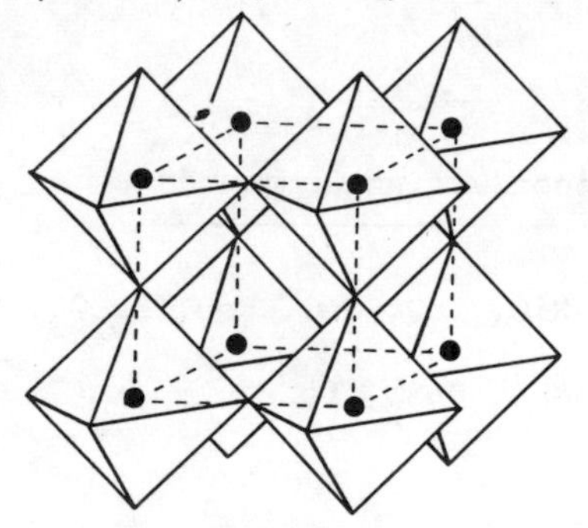

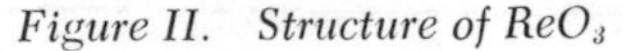
Figure II. Structure of ReO_3

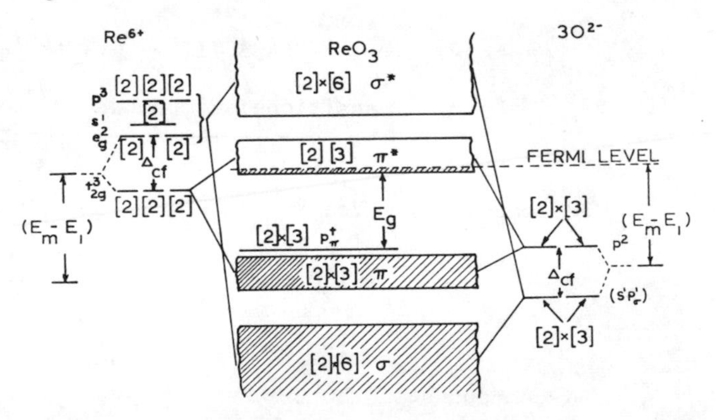

Figure III. Electron energy diagram for ReO_3

cubic bronze, Na_xWO_3, and the ordered perovskite $Sr_2M_gReO_6$ have been investigated by Goodenough and co-workers (9).

The one electron energy diagram for ReO_3 is shown in Figure III. Because of the octahedral crystal field the d-electron energy states of the rhenium are split into a less stable, doubly degenerate state (e_g) and a more stable triply degenerate (t_{2g}). The outer electron energy levels of the anion are similarly shown on the right side of the figure.

The overlap of cation t_{2g} and anion p_π orbitals results in the formation of bonding and antibonding bands extending throughout the crystal (10).

The Fermi level may be located by looking at the number of outer electrons per molecule. Consider the ReO_3 unit for rhenium $5d^5 6s^2$. There are [7] x [1] = 7 electrons. The oxygen $2s^2 2p^4$ contribute [6] x [3] = 18 electrons resulting in 25 electrons per molecule. Upon filling the available energy levels, the last electron occupies the π^*band. This band is partially filled and causes the observed metallic behavior.

The origin of the metallic conductivity of cubic Na_xWO_3 has been postulated by several models: 1. The direct overlap of sodium 3p orbitals (11) 2. The direct overlap of tungsten t_{2g} orbitals leading to tungsten-tungsten bonds (12) 3. The covalent mixing of tungsten t_{2g} orbitals and oxygen p_π orbitals to form partially filled bands (13).

Goodenough and his co-workers (9) performed a unique experiment which indicated that the third possibility was indeed the most probable mechanism for the observed metallic conduction in these materials. The compound $Sr_2M_gReO_6$ crystallizes in the ordered perovskite structure (see Figure IV). The B-site cations, Mg and Re, order such that each rhenium has only magnesium nearest cation neighbors and vice versa (9). Such an arrangement would still allow for the overlap of rhenium t_{2g} orbitals across a cube face. Consequently, metallic behavior would be expected if Model 2 applied. However, if the electrical properties are a result of cation t_{2g} and oxygen ρ_π interactions (Model 3), Sr_2MgReO_6 should be a semi-conductor, since Mg does not possess t_{2g} electrons. The observed semi-conductive behavior (E_a = 0.31eV) and temperature dependent paramagnetism confirm that the electrons are localized rather than occupying collective energy bands (9). Thus, model 3 is consistent with the interactions present in perovskite-like materials and best explains the observed properties.

II. Vanadium (IV) Oxide and Substituted Compounds. The high temperature, metallic VO_2 phase has a tetragonal rutile like (14) structure (space group $P4_2/mnm$). As can be seen from Figure V the structure may be described as strings of edge-shared octahedra joined by corners extending in the c direction. The V-V distances are equivalent within the strings (2.87 Å).

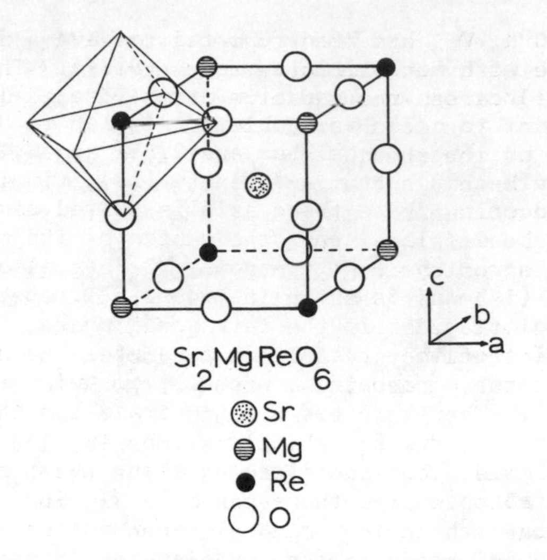

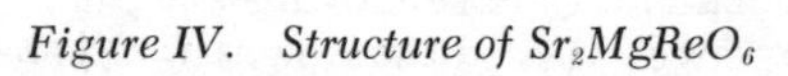

Figure IV. Structure of Sr_2MgReO_6

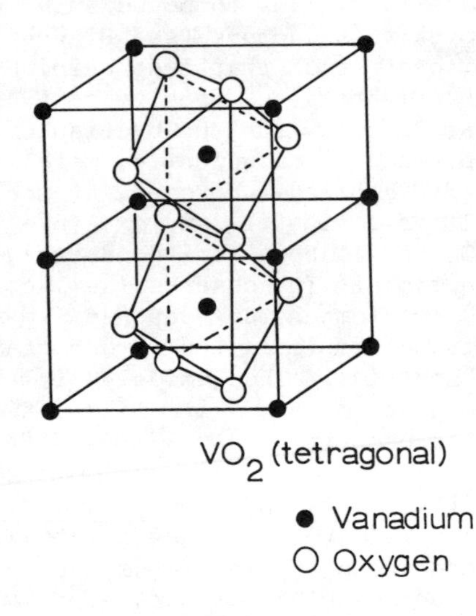

Figure V. Rutile structure

Below 340°K, VO_2 has been reported to have a distorted rutile structure with monoclinic symmetry ($P2_1/c$. This distortion (see Figure VI) causes the vanadium atoms located in the strings of VO_6 octahedra to occur as doublets (similar to MoO_2). The V-V distances are no longer equal but are 2.65Å and 3.12Å (14). As pointed out by Heckingbottom and Linett (15), in the low temperature semiconducting phase the c axis is tilted causing the vanadium atoms to be displaced from the center of its octahedra.

The band structure for tetragonal VO_2 has also been examined by Goodenough (14) and is shown in Figure VII. Because of edge sharing of VO_6 octahedra in the tetragonal phase, an orthorhombic component of the crystal field removes the d-state degeneracy. The two e_g orbitals, normally occuring in an octahedral field, are split into two $d\sigma$ orbitals and the three t_{2g} orbitals split into two d_π orbitals which mix with the anion p_π orbital and a $d_{||}$ orbital directed along the c rutile axis.

When the 17 outer electrons per molecule in VO_2 are placed into this energy scheme the final electron enters the overlapping π^* and $d_{||}$ bands. These overlapping, non-degenerate bands, being only partially filled, result in metallic conductivity in the tetragonal phase (14).

Below the transition temperature VO_2 has a monoclinically distorted rutile structure characterized by the formation of V-V pairs along the a monoclinic axis (c_{rutile} axis) and a consequent doubling of the crystalographic unit cell (16). This displacement of the V^{+4} ion from its center of symmetry is believed to be caused by a ferroelectric - type distortion (14). The effect of this distortion on the band structure of VO_2 is shown in Figure VIII. The $d_{||}$ band is split in two and the Fermi energy is lowered below the bottom of the π^* band. As a consequence of doubling the cell, the lower t_{2g} band is filled completely and semi-conducting behavior results (14).

III. Oxyfluorides. At low levels of fluorine, the oxyfluorides usually possess structures quite similar to the parent oxide. The cell parameters (orthorhombic indexing) a_o = 7.356 Å, b_o = 7.469 Å, c_o = 3.846 Å reported for $WO_{2.96}F_{0.04}$ are quite similar to those observed by Sleight (17) for monoclinic WO_3 a = 7.301 Å b = 7.538 Å, c = 3.844 Å, β = 90.89°.

Sleight did not rule out the possibility that $WO_{2.96}F_{0.04}$ could be monoclinic with the deviation of the monoclinic angle from 90° being too small to detect. Higher substitutions of fluorine in the $WO_{3-x}F_x$ systems stabilizes the cubic ReO_3 structure (17, 18). This structure is analogous to the perovskite type alkali metal bronzes with all A sites vacant. In the tungsten oxyfluoride system the cubic phase extended from x = 0.17 to 0.66. A one-electron energy diagram would be much simpler for $WO_{3-x}F_x$ than for the corresponding cubic tungsten bronzes Na_xWO_3. It is not necessary to consider interactions involving A site cations since these positions are empty.

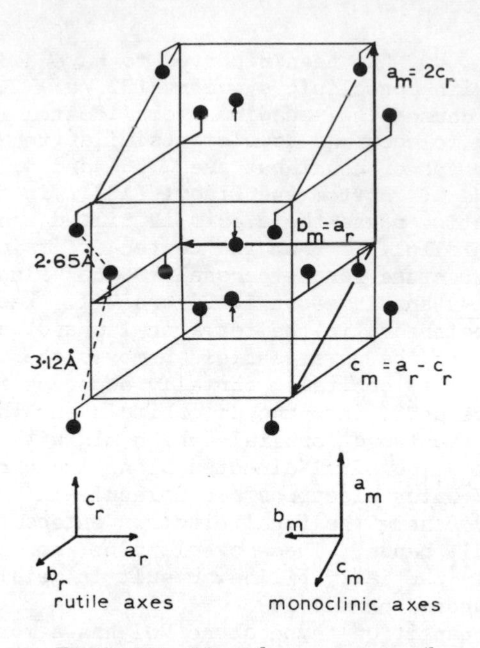

Figure VI. Monoclinic structure (low temperature) for VO_2

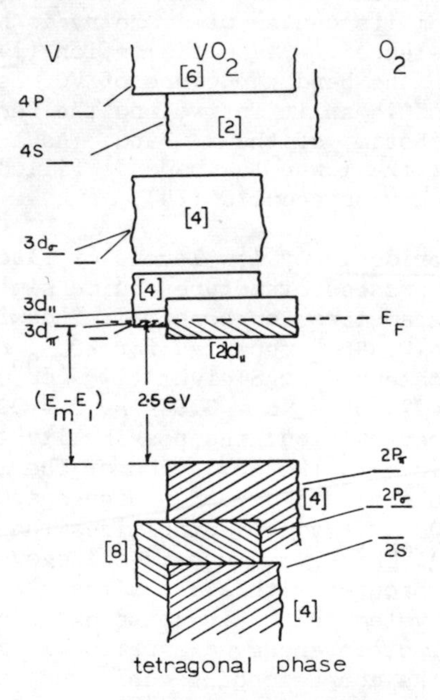

Figure VII. Electron energy diagram for tetragonal VO_2

The substitution of fluorine also results in additional electrons because of the formation of W^{5+} ions. Thus, the compounds $WO_{3-x}F_x$ have x additional electrons per molecule. Although it might be expected that the more electronegative fluoride ion would have a localizing effect on the additional electrons, metallic conductivity has been observed for the cubic oxyfluoride bronzes. By comparison with the $WO_{3-x}F_x$ system, in which large fluorine substitutions stabilize a high symmetry cubic phase, there may also be an analogous decrease in the semiconductor metal transition with increasing fluorine substitution in $VO_{2-x}F_x$. For VO_2, the monoclinic to tetragonal transition temperature decreases with increasing anion substitution i.e. the high temperature tetragonal phase becomes more stable (19).

For the system $VO_{2-x}F_x$ activation energies were calculated from the resistivity data (see Figure IX). the values are given in Table II for different values of x in $VO_{2-x}F_x$. It can be seen from these curves that a metallic to semiconductor transition occurs at a temperature which decreases with increasing values of x. This change in the electrical properties of $VO_{2-x}F_x$ can be explained by the corresponding transition from the monoclinic phase to the tetragonal phase which has been observed by means of low-temperature X-ray analysis. A linear relationship exists between the value of x and the transition temperature (T_t) shown in Figure X; this extrapolates to the correct transition temperature for pure VO_2. For the higher fluorine compounds, the transition region is considerably broadened and the transition point was chosen as the first deviation from log-linear behavior. A similar linear relationship exists between the volume of the tetragonal cell and the value of x, again extrapolating to the value of the pure VO_2 phase at x = 0 (see Figure XI). The same behavior has been observed in compounds corresponding to the formula $V_{1-x}W_xO_2$ ($0 \leq x \leq 0.067$) by Nygren and Israelsson (19).

It is seen, therefore, that the addition of fluorine tends to stabilize the high temperature, higher symmetry, rutile phase. The compositions containing larger amounts of substituted fluorine shows primarily metallic behavior. However, for all compositions studied there still appears to be a discontinuity in the resistivity at the expected transition temperature (T_t).

The metallic behavior observed in these materials may be explained on the model presented by Goodenough (13) and Rogers (20). In their model the band formed between the overlap of the $t_{2g}\sigma$ orbitals parallel to the crystallographic c direction splits into a more stable, pair-localized, bonding V-V state and a higher less stable σ^* state. The lower lying V-V level if filled with one electron per vanadium and hence the semiconducting properties of the monoclinic VO_2 may be explained. The substitution of fluorine for oxygen in VO_2 results in the creation of additional unpaired d-electrons. Despite the tendency for the more electronegative anion to localize d-electrons, it is apparent that for

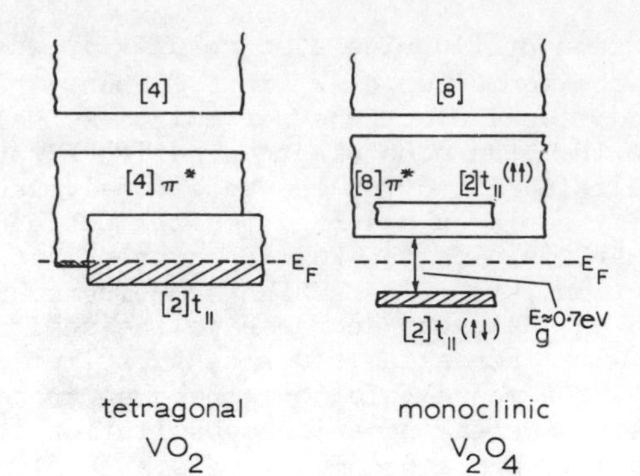

Figure VIII. Electron energy diagram for tetragonal and monoclinic VO_2

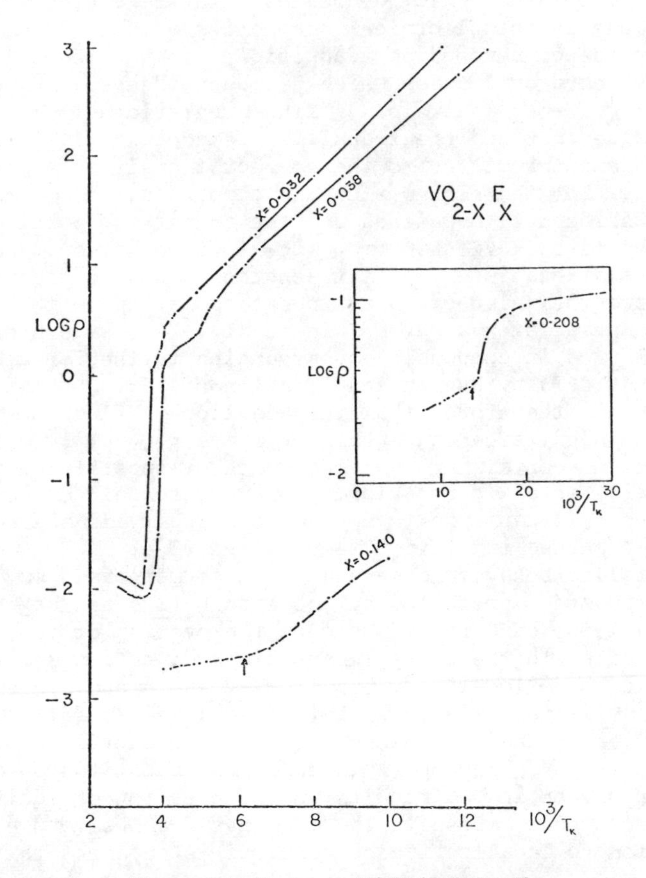

Figure IX. Log ρ vs. $10^3/T$ for $VO_{2-x}F_x$

Table 2. Cell Parameters of $VO_{2-x}F_x$ Compounds

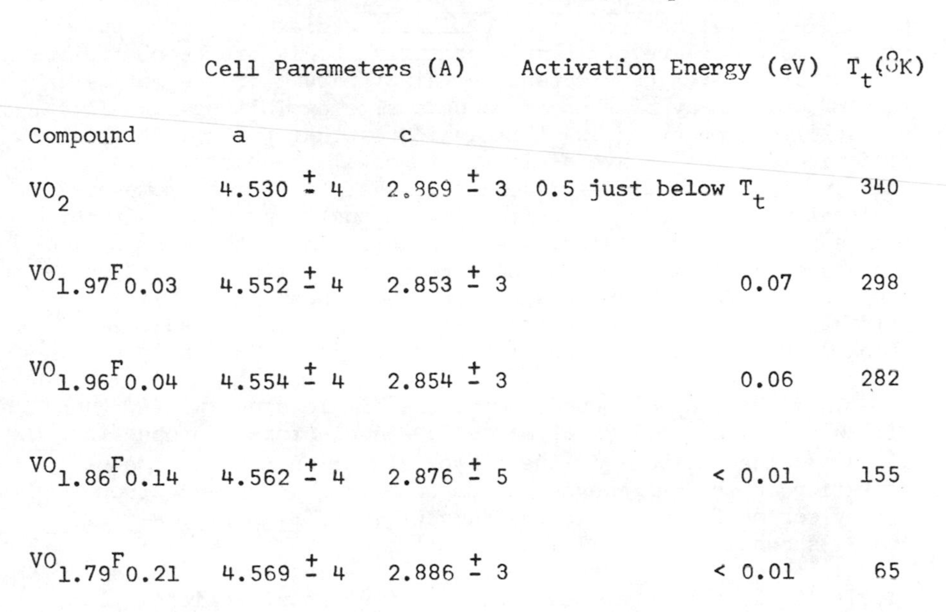

Compound	Cell Parameters (A) a	c	Activation Energy (eV)	T_t (°K)
VO_2	4.530 ± 4	2.869 ± 3	0.5 just below T_t	340
$VO_{1.97}F_{0.03}$	4.552 ± 4	2.853 ± 3	0.07	298
$VO_{1.96}F_{0.04}$	4.554 ± 4	2.854 ± 3	0.06	282
$VO_{1.86}F_{0.14}$	4.562 ± 4	2.876 ± 5	< 0.01	155
$VO_{1.79}F_{0.21}$	4.569 ± 4	2.886 ± 3	< 0.01	65

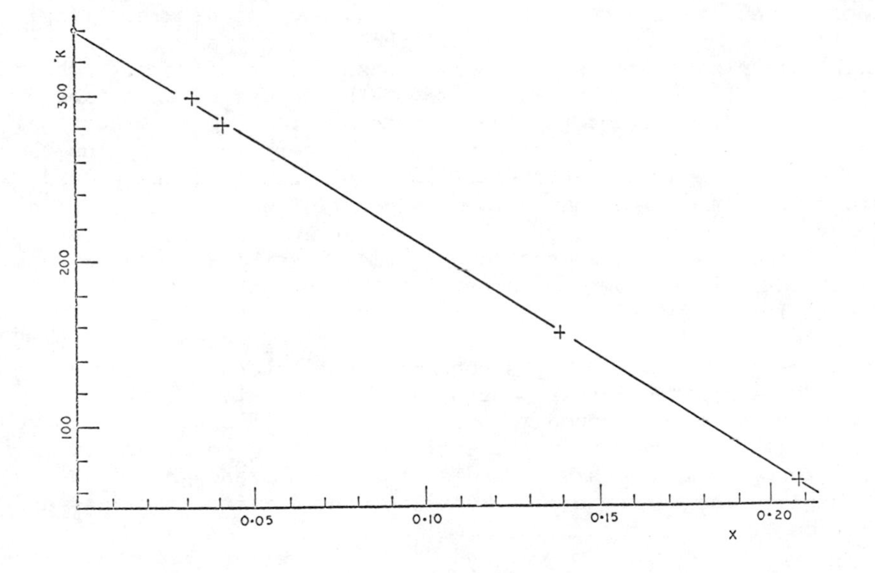

Figure X. Transition temperature vs. *composition for $VO_{2-x}F_x$*

the amount of fluorine which has been substituted, the conduction paths have not been substantially altered.

IV. Transition Metal Chalcogenides. Binary and ternary transition metal chalcogenides exhibit a variety of different structures which are often quite complex. This has been attributed in part to the considerable covalent nature of the metal-sulfur bonds. Many of these compounds also exhibit semimetallic or metallic properties and this indicates that not all of the bonding electrons behave as in simple covalent crystals. A characteristic of sulfide minerals is that partial or complete replacement of either the cations or anions is possible without a change in the crystal structure occuring. Among the many chalcogenides studied at Brown University the transition metal dichalcogenides, with the pyrite structure, have been particularly attractive since their electronic properties can be radically changed by substituting for either the cation or anion in the host compound.

This group of compounds combine a simple structure (Figure XII) with a wide variety of magnetic and electrical properties(21). Of particular interest is the effect of cation and anion substitution on the ferromagnetic compound CoS_2. A one-electron energy scheme for CoS_2 has been described by Bither et al (22) and is shown in Figure XIII.

In this model σ-d^2sp^3 orbitals on the metal atom and sp^3 orbitals on the anions are assumed. Sulfur has six valence electrons that are shared among four tetrahedral bonds. Each sulfur contributes one electron to the S-S bond and the remaining five to the three M-S bonds. Since all the M-S bonds are equivalent, each sulfur contributes 1 2/3 electrons to each of them. Cobalt is coordinated to six sulfur atoms at the apices of an octahedron and these sulfur atoms thus contribute 6 x 1 2/3 = 10 electrons to the bonding in each octahedron. These ten electrons plus the two 4 s electrons of the transition metal just fill the ground state σ(s-p) and $\sigma\ e_g$ manifold of states. The remaining d-electrons of the metal occupy the next-lowest available levels.

Since cobalt possesses seven d-electrons, the one unpaired electron indicated by the magnetic moment of CoS_2 suggests that the t_{2g} levels are filled with six electrons (spin-paired). The unpaired electron therefore occupies the $\sigma^*\ e_g$ state. In the presence of a sufficiently strong covalent interaction the $\sigma^*\ e_g$ level will broaden into a band of crystalline states; the electrons are no longer bound to specific sites but are free to move through the crystal under the influence of an electric field. This concept already has been discussed by Goodenough (23,24). to explain the occurrence of metallic conductivity in oxides and sulfides.

For both compounds CoS_2 and $CoSe_2$, as well as members of the

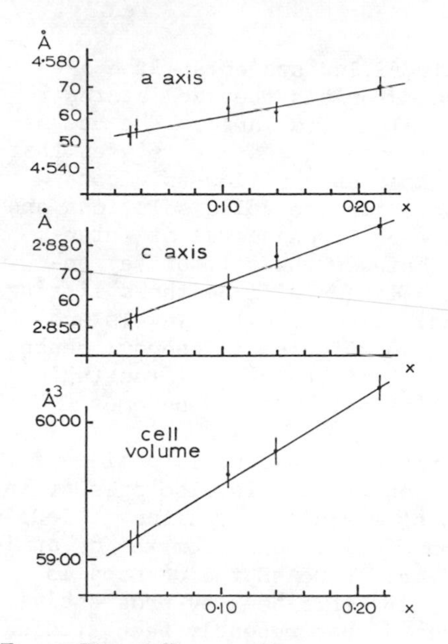

Figure XI. Cell constants vs. *composition for* $VO_{2-x}F_x$

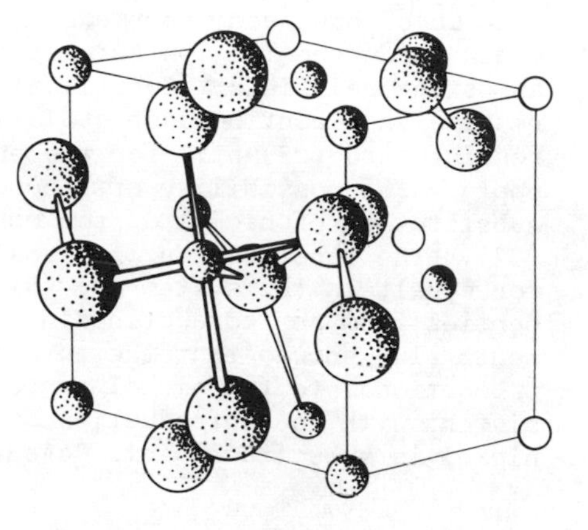

Figure XII. Pyrite structure. Large balls are sulfur; small balls, cobalt.

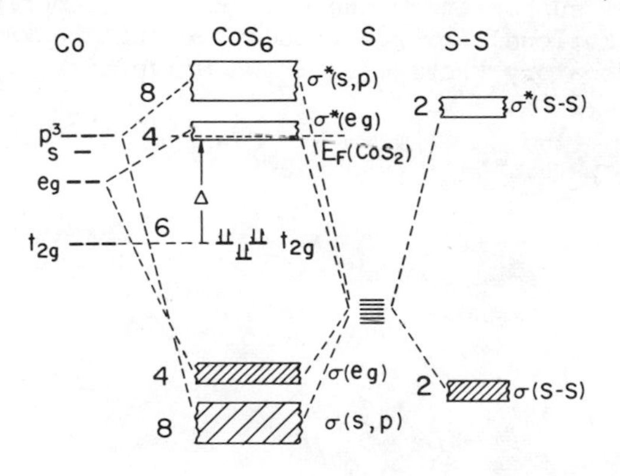

Figure XIII. Electron energy diagram for CoS_2

system $CoS_{2-x}Se_x$, the only partially filled states are the σ^*e_g states. This implies that in these materials the σ^*e_g states constitute a band; the degree of covalence is large. The occurrence of an energy difference between the t_{2g} and σ^*e_g states that is larger than the intraatomic exchange energy (low-spin configurations in ligand field parlance) for the solid solutions and CoS_2 is consistent with the presence of strong covalent interactions. The stronger the bonding between the cationic e_g orbitals and the anionic orbitals, the larger will be the splitting between the t_{2g} and the σ^*e_g states. Since the t_{2g} levels are essentially nonbonding they are little affected in energy; hence, the energy difference between the t_{2g} and the σ^*e_g states (which corresponds to the ligand field splitting, or 10Dq) becomes larger with increasing covalence.

CoAsS has been reported (25) to be a diamagnetic semiconductor where the low spin state cobalt (d^6) is also present in an octahedral field. Substitution of arsenic for sulfur in CoS_2 results in a continued depopulation of electrons from the σ* antibonding band (26)until for the end member CoAsS the σ* band is empty. The possibility of repopulating this band by progressive substitution of nickel d^7 for cobalt d^6 has recently been studied (27). The substitution of a small amount of nickel (x = 0.05) for cobalt in the system $Co_{1-x}Ni_xAsS$ changed the electrical properties from semiconducting to metallic. Hall-effect measurements also showed that the number of carriers (electrons) was proportional to the nickel concentration. These results are consistent with the introduction of electrons into the σ* band as nickel is substituted into CoAsS.

This work was supported by the U.S. Army Research Office, Durham, the National Science Foundation and the Materials Research Laboratory Program at Brown University.

Literature Cited

1 de Boer, J. H., and Verwey, E. J. W., Proc. Phys. Soc. London (1937) A49, 59
2 Block, F., Zeits, f., Phys. (1928) 52, 555
3 Morin, F. J., Phys. Rev. Letters (1959) 3, 34
4 Pearson, A. D., J. Phys. Chem Solids (1958) 5, 316
5 Adler, D., Rev. Mod. Phys. (1968) 40, 714
6 Goodenough, J. B., Mat. Res. Bull (1967) 2, 37
7 Goodenough, J. G., Mat. Res. Bull (1967) 2, 165
8 Braeken, Z. Krist., (1931) 78, 484
9 Ferretti, A., Rogers, D., Goodenough, J. B., J. Phys. Chem. Solids (1965) 26, 2007
10 Banks, E., and Wold, A., "Preparative Inorganic Reactions", 4, 237 Interscience Publishers, New York, 1968
11 MacKintosh, A. R., J. Phys. Chem., (1963) 38, 1991
12 Sienko, M. J., "Paper #21 in Non Stoichiometric Compounds", 39, 224 Adv. Chem., Amer. Chem. Soc., Wash., D. C. 1963
13 Goodenough, J. B., Bull Soc. Chim. Fr., (1965) 1200
14 Goodenough, J. B., Solid State Chem., (1971) 3, 490
15 Heckingbottom, Linett, J. V., Nature (London) (1962) 194, 678
16 Hagg, G., Z., Physik Chem. (1935) B29 , 192
17 Sleight, A. W., Inorg. Chem., (1969),8, 1764
18 Pierce, J. W., McKinzie, H. L., Vlasse, M., Wold, A., J. Solid State Chem., (1969) 1, 332
19 Nygren, M., Israelsson, Mat. Res. Bull., (1969) 4, 881
20 Rogers, D. B., Shannon, R. D., Sleight, A. W., Gilldon, J. L., Inorg. Chem., (1969) 8, 841
21 Hullinger, F., J. Phys. Chem. Solids (1965) 26, 639
22 Bither, T. A., Bouchard, R. J., Cloud, W. H., Donohue, P. A. and Siemond, W. J., Inorg. Chem. (1968) 7, 2208
23 Goodenough, J. B., "Magnetism and the Chemical Bond", Interscience Publishers, Inc., New York, (1963)
24 Goodenough, J. B., Proc. Colloque Int. Orsay 1965 (1967) C.N.R.S. No. 157, 263
25 Nahigian, H., Steger, J., McKinzie H., Arnott, R. J., Wold, A., to be published
26 Goodenough, J. B., J. Solid State Chem. (1971) 3, 26
27 Steger, J. J., Nahigian,H., Arnott, R. J., Wold, A., to be published

2

High Temperature Crystal Chemistry of V_2O_3 through the Metal-Insulator Transition and of $(Cr_{0.01}V_{0.99})_2O_3$, an Insulator

WILLIAM R. ROBINSON

Purdue University, West Lafayette, Ind. 47907

A variety of interesting structural and electrical phenomena are observed upon heating the chromium doped V_2O_3 system $(Cr_xV_{1-x})_2O_3$. At 155-170°K, those systems with $x \leq 0.018$ undergo an antiferromagnetic insulator to metal transition (1). This transition is accompanied by a change from a monoclinic structure to a rhombohedral structure isomorphous with α-corundum. Those systems with $x > 0.018$ exhibit an antiferromagnetic insulator to insulator transition at about 170°K and maintain their negative thermal coefficients of resistivity at elevated temperatures. The electrical behavior of systems containing less chromium ($x < 0.018$) is quite different at higher temperatures.

Pure V_2O_3 exhibits a continuous electrical transition in the range 225-325°C which results in an increase in resistivity of about one order of magnitude (1,2,3,4). The doped systems with $x < 0.018$ exhibit an electrical effect which was originally ascribed to a Mott transition (1) But which more recently has been described as an extrinsic effect resulting from the coexistence of two phases (4).

Above the transition at 155-170°K, all of the various electrical modifications of $(Cr_xV_{1-x})_2O_3$ are rhombohedral and isomorphous with corundum. However, the size of the rhombohedral cell varies between two extreme values. V_2O_3 at room temperature and the low temperature form of $(Cr_{0.01}V_{0.99})_2O_3$ (the α form) have unit cell parameters with a approximately 4.95A and c approximately 14.00A in the hexagonal indexing of the rhombohedral system. V_2O_3 at high temperature, the high temperature form of $(Cr_{0.01}V_{0.99})_2O_3$ (the β form), and $(Cr_{0.04}V_{0.96})_2O_3$ have a approximately 5.00A and c approximately 13.94A. The change in the V_2O_3 cell parameters is continuous with temperature while the primary change in the $(Cr_{0.01}V_{0.99})_2O_3$ cell parameters occurs with the α-β structural transition.

In view of the structural changes accompanying the semiconductor to metal transition upon heating Ti_2O_3(5,6) or doping with vanadium to give $(V_xTi_{1-x})_2O_3$ where $x \leq 0.1$ (7,8), we thought it interesting to follow the structural changes accompanying the electrical transition in V_2O_3 in the 225°-325°C range. Since

β-$(Cr_{0.01}V_{0.99})_2O_3$ does not change much in this range, it was selected as a reference material.

Experimental.

Samples of V_2O_3 and $(Cr_{0.01}V_{0.99})_2O_3$ (4) were provided by Professor J. M. Honig, Department of Chemistry, Purdue University. A small single crystal of each material was cut from the boule and mounted along the 210 direction in evacuated silica capillaries using the procedure described by Brown, Sueno, and Prewitt (9). These crystals were used in all subsequent studies.

The unit cell dimensions reported in Table I and Figure 1 were determined by least-squares refinement of the 2Θ values of 15-20 reflections, principally of the type hkl, lying in the region $50° < 2\Theta < 55°$. The 2Θ values were determined on a Picker diffractometer equipped with a crystal heater (9). The average peak position at positive and negative 2Θ was determined using the quarter height technique and graphite monochromated MoKα radiation.

At each temperature indicated in Table I, about 360 intensities lying in a quadrant of reciprocal space with $5° < 2\Theta < 62°$ were collected. The diffractometer was operated in a Θ - 2Θ mode using a graphite monochromator and MoKα radiation. Following application of absorption corrections and the Lp correction, equivalent reflections were averaged giving about 100 independent reflections at each temperature.

Both isotropic and anisotropic refinements were carried out for each set of intensity data in space group $R\bar{3}c$ using the V and O positions reported by Newnham and deHaan (10) as starting parameters. Three cycles of refinement in each case resulted in convergence and gave final R values of 0.02-0.04. Selected structural parameters are reported in Table I and Figures 2 and 3.

Results

The crystal structure of V_2O_3 at 23° was found to be identical within experimental error to that reported by Dernier (11) and Newnham and deHaan (10). The structure of V_2O_3 at 600°C and the structure of β-$(Cr_{0.01}V_{0.99})_2O_3$ were found to be identical to that of $(Cr_{0.04}C_{0.96})_2O_3$ (11).

The structures of all of these systems, like that of corundum, consists of an approximately hexagonal closest packed array of oxide ions with metal ions in two-thirds of the octahedral holes (Figure 4). Each metal ion has four near metal neighbors; one sharing an octahedral face of the coordination polyhedron (M_1-M_2 in Figure 4) and three sharing edges of the octahedron (M_1-M_3 in Figure 4).

The nonlinear variation of the unit cell parameters of V_2O_3 upon heating (Table 1 and Figure 1) is similar to that previously reported (1,12). The linear increase with temperature of the cell parameters of β-$(Cr_{0.01}V_{0.99})_2O_3$ parallels that of

Table I: Crystallographic Data for V_2O_3 and $(Cr_{0.01}V_{0.99})_2O_3$

V_2O_3

	23°	115°	212°	300°	400°	600°
$\underline{a}_{hex}$*	4.9492(2)	4.9605(1)	4.9776(2)	4.9912(2)	5.0013(1)	5.0140(2)
$\underline{c}_{hex}$	13.998(1)	13.9857(6)	13.9647(7)	13.9489(6)	13.9416(5)	13.9427(7)
M_1-M_2	2.697(1)	2.705(1)	2.718(1)	2.728(1)	2.734(1)	2.738(1)
M_1-M_3	2.880(1)	2.888(1)	2.900(1)	2.910(1)	2.916(1)	2.924(1)
M_1-O_1	2.051(1)	2.053(1)	2.057(2)	2.061(1)	2.064(1)	2.066(1)
M_1-O_4	1.968(1)	1.970(1)	1.973(1)	1.975(2)	1.977(2)	1.981(1)
O_1-O_2	2.676(3)	2.675(3)	2.675(4)	2.676(3)	2.680(3)	2.681(3)
O_1-O_4	2.804(1)	2.802(1)	2.800(1)	2.799(2)	2.799(2)	2.800(1)
O_1-O_5	2.889(1)	2.891(2)	2.893(1)	2.895(2)	2.897(1)	2.901(1)
O_4-O_5	2.952(1)	2.962(2)	2.978(1)	2.990(1)	2.997(2)	3.007(1)

$(Cr_{0.01}V_{0.99})_2O_3$

	23°	113°	310°
$\underline{a}_{hex}$	4.9974(1)	5.0018(1)	5.0052(1)
$\underline{c}_{hex}$	13.9260(6)	13.9238(7)	13.9318(6)
M_1-M_2	2.747(1)	2.746(1)	2.739(1)
M_1-M_3	2.917(1)	2.919(1)	2.920(1)
M_1-O_1	2.061(1)	2.061(1)	2.062(2)
M_1-O_4	1.976(1)	1.977(1)	1.979(1)
O_1-O_2	2.661(2)	2.663(2)	2.670(2)
O_1-O_4	2.792(1)	2.793(1)	2.796(1)
O_1-O_5	2.897(1)	2.898(1)	2.899(1)
O_4-O_5	3.004(1)	3.006(1)	3.006(1)

*All entries in Å.

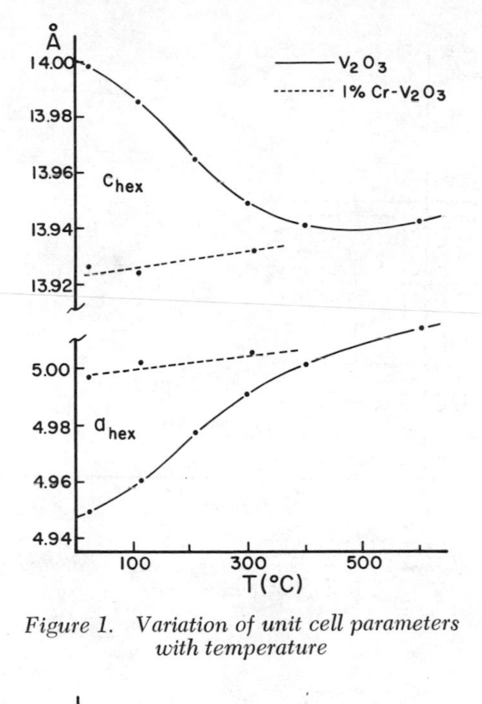

Figure 1. Variation of unit cell parameters with temperature

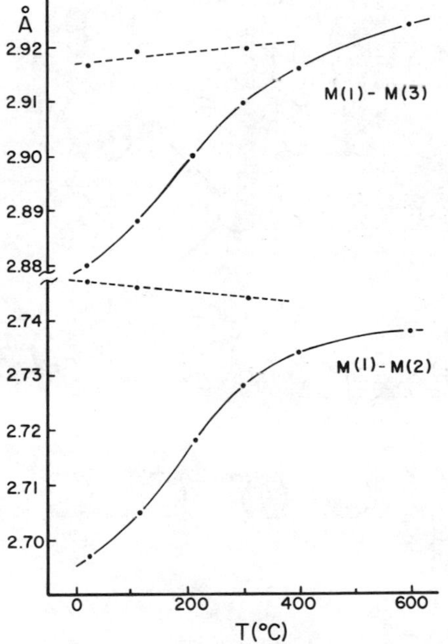

Figure 2. Variation of metal–metal distances with temperature

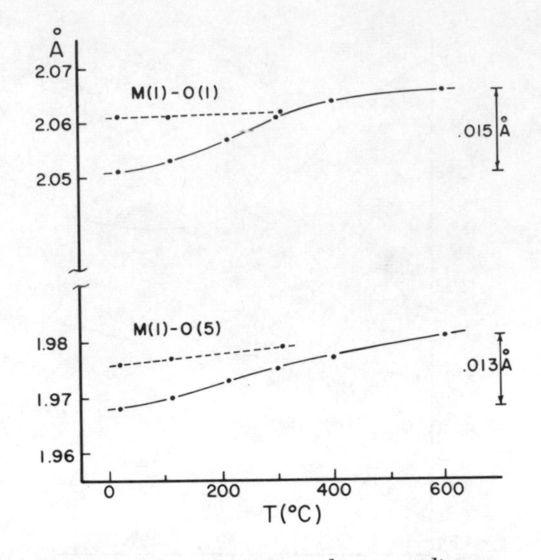

Figure 3. Variation of metal–oxygen distances with temperature

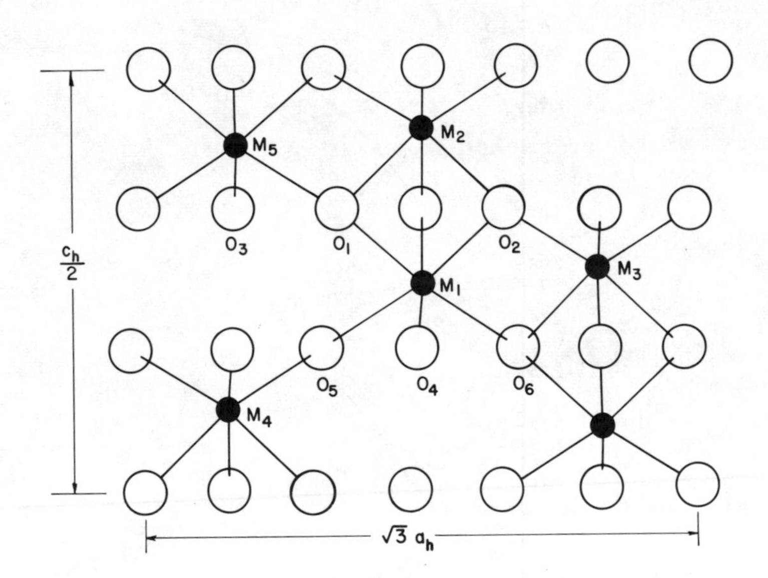

Figure 4. Projection of the corundum structure

$(Cr_{.04}V_{.96})_2O_3$ (1) although the cell dimensions differ slightly. It is apparent from the data that the unit cell dimensions of the pure and doped systems converge at elevated temperatures.

The coordinates of the metal and oxygen atoms of V_2O_3 and β-$(Cr_{.01}V_{.99})_2O_3$ also converge with increasing temperature. The changes in metal-metal, metal-oxygen, and oxygen-oxygen distances which result are given in Table 1. Plots of metal-metal and metal-oxygen distances are presented in Figures 2 and 3.

Both the M_1-M_2 and M_1-M_3 equilibrium distances increase with temperature in V_2O_3. The M_1-M_2 change from 2.697(1)A at 23° to 2.738(1)A at 600° represents a change of about 40 e.s.d.'s as does the increase from 2.880(1)A at 23° to 2.924(1)A at 600° in the M_1-M_3 distance. Over the temperature range 23° to 310°C the M_1-M_3 equilibrium distance of β-$(Cr_{.01}V_{.99})_2O_3$ increases slightly from 2.917(1)A to 2.920(1)A while the M_1-M_2 distance decreases from 2.747(1)A to 2.739(1)A. This latter change occurs even though the c axis which is parallel to the M_1-M_2 direction is increasing.

No significant changes in the metal-oxygen distances in the doped system are observed upon heating. The maximum change is three e.s.d.'s. Small changes in the metal-oxygen distances in V_2O_3 are observed. The M_1-O_1 distance varies by 0.015A from 2.051(1)A to 2.066(1)A while M_1-O_4 varies from 1.968(1)A to 1.981(1)A, a change of 0.013A.

The only significant change in oxygen-oxygen distances is in the O_4-O_5 distance in V_2O_3. This distance changes from 2.952(1)A at 23° to 3.007(1)A at 600°C. The other changes are 0.009A or less.

In general, the structural changes in V_2O_3 upon heating involve a slight reduction of the spacing between the approximately hexagonally closest packed planes of oxygen atoms with a concommitant increase in the average oxygen-oxygen separation in the planes. One can distinguish two types of oxygen-oxygen distances within a given plane; distances of the type O_1-O_2 along the edge of a shared octahedral face and distances of the type O_1-O_3 or O_4-O_5 along the edges of an unshared face. It is expansion of the oxygen-oxygen distances along the edges of the unshared face which is reflected in the increase in the a axis. Paralleling the increase in the a axis is the M_1-M_3 separation. The M_1-M_2 distance increases at the same rate as the M_1-M_3 distance. The net result of these changes is that at high temperatures V_2O_3, which is isostructural with α-$(Cr_{.01}V_{.99})_2O_3$ (13) at room temperature, becomes isostructural with β-$(Cr_{.01}V_{.99})_2O_3$ and $(Cr_{.04}V_{.96})_2O_3$.

Acknowledgements.

This work was done in the Department of Earth and Space Sciences, SUNY, Stony Brook, N.Y. We wish to thank Professor C. T. Prewitt for his assistance with this project and the

National Science Foundation (Grant No. GP-17554) and Purdue University for support of the work.

Literature Cited.

1. McWhan, D. B. and Remeika, J. P., Phys. Rev. B, (1970), 2, 3734.
2. Foex, M., Compt. Rend., (1946), 223, 1126.
3. Morin, F. J., Phys. Rev. Lett., (1959), 3, 34.
4. Honig, J. M., Chandrashekhar, G. V., Sinha, A. P. B., Phys. Rev. Lett., (1974), 32, 13.
5. Honig, J. M., and Reed, T. B., Phys. Rev., (1968), 174, 1020.
6. Simonyi, E., and Raccah, P. M., Bull. Am. Phys. Soc., (1973), 11, 339.
7. Chandrashekhar, G. V., Won Choi, Q., Moyo, J., and Honig, J. M., Mat. Res. Bull., (1970), 5, 999.
8. Robinson, W. R., J. Solid State Chem., (1974), 9, 255.
9. Brown, G. E., Sueno, S., and Prewitt, C. T., Amer. Mineral., (1973), 58, 698.
10. Newnham, R. E. and deHaan, Y. M., Z. Kristallog., (1962), 117, 235.
11. Dernier, P. D., J. Phys. Chem. Solids, (1970), 31, 2569.
12. Eckert, L. J. and Bradt, R. C., J. Appl. Phys., (1973), 44, 8.
13. Robinson, W. R., Unreported Observations.

3

Synthesis and Properties of Some New Organic Intercalation Complexes of Tantalum Sulfide

R. L. HARTLESS and A. M. TROZZOLO

Bell Laboratories, Murray Hill, N.J. 07974

Intercalation of layered compounds by metals has been previously reported (1-5). A number of inorganic and biological molecules have been found to intercalate clays and graphite (6,7). The first intercalation of layered transition metal chalcogenides by organic molecules was reported by Weiss and Rudhardt (8). They intercalated TiS_2 with hydrazine and aliphatic amides and observed increases of several angstroms in the interlayer spacing of the chalcogenide. More recent studies show that a variety of organic molecules intercalate layered transition metal chalcogenides (9-11). The organic molecules penetrate the interlayer planes and form a periodic crystalline structure. In addition to increased interplanar spacing, the intercalated complexes showed enhancement of the critical superconducting temperature (T_c) above that observed for the unintercalated chalcogenide.

Tantalum sulfide crystallizes in several polytypes which are characterized by layer-like structures in which the metal atoms are located between alternate sulfur layers. The bonds forming the intraplanar layers are strong (primarily covalent) bonds. Each sulfur-tantalum-sulfur layer (interplanar) is bonded to adjacent layers by weak van der Waals attraction. Intercalation (insertion) of atoms or molecules between the layers occurs by cleavage along the planes which have the low energy interaction. Figure 1 gives a schematic description of the structure of the layered transition metal chalcogenide prior to intercalation where M is the transition metal atom and X is the chalcogen atom. Shown are three intraplanar layers separated by the van der Waals gap. The intraplanar layers are approximately 6Å thick and the interlayer distance between transition metal atoms is approximately 6Å. Upon intercalation of the organic molecule, the van der Waals gap expands as schematically represented in Figure 2. Table 1 shows the synthesis conditions and parameters for some previously intercalated complexes (9). The interlayer distance for the pyridine-TaS_2 complex increases ~6Å to give a metal-metal distance of ~12Å. In the case of the octadecylamine-TaS_2 complex, the enhanced interlayer distance was ~50Å which was attributed to two layers of the amine oriented end-to-end and perpendicular to the TaS_2 planes (Figure 3). In general, organic molecules capable of intercalating TaS_2 are Lewis bases containing an sp^3 or sp^2 hybridized nitrogen.

This study involves intercalation of the 2H-polytype of tantalum sulfide (TaS_2) with organic molecules of various structures to form complexes with increased interlayer spacing and enhanced critical superconducting temperatures. The extent of interplanar layer expansion and T_c enhancement varies with the particular intercalate, thus allowing possible chemical control of T_c.

Experimental

The powdered tantalum sulfide was placed in a 4mm pyrex tube with the liquid intercalate or a benzene solution of the solid intercalate. The tube was evacuated and sealed and the reactions carried out in isothermal baths at 25, 100, 150 and 200°. Intercalation was evident when the volume of the chalcogenide increased. After intercalation, the excess organic reactants were removed by washing with benzene followed by drying at reduced pressure. Supporting evidence for intercalation was obtained by powder x-ray diffraction with a Guinier camera using CuK_α radiation. The critical superconducting temperatures were determined on powdered complexes by AC susceptibility measurements.

Results and Discussion

Several mono- and disubstituted amines form complexes with TaS_2 (Table 2). In general the amines which intercalated had $pK_a > 9$. Diethylamine intercalates in two hours at ambient temperatures to yield a complex with an interlayer distance increase (σ) of 3.69 Å and an onset of superconductivity (T_o) of 3.0°K. The dibenzyl amine complex was superconducting at 2.7°K and the interlayer spacing was increased by 6.39Å. Piperidine intercalates TaS_2 and increases the interlayer distance by 4.20Å, but the complex is not

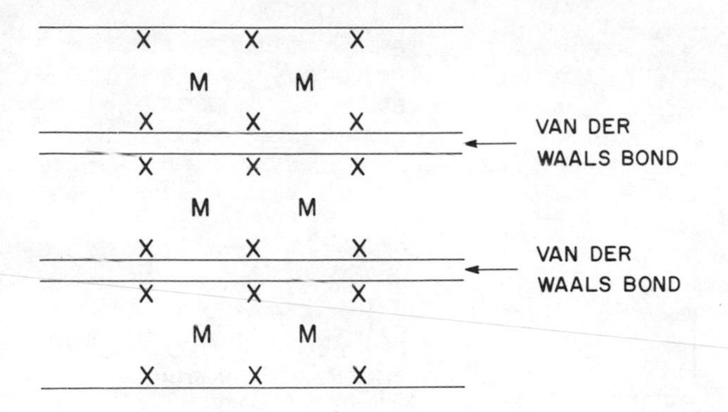

Figure 1. Schematic of structure of TaS_2

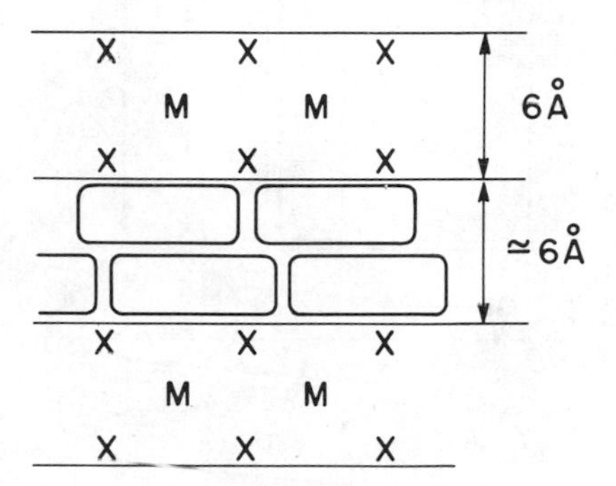

Figure 2. Schematic of structure of TaS_2 substituted pyridine complexes

TABLE I

SYNTHESIS OF TaS_2 COMPLEXES

	TEMP. (°C)	TIME (DAYS)	δ(Å)	T_0 (°K)
TaS_2	—	—	6.05	0.8
(pyridine)	200	1	5.81	3.5
$CH_3-(CH_2)_{17}-NH_2$	40	30	50	3.0

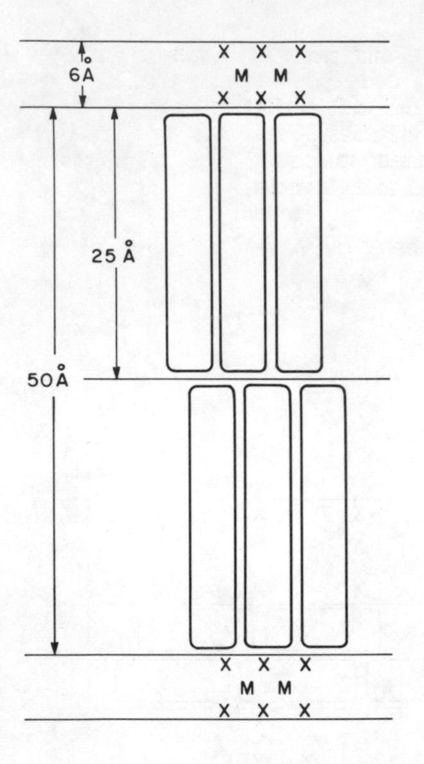

Figure 3. Schematic of structure of TaS_2 (octadecylamine)$_{2/3}$

TABLE 2
TaS_2 – AMINE COMPLEXES

INTERCALATE	SYNTHESIS TEMP. (°C)	SYNTHESIS TIME (DAYS)	δ(Å)	T_0 (°K)
CH_3CH_2 N–H CH_3CH_2	25	2	3.69	3.0
$CH_3CH_2CH_2CH_2$ N–H $CH_3CH_2CH_2CH_2$	150	3	4.02	
–CH_2 N–H –CH_2	150	4	6.39	2.7
–CH_2–N(H)(H)	25	2	6.15	2.4
N–H	150	3	4.20	NS
N N	200	8	3.65	1.8

superconducting down to 1°K. 2-Aminoanthracene failed to form a complex even at 200° for 30 days. This contrasts with the intercalation of aniline reported earlier. The inability of 2-aminoanthracene to intercalate may be related to a lesser capability to donate electrons to the unfilled metallic bands.

Substituted hydrazines readily intercalate the 2H-polytype of TaS_2 to form complexes (Table 3). The electron donor ability of the hydrazines varies sufficiently such that correlations with the critical superconducting temperature can be made. The charge density potentially available for transfer to the unfilled metallic bands is related to the asymmetric charge distribution represented by the quadrupole moment of a nucleus.

Gamble and others have suggested that the bonding of electron donors in intercalated layered-transition-metal chalcogenides involves some charge transfer from the intercalated molecule to the incompletely filled bands of the chalcogenide (9,10). Ehrenfreund, Gossard and Gamble have observed an alteration in the transition-metal Knight shift and electric field gradient which indicates that occupancy of the conduction band is changed upon intercalation (12). Transmission spectra (0.4-5.0 eV) of layered transition metal chalcogenides show shifts and broadenings upon intercalation which have been interpreted in terms of increased occupation of d bands (13). According to the Bardeen-Cooper-Schrieffer theory of superconductivity, the critical superconducting temperature can be increased by electron density enhancement at the Fermi level (14). The formation of intercalated complexes from intercalates of varying electron donor ability appears to offer a method of enhancing the electron density at the Fermi level thereby increasing the temperature for the onset of superconductivity. There appears to be a direct correlation between the enhancement of the superconducting temperature of the hydrazine-TaS_2 complexes and the atomic-orbital occupation numbers of the unsubstituted amino nitrogen atoms (Table 4). The occupation numbers are from Sauer and Bray and were determined prior to intercalation (15). The occupation number (σ) was derived from nuclear quadrupole resonance data and is expressed as the difference between the σ-charge densities of the nitrogen-hydrogen and nitrogen-nitrogen bonds. As the occupation numbers increase from 0.3429 to 0.3795, the temperature for the onset of superconductivity species increases from 2.0 to 4.9°K. Interestingly, there appears to be no correlation between T_o and pK_a of the

TABLE 3

TaS_2 – HYDRAZINE COMPLEXES

INTERCALATE	SYNTHESIS TEMP. (°C)	SYNTHESIS TIME (DAYS)	δ(Å)	T_0(°K)
H_2N-NH_2	25	2 HOURS	3.15	4.9
$(CH_3)(H)N-NH_2$	25	2 HOURS	3.53	3.3
$(CH_3)_2N-NH_2$	25	2 HOURS	3.53	4.0
$(C_6H_5)(H)N-NH_2$	25	2 HOURS	3.14	2.0
$(C_6H_5)(CH_3)N-NH_2$	25 100	14 5	13.00,6.26 6.26,4.08	2.9 2.9,2.4
(cyclic)N–NH_2	25	10	3.14,1.66	2.4,1.8
$H-C(=O)-NH-NH-C(=O)-H$	150	5	3.48,2.09	2.5,1.6

TABLE 4

CORRELATION OF PKa, CHARGE DENSITY (Δσ) AND THE TEMPERATURE FOR THE ONSET OF SUPERCONDUCTIVITY (To)

INTERCALATE	pKa	Δσ	T_0(°K)
H_2N-NH_2	7.93	.3795	4.9
$(CH_3)_2N-NH_2$	7.21	.3608	4.0
$(CH_3)(H)N-NH_2$	7.87	.3485	3.3
$(C_6H_5)(H)N-NH_2$	5.2	.3429	2.0

intercalate. The critical superconducting temperatures for the hydrazine, 1,1-dimethylhydrazine and methylhydrazine complexes are 4.9, 4.0 and 3.3°K, whereas the pK_as are 7.93, 7.21 and 7.87 respectively. This has been observed by DiSalvo for amine-TaS_2 complexes in which an increase in pK_a may produce an increase or decrease in T_o. This is confirmed by nuclear quadrupole resonance data where the electron density in the nonbonding nitrogen orbitals of hydrazones does not correlate with T_o of the intercalated complexes. Bray and Sauer have observed the same effect for pyridine-TaS_2 complexes (16).

The correlation of nuclear quadrupole resonance derived charge densities with To is complicated by the formation of two phase complexes-complexes which have two distinct interlayer distances. For example, the complex formed between TaS_2 and 1-phenyl-1-methylhydrazine has interlayer distance increases of 13.00 and 6.26 Å and the Sauer and Bray occupation number for the hydrazine is 0.388. The temperature for the onset of superconductivity for the complex should be > 4.9°K, but the transition occurs at 2.9°K.

The synthesis condition, interlayer distance increase, and the temperature for the onset of superconductivity of some TaS_2 amide complexes are included in Table 5. The intercalation of N-methylformamide under various reaction conditions produces different complexes. The complex formed at ambient temperatures had an interlayer spacing of 3.65 Å and T_o = 4.2°K. The complex formed at 100° had an interlayer spacing of 3.00 Å and was superconducting at 2.4°K. If the reaction is carried out at 90° with ultrasonic mixing of the reactants, the interlayer distance increases to ~50 Å and the complex is superconducting at 2.7°K. This evidence supports the previous observation that there is no correlation between the interlayer distance and the temperature for the onset of superconductivity (9). Di-n-butyl formamide, N-methylformanilide, and diformylhydrazine also form complexes with TaS_2. Diformylhydrazine formed a two phase complex with interlayer spacing increases of 3.48 and 2.09 Å and showed two transitions for the onset of superconductivity at 2.5 and 1.6°K. Interpretation of the results of the amide complexes is complicated by the tendency of these molecules to decompose under the reaction conditions. Elemental analysis indicates the possibility of the intercalation of more than one specie for the N-methylformamide complex. The tendency of amide intercalates to decompose has previously been observed by DiSalvo.

TABLE 5

TaS_2 – AMIDE COMPLEXES

INTERCALATE	SYNTHESIS TEMP. (°C)	SYNTHESIS TIME (DAYS)	δ(Å)	T_0 (°K)
$CH_3-N(H)-C(=O)-H$	25	30 MINS.	3.65	4.2
	100	14	3.00	2.4
	90	14	~50	2.7
$(CH_3CH_2CH_2CH_2)_2N-C(=O)-H$	150	4	3.97	
$H-C(=O)-NH-NH-C(=O)-H$	150	5	3.48, 2.09	2.5, 1.6
$C_6H_5-N(CH_3)-C(=O)-H$	200	30	6.05	

TABLE 6

TaS_2 COMLEXES EXHIBITING TWO PHASES

INTERCALATE	SYNTHESIS TEMP. (°C)	SYNTHESIS TIME (DAYS)	δ(Å)	T_0 (°K)
$C_6H_5-N(CH_3)-NH_2$	25	14	13.00, 6.26	2.9
	100	5	6.26, 4.08	2.9, 2.4
$CH_3CH_2O-P(=O)(NHCH_3)-OCH_2CH_3$	100	30	4.85	2.7
	150	30	4.85, 3.95	2.7, 1.5
	150	60	3.95	1.5
$[(CH_3)_2N]_3P=O$	100	30	13.63, 7.58	3.5

The preceding paragraph referred to the formation of a two-phase complex between TaS_2 and diformylhydrazine in which the complex had two interlayer spacings and two temperatures for the onset of superconductivity. The formation of two phase complexes appears to be fairly general and there are interesting relationships between σ, To, and reaction conditions (Table 6). Reaction of 1-methyl-1-phenylhydrazine with TaS_2 at ambient temperatures yielded a two-phase complex with interlayer distance increases of 13.00 and 6.26 Å which was superconducting at 2.9°K. Reaction at higher temperatures yielded a complex in which the 13.00 Å distance disappeared and a new distance (4.08 Å) appeared. The 6.26 Å interlayer distance increase was present in both complexes. The same appearance and disappearance of spacings was observed in the formation of diethyl methylamidophosphate complexes. The complex formed at 100° had one interlayer distance of 4.85 Å and one temperature for the onset of superconductivity. The complex formed at 150° had 3.95 Å and 1.5°K for the interlayer distance increase and superconducting transition, respectively. At intermediate reaction conditions, the complex formed had both interlayer distance increases and both superconducting transitions. From previous results the decrease in the transition temperature with a decrease in interlayer spacing cannot be attributed solely to the interlayer distance change.

It seems reasonable that the increased temperatures and reaction times allow the intercalate to reach an equilibrium orientation where electron donation to the unfilled metallic bands is less efficient, thereby causing a decrease in the transition temperature.

Summary

A number of organic intercalated complexes of the 2H-phase of TaS_2 have been prepared which have increased interlayer spacings and enhanced critical superconducting temperatures. There is a direct correlation between the nuclear quadrupole resonance derived charge densities of the intercalate with the enhanced superconducting temperatures of hydrazine complexes. Some organic intercalates form two phase complexes which may have one or two temperatures for the onset of superconductivity.

Literature Cited

1. Voorhoeve, J. M. and Robbins, M., J. Solid State Chem., (1970), 1, 134.
2. Huisman, R., Kadijk, F., and Jellinek, F., J. Less-Common Metals, (1970), 21, 187.
3. Omloo, W. P. F. A. M., and Jellinek, F., J. Less-Common Metals, (1970), 20, 121.
4. Rudorff, W., Chimia (1965), 19, 489.
5. Rudorff, W. and Sich, H. H., Angew. Chem., (1959), 71, 724.
6. Freeman, A. G. and Johnston, J. H., Carbon (Oxford), (1971), 9, 667.
7. Grim, R. E., "Clay Minerology," McGraw-Hill, New York, N.Y., 1968.
8. Weiss, A. and Rudhardt, R., Z. Naturforsch., B (1969), 24, 265, 355, 1066.
9. Gamble, F. R., DiSalvo, F. J., Klemm, R. A., and Geballe, T. H., Science (1970), 168, 568.
10. Gamble, F. R., Osiecki, J. H., Cais, M., Pisharody, R., DiSalvo, F. J., and Geballe, T. H., Science, (1971), 174, 493.
11. DiSalvo, F. J., Schwall, R., Geballe, T. H., Gamble, F. R., and Osiecki, J. H., Phys. Rev. Lett., (1971), 27, 310.
12. Ehrenfreund, E., Gossard, A. C., and Gamble, F. R., J. Phys. Rev. B (1972) 5, 1708.
13. Beal, A. R. and Liang, W. Y., Phil. Mag. (1973), 27, 1397.
14. Bardeen, J., Cooper, L. N. and Schrieffer, J. R., Phys. Rev., (1957), 108, 1175.
15. Sauer, Enrique G. and Bray, P. J., J. Chem. Phys., (1972), 56, 820.

16. Bray, P. J. and Sauer, Enrique G., Solid State Communications, (1972), 11, 1239.

Acknowledgment

The authors acknowledge helpful discussions with F. J. DiSalvo, Jr., G. W. Hull, T. H. Geballe, D. W. Murphy and F. J. Padden, Jr. We particularly thank G. W. Hull for the superconductivity data.

4

Intermolecular Magnetic Exchange Interactions in Molecular Crystals

J. M. GROW and H. H. WICKMAN

Oregon State University, Corvallis, Ore. 97331

Abstract

Intermolecular exchange interactions have been studied in a representative, homologous series of metal complexes: iron(III) bis (dithiocarbamates). Several of these molecular crystals exhibit cooperative magnetic order at temperatures below approximately 4 °K. Exchange paths include weak interactions between formally non-bonded molecular units. The magnetic phenomena are sensitive to zero-field splittings within the orbital-singlet, spin-quartet ground term of iron(III).

Discussion

Molecules containing a single paramagnetic ion surrounded by bulky, organic ligands often form magnetically dilute molecular crystals. For metal-metal distances greater than about 7 Å, the dipolar fields are significant as an origin of magnetic order only at temperatures below ~0.01 °K. Hence, most such metal complexes remain simple paramagnets to quite low temperatures. However, several homologues of the iron(III) halo-bis-dithiocarbamates order magnetically at temperatures in the range 1-4 °K (1-3). We summarize here certain structural or crystallographic features of these complexes which appear to be important determinants for exchange paths in these molecular crystals. In addition, we note that variable ground term crystal field interactions may enhance or depress the cooperative transitions for a fixed crystal structure or set of exchange paths.

The structural formula for an iron(III) bis-dtc is given in Fig. 1. Much of the crystallographic information on bis-dtcs has been reported by Hoskins,

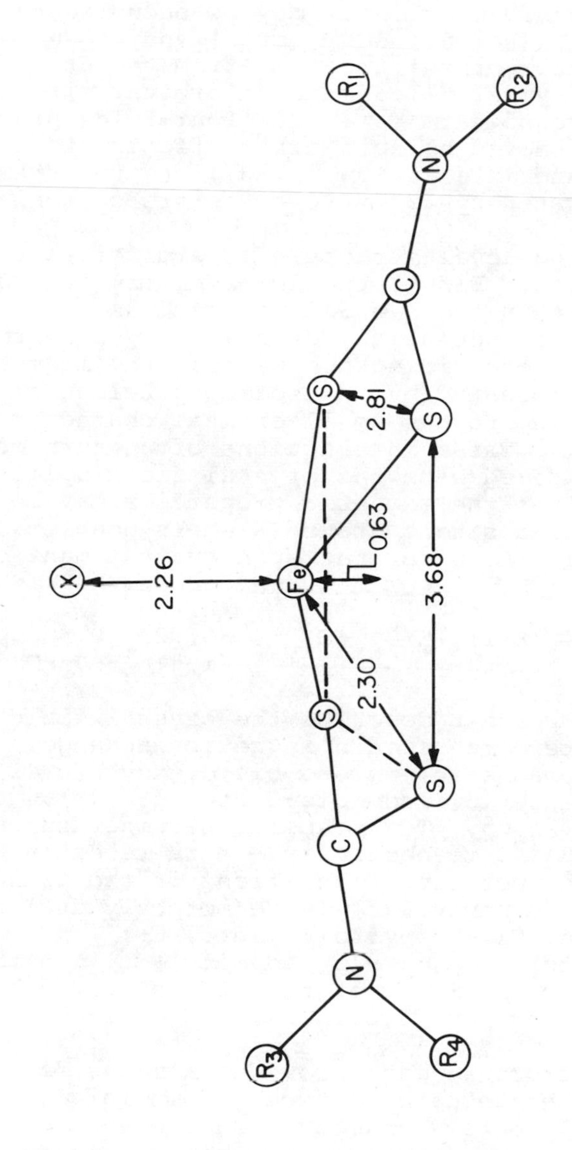

Figure 1. Structural formula of a halobisdithiocarbamatoiron(III) complex

Martin, White, and co-workers (4-6). For example, the dimensions given in the Figure are for the diethyl-chloro derivative (R_i = ethyl, i=1,4; X = Cl) (5). The iron(III) is situated near the centroid of a rectangular pyramid whose apex may be occupied by a monovalent anion (halide, SCN^-, etc.) and whose base consists of 2 pair of sulfur atoms from the dtc ligands. This symmetry leads to an orbital-singlet, spin-quartet ground term (4). The four-fold spin degeneracy is removed by spin-orbit effects, together with metal-ligand interactions (7-8), and two Kramers doublets, separated by ~2 to 20 cm^{-1}, are commonly observed (9).

The foregoing level structure is attractive for magnetism studies. First, the low-symmetry environment is felt within the S = 3/2 manifold as a second-rank crystal field potential, described by only two spin-Hamiltonian coefficients. Second, the magnetic exchange is represented by an isotropic Heisenberg interaction, owing to the small orbital character of the ground term. These interactions often turn out to be small compared with the crystal field splittings; the result is that the magnetic properties may be discussed within a single ground Kramers doublet. The well-known Hamiltonian for the spin-quartet manifold is

$$\hat{\mathcal{H}} = \sum_i \{D[\hat{S}_{iz}^2 - \tfrac{1}{3}S(S+1)] + E(\hat{S}_{ix}^2 - \hat{S}_{iy}^2)\} - \sum_{(i,j)} 2J_{ij}\vec{S}_i \cdot \vec{S}_i . \quad (1)$$

The parameters D and E describe the crystal field and the remaining term represents magnetic exchange.

Both ferromagnetic, chloro-bis-diethyldtc (1), and antiferromagnetic, iodo-bis-diethyldtc (3), homologues are known. The transition temperatures are 2.48 °K and 1.9 °K, respectively. More recently we have observed cooperative transitions in the following complexes: $Fe(morphylyldtc)_2I$, $Fe(morphylyldtc)_2Br$, $Fe(morphylyldtc)_2Cl$, $Fe(pyrrolidyldtc)_2Cl$, $Fe(pyriolidyldtc)_2I$, and $Fe(pyridyldtc)_2Cl$. In these complexes, the magnetic order has been observed by Mossbauer spectroscopy (1,2). All transitions occur below 4 °K.

Crystal structure data provides a means for identifying exchange paths between atoms in adjacent molecules. Only a few structures have been reported, but included are the ferromagnetic and antiferromagnetic derivatives mentioned above (5-6). A detailed discussion of the orientations of the molecules in connection with exchange paths is available elsewhere (3,5,6). We have recently obtained structural information on the SCN-bis-diethyldtc, bromo-bis-diethyldtc, and isodo-bis-morpholyldtc-toluene adduct. The latter exhibits a magnetic transition at approximately 2 °K.

The following picture emerges from a study of the structural information. As expected, iron-iron distances in all complexes are in excess of 7 Å. In those complexes whose structure is known and which show a magnetic transition, or exchange broadened epr spectra, there exists a juxtaposition of molecules such that there is an intermolecular S•••S linkage. That is, we may write the intermolecular interaction as Fe-S•••S-Fe, with the Fe-S units belonging to distinct molecules. The S•••S distance is typically 3.6 to 4.5 Å. The exchange often propagates through the lattice in rather complicated paths: however, simple dimer interactions may also occur. Because of the packing characteristics of the particular lattice determined thus far, we have not observed molecules arranged so that the potential strong exchange path Fe-X•••Fe-X is favored. We have also not observed packing which produces sandwiched dtc ligands, such as overlapping aromatic ring structures. Our pmr shifts (3) show a delocalization of the iron spin density to the dtc periphery, but the major delocalization is to the immediate iron ligands. This is of course consistent with the Fe-S•••S-Fe type exchange paths.

It is interesting that the existence of substantial exchange interactions is a necessary but not sufficient condition for magnetic order in the temperature range of our experiments, $T \gtrsim 1°K$. This may be seen by considering the effect of ground Kramers level anisotropies on the collective magnetic properties of the complexes. That is, we consider a situation where a fixed intermolecular exchange exists between molecules in two isomorphons lattices. The molecular lattices are assumed to differ only in the type of splitting that occurs in the ground $S = 3/2$ manifold. In Eq. (1), if D is large in magnitude and dominates the rhombic crystal field and the exchange term, we may rewrite the exchange interaction within the ground Kramers doublet using an effective spin $S = 1/2$ (low temperature approximation). This illustrates the magnetic anisotropy of the original $|S = 3/2, M_S = \pm 3/2\rangle$ Kramers doublet. The result is

$$\hat{\mathcal{H}}_I = -2 \sum_{(i,j)} 9J_{ij}\hat{S}_{iz}\hat{S}_{iz} \qquad (2)$$

This is the usual Ising Hamiltonian (E is neglected). Upon changing the sign of D, but maintaining the relative magnitudes of the interactions, the alternative ground doublet yields the anisotropic Heisenberg interaction

$$\hat{\mathcal{H}}_A = -2 \sum_{(i,j)} J_{ij}[\hat{S}_{iz}\hat{S}_{jz}+4(\hat{S}_{ix}\hat{S}_{jx}+\hat{S}_{iy}\hat{S}_{jy})]. \quad (3)$$

It is easily shown (10) that in the molecular field approximation the ratio of magnetic transition temperatures with and without the crystal field D term (exchange forces constant), is given by the expression

$$T_c(D)/T_c(O) = 3 \langle S_z^2 \rangle_{T_c} /S(S+1) \quad (4)$$

In Eq. (4), $\langle S_z^2 \rangle_{T_c}$ is the ionic expectation value of $\hat{S}_z^2$ at the magnetic transition temperature. For $D \to +\infty$ and our low temperature approximation with H_{exch} parallel to the z-axis of Eqs. (1)-(3), the ratio is

$$T_c(D)/T_c(O) = 1/5. \quad (5)$$

This represents a depression of the ordering temperature due to crystal field effects. Conversely, as $D \to -\infty$, we find

$$T_c(D)/T_c(O) = 9/5. \quad (6)$$

Thus the transition temperature is enhanced. In either case, however, the crystal field induced anisotropies are important factors in the occurence of magnetic order in a bis-dtc. This point becomes important when it is necessary to assess accurately the magnitude of exchange constants in Eq. (1) while working with systems in which only a single Kramers level is occupied (3).

Ideally, it is desirable to know in some detail the spatial distribution of spin density in the periphery of a molecule, since this in large measure dominates intermolecular exchange. A single, static orientation of molecules in a crystal lattice does not provide this information. Fortunately, molecular crystals often exhibit polymorphism in their crystal habits, or they may form several solvent adduct structures. We have recently found that such is the case in $Fe(morphylyldtc)_2I$. At least five polymorphs and/or solvent adducts have been isolated; each exhibits somewhat different magnetic properties. Determination of the crystal structures is in progress. As the relative orientations of molecules are correlated with magnetic properties, we expect a further extension of our understanding of factors important for magnetic exchange in Van der Waals' lattices.

Literature Cited

1. Wickman, H. H., Trozzolo, A. M., Williams, H. J., Hull, G. W., and Merritt, F. R., Phys. Rev. (1967) 155, 563.
2. Kostikas, A., Petridis, D., Simopoulos, A., and Pasternak, M., Solid State Comm. (1973) 13, 1661.
3. Chapps, G. E., McCann, S. W., Wickman, H. H., and Sherwood, R. C., J. Chem. Phys. (1974) 60, 990.
4. Hoskins, B. F., Martin, R. L., and White, A. H., Nature (1966) 211, 627.
5. Hoskins, B. F., and White, A. H., J. Chem. Soc. (A) (1970), 1668.
6. Healy, P. C., White, A. H., and Hoskins, B. F., J. Chem. Soc. (A) (1972), 1369.
7. Ake, R. L., and Harris-Loew, G. M., J. Chem. Phys. (1970) 52, 1098.
8. de Vries, J. L. K. E., Keijzers, C. P., and de Boer, E., Inorg. Chem. (1972) 11, 1343.
9. Brackett, G. C., Richards, P. L., and Caughy, W. S. Jr., J. Chem. Phys. (1971) 54, 4383.
10. Lines, M. E., Phys. Rev. (1967) 156, 534.

5

Computation of Field-Swept EPR Spectra for Systems with Large Interelectronic Interactions

R. L. BELFORD, P. H. DAVIS, G. G. BELFORD, and T. M. LENHARDT

Center for Advanced Computation, University of Illinois, Urbana-Champaign, Ill. 61801

Abstract. The computation of field-swept epr spectra from a physical model often is a particular problem for metal systems having large interelectronic interactions, because of the large field range covered. This contribution points out that theoretical formulations can be devised specifically for this purpose and sketches the development and use of two such methods - one exact and one approximate.

1. Introduction

As several other contributions to this volume attest, electron paramagnetic resonance is among the most important tools for studying electronic interactions between metal centers. Large zero-field splittings, leading to epr transitions over a wide range of magnetic field in the typical fixed-frequency spectrum, characterize many high-spin multicenter systems. In these cases, computation of predicted epr spectra from trial Hamiltonian parameters is complicated by the fact that the Hamiltonian is a function of the spectroscopic sweep variable, the magnetic field. Suppose all the Hamiltonian parameters which characterize a system are given. Then the problem is to find all values of the magnetic field, x, such that two eigenvalues of the Hamiltonian differ by the spectrometer energy, w. In the usual case of linear Zeeman terms, a system of n basis states may give rise to as many as $n(n-1)/2$ different transition fields. Although the two energy eigenvectors associated with each transition field are orthogonal, neither need be orthogonal to any of the other n^2-n-2 eigenvectors involved in transitions at the other fields. Various methods to treat this problem are commonly employed (1). One example is construction of a map of energy levels as functions of field by eigensystem solution on a fine magnetic field grid (2). Another is the conventional perturbation treatment in which the magnetic field (or field shift) is the perturbation parameter and the energy levels and thus frequency are developed as power series in the field (shift); the frequency series must be terminated and the correct root of the polynomial w(x)-w extracted (3,4). While all these methods can be made to work, and some do very well in some cases, they are generally adapted with minimal modification from techniques appropriate to calculation of energy levels for systems characterized by fixed Hamiltonians. As such, one might expect them to be generally less efficient or less precise and direct than methods especially developed to handle the fixed-frequency field-swept problem. We have begun to look into the possibility of devising such methods, and here we outline two of them. The first, the eigenfield formulation (1), displays all the transition fields as the real eigenvalues, x, of a generalized eigenvalue equation AZ=xBZ. Section 2 discusses the eigenfield technique briefly; it has been described in some detail recently (1) as has its application to time-dependent slow-motion or exciton hopping problems (5). The eigenfield approach is exact and straightforward but slow for problems with many basis states. The second method, developed in Section 3, is a perturbation formulation devised to predict transition fields.

2. Eigenfield Equations for Exact Calculation

A recent paper (1) describes a new (generalized eigensystem) formulation for exact calculation of resonance fields and intensities for a molecule or lattice site described by a time-independent Hamiltonian which is a polynomial function of field; it also describes some properties of this formulation, which we call the eigenfield equation. The eigenfield concept can also be used to facilitate solution of relaxation master equations, which provide intensity as a function of field for molecules in slow motion or for exciton hopping among lattice sites in a crystal (5). Here, we restrict our discussion to a simplified development of the static eigenfield equation for systems with linear Zeeman terms and to a practical application to a system of interacting transition metal ions. As we shall see, the eigenfield approach is very attractive because it is a direct, straightforward, exact method, but at present it has the drawback that computation time increases much too rapidly with the number of basis states. Accordingly, until better numerical techniques are developed for computer solution of the eigenfield equations, the method is of practical use only for systems having just a few (~10) basis states. It does provide resonance fields and transition vectors of high accuracy which we have employed as test standards of excellence for approximate methods.

2.1 Development of Eigenfield Equation. For convenience, we use the matrix representation throughout. Matrix operators, vectors, and scalars are distinguished by context, not by special symbols. The Hamiltonian, H, for the problem is the sum of two parts -- a zero-field term, F, (exchange coupling, dipolar and pseudodipolar coupling, hyperfine coupling, etc.) and a Zeeman term, xG (electronic and nuclear Zeeman energies). Here x denotes the magnitude of the magnetic field. For example, the coupling of two Cr^{+++} ions ($S=3/2$) may be described by a spin-Hamiltonian $H=F+xG$ in which $F=D(S_{1z}^2+S_{2z}^2-5/2)+\vec{S}_{1z}\cdot\overleftrightarrow{d}\cdot\vec{S}_{2z}+J\vec{S}_1\cdot\vec{S}_2$ and $G=\beta\vec{e}\cdot\overleftrightarrow{g}\cdot\vec{S}$, where $\vec{e}$ denotes a unit vector in the direction of the magnetic field.

Now for a fixed spectrometer energy, w, resonance lines will only occur between two states, $u\rightarrow v$, when the following two conditions are satisfied together:

$$\text{[2-1]} \qquad \begin{aligned} (F+xG-\epsilon)u&=0 \ , \\ (F+xG-\epsilon-w)v&=0 \end{aligned}$$

That is, if ϵ is the energy of the lower state, $\epsilon+w$ must be the energy of the upper state. The two coupled equations can be converted into a single one which contains only one eigenvalue, x, by means of outer-product algebra. The outer product, $A\otimes B$, of an $m\text{X}n$ matrix A and an $m'\text{X}n'$ matrix B is an $mm'\text{X}nn'$ matrix: $(A\otimes B)_{i,i';k,k'}=(A\otimes B)_{m'i-m'+i',n'k-n'+k'}=A_{ik}B_{i'k'}$. The outer product operation has the following useful properties: (1) it is associative; (2) $A\otimes B$ is Hermitian if and only if A and B are both Hermitian; and (3) $(A\otimes B)(C\otimes D)=(AC)\otimes(BD)$ (6). The two equations [2-1] can be outer-multiplied on the right and left, respectively, by Iv and Iu (I denotes a unit matrix), converted by use of property (3) of the outer product, and subtracted to yield the eigenfield equation:

$$\text{[2-2]} \qquad (F\otimes I-I\otimes F+wI\otimes I)(u\otimes v) = x(I\otimes G-G\otimes I)(u\otimes v) \quad .$$

2.2 Properties of Eigenfield Equation. Here we summarize some of the more important properties of [2-2] and its solutions; a more detailed description is found in ref. 1. Equation [2-2] is a generalized eigenvalue equation of the form $AZ=xBZ$, where the operators A and B are Hermitian matrices of order n^2 (for H of order n) and the eigenvectors, Z, are transition vectors $u\otimes v$ of dimension n^2. Accordingly, there are n^2 eigenfield values. The form of B insures that it is at least n-fold singular; therefore, n of the eigenvalues are infinite. They correspond to unrealizable transitions from an energy level to itself. In general, if the eigenvalues of the Zeeman operator, G, are grouped into degenerate sets, B will have singularity equal to the sum of squares of those degeneracies (Σm^2); therefore there will be ($n^2-\Sigma m^2$) noninfinite eigenfields (1). Normally, these will occur in pairs of equal magnitude but opposite sign, reflecting the physical fact that reversal of the magnetic field direction does not change the resonance spectrum (cf. ref. 1 concerning the time-reversal symmetry of F and G). Thus, there are

$(n^2-\Sigma m^2)/2$ physically distinct transitions. Moreover, A is positive-definite (i.e., all of its eigenvalues are greater than 0) if and only if the spectrometer energy is greater than the total span of the zero-field energy levels. Only in this case are all eigenfield values guaranteed to be real. Obviously, complex eigenfield solutions, like the infinite ones, are physically spurious; thus there actually may be fewer transitions than $(n^2-\Sigma m^2)/2$. Intensities are readily obtained from eigenfield transition vectors (1).

2.3 Example of Eigenfield Equation. The eigenfield method yields all transition fields and intensities with no substantial difficulties; it is effectively automatic, and we have been using it routinely for small problems (e.g., triplet states, S=1; coupled doublets -- i.e., binuclear Cu(II) sites, $S_1=S_2=1/2$; quartets, S=3/2) both for generation of fixed-orientation spectra and for simulation of powder spectra. Since the work and storage requirements increase very rapidly with size of basis set, the largest problem which we have yet handled by direct solution of the eigenfield equations is that of a coupled Cr^{+++}-Cr^{+++} pair, for which F and G are 16x16 matrices and A and B consequently are 256x256 matrices. That will be the example described here. We take two identical S=3/2 ions on the z axis with isotropic Zeeman terms and axial inter- and intra- ion zero-field terms. The pair basis set can be constructed as the outer product of basis sets for the two S=3/2 particles, and the 16x16 pair Hamiltonian as the outer product of 4x4 single-particle factors. Therefore, we can set up the entire problem in terms of 4x4 matrices. Our example is as follows:

[2-3]

$$F=D(S_z{}^2 \otimes I+I \otimes S_z{}^2-2.5I \otimes I)+d(2S_z \otimes S_z-S_x \otimes S_x-S_y \otimes S_y)$$

[2-4] $$G=g\beta(\cos\theta[S_z \otimes I+I \otimes S_z]+\sin\theta[S_x \otimes I+I \otimes S_x])$$

The magnetic field is in the xz plane at an angle θ to the z axis. Here I is the 4x4 unit matrix, and S_z is a diagonal matrix with values (-3/2,-1/2,+1/2,+3/2), and S_x and S_y are as follows:

[2-5] $$S_x=1/2\begin{pmatrix} 0 & \sqrt{3} & 0 & 0 \\ \sqrt{3} & 0 & 2 & 0 \\ 0 & 2 & 0 & \sqrt{3} \\ 0 & 0 & \sqrt{3} & 0 \end{pmatrix}; \quad S_y=\frac{\sqrt{-1}}{2}\begin{pmatrix} 0 & \sqrt{3} & 0 & 0 \\ -\sqrt{3} & 0 & 2 & 0 \\ 0 & -2 & 0 & \sqrt{3} \\ 0 & 0 & -\sqrt{3} & 0 \end{pmatrix}$$

Parameters for the three tests at θ=1.5°, 51°, and 88.5° were

w/β=6802.6523 gauss, g=1.9759, D/β=1006.25 gauss, and d/β= -107.74 gauss.

Now F and G are readily (and in a computer program, quite easily) constructed, as are A and B. For example, the first term of F leads to a diagonal matrix contribution to A, $D(S_Z{}^2 \otimes I \otimes I \otimes I - I \otimes I \otimes S_Z{}^2 \otimes I)$, its diagonal elements being (D/4) (64 9's, 128 1's, 64 9's) - (D/4) (16 repeats of the pattern 4 9's, 8 1's, 4 9's).

Note here that G has 2 nondegenerate eigenvalues ($\pm 3g\beta$), 2 twofold ($\pm 2g\beta$), 2 threefold ($\pm g\beta$), and 1 fourfold (0) degenerate sets. Thus there must be exactly 2+8+18+16=44 infinite fields. The computer program yielded 44 fields between 3×10^{17} and 3×10^{19}in absolute value. In no case was the computed intensity for one of these 'infinite-field' transitions as great as 10^{-28} (strongest line=1). Also, since for our test cases D and d are sufficiently small compared with w ($w > |9D/2| + |9d|$) so that F is positive-definite, there should be exactly (256-44)/2=106 real positive eigenfields and 106 negative images. The computer results also showed this feature, with the plus-minus pairs generally agreeing to 14 digits. At $\theta = 51^0$ the 106 eigenfields ranged from 404 to 3716 gauss.

A useful feature of the eigenfield program is that it produces field values and intensities for all lines at once. Here, it is interesting to see that several lines are predicted in the 400-900 gauss region with intensities ~.01 to .03% of those of the most strongly allowed lines and ~1-3% of those of the weaker lines in the 1000 gauss region. These are essentially double-spin-flip transitions involving excitation of one ion and de-excitation of its partner; under some circumstances they could be observed experimentally and would be sensitive to the intermolecular coupling. Most approximation methods would provide only the 'normal' lines, with the result that experimentalists might ignore the abnormal ones altogether.

The computer program used to solve this generalized eigenvalue problem was written by A. Sameh and C. Chang of the University of Illinois' Center for Advanced Computation. Their method ("SQZ") is a version of the QZ algorithm of Moler and Stewart (7), which we have used (1) for smaller problems. The SQZ algorithm (8) requires that A and B both be real, symmetric matrices (with one of them positive definite) and then takes advantage of symmetry to generate solutions more rapidly (and often more accurately) than QZ. Even then, the example discussed here required about ten minutes of processing time on an IBM 360/91 computer, whereas a system of two coupled ions of S=1/2 each would require only a small fraction of a second. The space requirements were also formidable, the example given requiring more than one million bytes. The reason for the time and space difficulties is that the current state of numerical analysis of generalized eigenvalue problems is as yet quite primitive; neither QZ nor SQZ takes any advantage of the sparseness or special form of A and B. Advent of a numerical method which does so will make the eigenfield formulation widely practical for reasonably large systems.

A measure of the goodness of the eigensystem is the residual R_k associated with each eigenfield x_k. The program calculated for each field a residual $R_k \equiv \frac{\max_j [(A-x_k B)Z_k]_j}{||A|| + |x_k|\ ||B||}$, where $||A|| \equiv \max_\ell \Sigma_p |A_{\ell p}|$, and the normalization of Z_k is chosen so that $0 \leq R_k \leq 1$. An exact solution must give a zero residual, and a good solution cannot give a large residual. The residuals were all less than 10^{-10}, many of them less than 10^{-17}. We suggest that these residuals could be calculated to check on accuracy even when some other method (e.g., perturbation technique) is used to obtain the transition fields and associated vectors.

3. Frequency-Shift Perturbation

3.1 Approach. The Hamiltonian, $H(x)$, is a prescribed function of the field, x. The spectrometer frequency, w, is given. We choose a starting approximation x^0 and find the eigensystem of $H(x^0)$ to obtain zero'th order states of origin and termination, u^0 and v^0, energies, μ and ν, and frequencies, $w^0=\nu-\mu$. Then we regard the frequency shifts, $\delta w \equiv w-w^0$, as perturbations. That is, to any initially calculated frequency we add a perturbation $\lambda\delta w$ where λ is a perturbation parameter and develop a perturbation series for the field: $x=x^0+\lambda x'+\lambda^2 x''+...$ As in the standard versions of perturbation theory, initial degeneracies cause complications and must be handled carefully. The approach described here allows perturbations in other parts of the Hamiltonian -- for example, nuclear hyperfine, quadrupole, and Zeeman terms -- to be treated along with the frequency shift.

3.2 Nondegenerate Case. As before, let $H=F+xG$, where F is the zero-field operator and G the Zeeman operator. Let the n'th prime on any symbol denote its n'th order contribution. Then the following relations hold.

$$H=\sum_0^\infty \lambda^n H^{(n)};\ H^0=F^0+x^0G^0,\ H'=F'+x^0G'+x'G^0, \qquad [3\text{-}1]$$
$$H^{(n)}=x^{(n)}G^0+x^{(n-1)}G',\ n\geq 2$$

Notice that we have assumed $F=F^0+\lambda F'$ and $G=G^0+\lambda G'$; higher order terms may be included if desired.

$$Hu=\epsilon_u u \text{ and } Hv=\epsilon_v v;\ \epsilon_v-\epsilon_u=w-(1-\lambda)\delta w \qquad [3\text{-}2]$$

$$u=\sum_n u^{(n)}\lambda^n,\ v=\sum_n v^{(n)}\lambda^n \qquad [3\text{-}3]$$

[3-4] $$\varepsilon_u = \mu + \lambda\mu' + \sum_{n>1} \lambda^n \varepsilon^{(n)}$$

[3-5] $$\varepsilon_v = (\mu + w^0) + \lambda(\mu' + \delta w) + \sum_{n>1} \lambda^n \varepsilon^{(n)}$$

The difference between the usual sort of perturbation treatment and this one is simply that here we let x have corrections to all orders as forced by the constraint that w has only a first order correction, while normally one introduces all the corrections to x as a first-order term, thereby causing w to have contributions in all orders.

Now expansion of [3-2] and isolation of the coefficient of λ^n for each n leads to the following.

[3-6] $$\begin{aligned}(H^0-\mu)u' + (H'-\mu')u^0 &= 0\\ (H^0-\mu-w^0)v' + (H'-\mu'-\delta w)v^0 &= 0\end{aligned}$$

[3-7] $$\begin{aligned}(H^0-\mu)u'' + (H'-\mu')u' + (H''-\varepsilon'')u^0 &= 0\\ (H^0-\mu-w^0)v'' + (H'-\mu'-\delta w)v' + (H''-\varepsilon'')v^0 &= 0\end{aligned}$$

In general, for $n \geq 2$, we have

[3-8] $$\begin{aligned}(H^0-\mu)u^{(n)} + (H'-\mu')u^{(n-1)} + \sum_{j=2}^{n}(H^{(j)}-\varepsilon^{(j)})u^{(n-j)} &= 0\\ (H^0-\mu-w^0)v^{(n)} + (H'-\mu'-\delta w)v^{(n-1)} + \sum_{j=2}^{n}(H^{(j)}-\varepsilon^{(j)})v^{(n-j)} &= 0\end{aligned}$$

Each order is represented by a coupled pair of equations. Taking the inner product of the first equation in each pair with u^0 and the second with v^0, subtracting, and rearranging with the aid of [3-1], provide equations for the successive corrections to the field. For convenience, we let $(u^{(n)},u^0) = (v^{(n)},v^0) = \delta_{no}$.

[3-9] $$x' = [\delta w + (F'_{uu} - F'_{vv}) + x^0(G'_{uu} - G'_{vv})]/(G^0_{vv} - G^0_{uu})$$

[3-10]
$$x'' = \frac{F'_{u'u} - F'_{v'v} + x^0(G'_{u'u} - G'_{v'v}) + x'(G^0_{u'u} - G^0_{v'v} + G'_{uu} - G'_{vv})}{(G^0_{vv} - G^0_{uu})}$$

etc. Here we use the condensed notation $(u', Gu^0) = G_{u'u}$, etc.

The complete prescription for a perturbation solution is now clear. One first calculates $(u^0, v^0, \mu, w^0, \delta w)$ by exact diagonalization of H^0. Then one finds x' by [3-9], μ' from one of the equations which is a precursor to [3-9], and u' and v' from [3-6]. The process can continue to any order. It is also possible, as in the standard perturbation developments, to derive the first 2n+1 corrections to the eigenfield from the first n corrections to the eigenvectors.

A good strategy of calculation is to work in the basis in which H^0 is diagonal; then many of the relationships take a simple form. In our first programs using frequency-shift perturbations, we start by applying to all matrices in the problem the similarity transformation which diagonalizes H^0. From this point until the end (where eigenvectors are transformed back into the original basis for output) we work in this basis and find that the programs thereby can be made very economical. We have programmed the method up to fourth order in values and third order in vectors.

<u>3.3 Example of Frequency - Shift Perturbation and Comparison with Eigenfield Results.</u> We have used a frequency-shift perturbation program to analyze the Cr^{+++}-Cr^{+++} pair spectra described in the next paper in this volume (9). The eigenfield examples described in Section 2.3 provided standards of comparison for the frequency-shift perturbation program. For these studies, only the frequency shift was taken as the perturbation -- i.e., $F'=G'=0$. No actual degeneracies of H^0 caused trouble. The diagonalizations of H^0 used the widely available EISPAC matrix eigenvalue package. Generally, the perturbation program, up through fourth-order field corrections, requires only about 1% of the processing time expended by the eigenfield program, and it can be made more efficient. It also requires only a fraction (~10%) of the storage space used by the eigenfield program.

For the most part, agreement between the exact (eigenfield) and the perturbation fields was good to 6, 7, or 8 significant digits (i.e., the differences were ~10^{-3} gauss or less). A typical result is given below; $\theta=51^\circ$:

$x^0\equiv 1022.94$, $x'=59.5029$, $x''=-0.1604$, $x'''=0.0066$, $x''''=-0.0003$
$x=1082.288879$. (Compare exact=1082.288891)
Intensity: 0.1013 (Compare exact=0.1013)

Final vectors (all in agreement with the exact result to the four places printed):

u=(.0720, .0672, .1350, .2371, .0672, -.0938, -.1747, -.4974, .1350, -.1747, .1826, .3233, .2371, -.4974, .3233, -.1728)

v=(.0026, .0086, .0256, .0612, .0086, .0259, .0707, .1571, .0256, .0707, .1812, .3827, .0612, .1571, .3827, .7780)

In almost all cases, the results could be improved by a very simple, rapid extrapolation of the third and fourth order field corrections toward infinite order. The idea behind the extrapolation is that the successive perturbations may approximate a geometric sequence, i.e., $x^{(j+1)}=rx^{(j)}$ where r is nearly constant for $j \geq 3$. By taking x''''/x''' to be this ratio, the perturbation series $x^{(5)}+x^{(6)}+\ldots$ sums to $(x'''')^2/(x'''-x'''')$. Some examples follow:

$$\Sigma_0^4 x^{(j)} = 1022.94+173.0747-2.0890+0.7314-0.2216=1194.4355$$
$$x(\text{extrapolated})=1194.4870 \qquad x(\text{exact})=1194.4857.$$

$$\Sigma_0^4 x^{(j)} = 1022.94+97.7430-0.6986+0.0915-0.0157=1120.0602$$
$$x(\text{extrapolated})=1120.0625 \qquad x(\text{exact})=1120.0626.$$

$$\Sigma_0^4 x^{(j)} = 1022.940+407.033+3.635-15.536+10.061=1428.133$$
$$x(\text{extrapolated})=1424.178 \qquad x(\text{exact})=1425.149.$$

The last case in the previous paragraph illustrates a slow-convergence problem which occurred occasionally and which was in a few cases manifested by divergence of the perturbation series. These cases occurred when the character of one or both of the state vectors involved in a transition was changing very rapidly with field, so that the range of convergence of the perturbation method was pathologically small -- i.e., only a small δw could be tolerated. This tends to happen close to the crossing point of two interacting states. We suffer no severe difficulties from this problem, because it is easy to recognize such cases and to correct them by trying a new x^0. However, for a practical self-contained program, one would need to devise an algorithm for automatic recognition and adjustment. Incidentally, in the pathological cases the vectors and intensities are far worse than the fields. As expected, here the approximate method, though faster, is much less foolproof than the exact (eigenfield) method.

In some cases, the same transition was computed starting from two or three rather distant values of x^0. In most cases, the agreement was reasonable. Of course, large field shifts from x^0 produced less accurate results than small shifts, the pathological cases mentioned above excepted. Some comparisons follow: For $\theta=88.5^0$, $x(\text{exact})=2858.5430$, intensity=1.0007. Three perturbation trials from different starting fields produced

$$\Sigma_0^4 x^{(j)} = 2200+655.8418+3.7346-1.3842+0.4590=2858.6512$$
$$x(\text{extrapolated})=2858.5269; \text{ Intensity}=0.9985$$

$$\Sigma_0^4 x^{(j)} = 2900-41.4622+0.0051+0.0001+0.0000=2858.5430$$
$$x(\text{extrapolated})=2858.5430; \text{ Intensity}=1.0000$$

$$\Sigma_0^4 x^{(j)} = 3600-742.3981+0.6875+0.1904+0.0483=2858.5281$$
$$x(\text{extrapolated})=2858.5445; \text{ Intensity}=0.9990$$

All the third-order vectors agreed to within a few parts in the fourth decimal place for each vector component; in each case the transition arose from the 5th→10th levels of the starting Hamiltonian.

3.4 Degenerate States. Suppose that a field position must be computed from an initially degenerate set of transitions. Consider $\{u_1^0,\ldots,u_p^0\}\rightarrow\{v_1^0,\ldots,v_q^0\}$, where the set $\{u^0\}$ is a p-fold degenerate set of eigenstates of H^0 and $\{v^0\}$ a q-fold set.

$$H^0u_i^0=\mu u_i^0\ ;\ H^0v_j^0=(\mu+w^0)v_j^0 \qquad [3\text{-}11]$$

There are several ways to proceed. Normally, a degenerate perturbation treatment would call for diagonalization of H' within degenerate blocks of H^0. However, that is not always possible here, a priori, because H' is $F'+x^0G'+x'G^0$, in which x' is not yet known. The desired separation can be obtained by diagonalizing both the $\{u^0\}$ and the $\{v^0\}$ sub-blocks of $F'+x^0G'$ and then diagonalizing G^0 within any remaining degenerate blocks. If any degeneracies persist, they must be attended to in higher order.

An alternative for degenerate zeroth-order states leads to a small eigenfield problem. Beginning with [3-11], one gets the following pair of first-order equations analogous to those for the nondegenerate case.

$$\begin{aligned}&\Sigma_i b_i(H'-\mu')u_i^0+(H^0-\mu)u'=0\\&\Sigma_j c_j\ (H'-\mu'-\delta w)v_j^0+(H^0-\mu-w^0)v'=0\end{aligned} \qquad [3\text{-}12]$$

Taking the inner product of the first equation with u_k^0 on the left, and the second by v_ℓ^0, and denoting the submatrices of H' within the $\{u^0\}$ and $\{v^0\}$ blocks as h^u and h^v and the vectors of coefficients b_i and c_j as b and c respectively, we get the following matrix equations.

$$(h^u-\mu'I^u)b=0;\ (h^v-(\mu'+\delta w)I^v)c=0 \qquad [3\text{-}13]$$

(Here I^u denotes the unit matrix having the same order as h^u -- i.e., p.) Applying $\otimes I^v c$ to the first and $I^u b\otimes$ to the second of these equations and subtracting yields [3-14].

$$(h^u\otimes I^v-I^u\otimes h^v+\delta wI^u\otimes I^v)(b\otimes c)=0 \qquad [3\text{-}14]$$

The form of h^u is now $f^u+x'g^u$, where f^u and g^u are submatrices within the $\{u^0\}$ block of $F'+x^0G'$ and G^0, respectively. Thus [3-14] can be rearranged to [3-15].

[3-15]

$$(f^u\otimes I^v-I^u\otimes f^v+\delta wI^u\otimes I^v)(b\otimes c)=x'(I^u\otimes g^v-g^u\otimes I^v)(b\otimes c)$$

This is a small eigenfield equation based on the transitions from the $\{u^0\}$ to $\{v^0\}$ blocks; the eigenfields are the first-order transition fields for these transitions, and the eigenvectors are the correct zero'th order transition vectors.

Acknowledgements.

This research was supported by the Petroleum Research Fund, the National Science Foundation, and the Advanced Research Projects Agency.

Literature Cited.

1. Belford, G. G., Belford, R. L., and Burkhalter, J. F., J. Magn. Resonance (1973) 11, 251.
2. Hempel, J. C., Morgan, L. O., and Lewis, W. B., Inorg. Chem. (1970) 9, 2064.
3. Byfleet, C. R., Chong, D. P., Hebden, J. A., and McDowell, C. A., J. Magn. Resonance (1970) 2, 69.
4. Abragam, A. and Bleaney, B., "Electron Paramagnetic Resonance of Transition Ions," Clarendon Press, Oxford, 1970.
5. Belford, R. L. and Belford, G. G., J. Chem. Phys. (1973) 59, 853.
6. Korn, G. A. and Korn, T. M., "Mathematical Handbook for Scientists and Engineers," §13.2-10, McGraw-Hill, New York, 1961.
7. Moler, C. B. and Stewart, G. W., SIAM J. Numer. Anal. (1973) 10, 241.
8. Chang, C.-C., M.S. Thesis in Computer Science, Univ. of Illinois at Urbana-Champaign, 1974.
9. Davis, P. H. and Belford, R. L., this volume.

6

Multisite Magnetic Interactions in Hexaurea-Metal(III) Halide Lattices

PHILLIP H. DAVIS and R. L. BELFORD

University of Illinois, Urbana, Ill. 61801

Introduction

X-ray crystallographic studies carried out in our laboratory and others on a number of salts of the type $M(urea)_6X_3$, where M = Al, Ti, V, Cr, or Fe and X = halide or perchlorate, have shown them to be isomorphous (1-7). Furthermore, where complete x-ray structure determinations were done, the molecular geometries of the various $M(urea)_6^{3+}$ complex ions were found to be virtually identical (2-7). The crystal structure is such as to suggest the possibility of observing weak one-dimensional interactions between the metal ions which occur in infinite linear chains parallel to the crystallographic c-axis. The existence of the isomorphous series provides an opportunity for a variety of comparative studies. For example, by varying the concentration of the paramagnetic species in binary systems involving one of the paramagnetic ions doped into one of the diamagnetic aluminum salts, it would be possible to vary the effective length of the chain along which interactions might take place. Also of interest are studies of interactions among different paramagnetic ions in binary and ternary systems (e.g. $Cr:Ti(urea)_6X_3$ and $Cr:Ti:Al(urea)_6X_3$).

Structural Background

The above salts crystallize in the trigonal (rhombohedral) space group $R\bar{3}C$ with approximate unit cell parameters a = 17.5Å and c = 14.0Å. Table I provides a comparison of unit cell parameters for the various salts.

The basic structure of the complex ion is shown in Figure 1. Each triply primitive unit cell contains six formula units. The metal atoms occupy the six-fold set of special positions a of 32 symmetry. While the complex is thus crystallographically constrained to possess D_3 symmetry, the coordination polyhedron is very nearly a regular octahedron. Two types of distortion of an

Table I

Comparative Unit Cell Parameters for Salts $M(urea)_6X_3$[a]

Compound	a (Å)	c (Å)	Ref.
$Al(urea)_6I_3$	17.539(3)	13.855(8)	6
$Ti(urea)_6I_3$	17.67(2)	14.15(2)	3
$Ti(urea)_6(ClO_4)_3$ (300°K)	18.132(5)	14.149(5)	4
$Ti(urea)_6(ClO_4)_3$ (90°K)	17.57	13.84	5
$Cr(urea)_6I_3$	17.53	13.87	6
$V(urea)_6I_3$ (300°K)	17.49(2)	14.38(3)	7
$V(urea)_6I_3$ (90°K)	16.74(3)	14.22(3)	7

[a] Where known, estimated standard deviation in last significant digit given in parentheses.

octahedron are possible which maintain at least D_3 symmetry. The first involves the compression or elongation of the octahedron along a three-fold axis and is manifested in a deviation of the angle (commonly denoted θ) between an M-O bond and the three-fold axis from the octahedral value of 54.74°. The second distortion involves a twisting about the three-fold axis of opposite faces of the octahedron resulting in a deviation from 60° of the angle (usually denoted φ) between projections onto the plane of the metal atom of successive M-O bonds. Distortions of the former type are virtually absent in these complexes (note, however, that the symmetry of the electron distribution about the metal atom may differ significantly from that of the positions of the atomic centers of the coordinating atoms), while those of the latter type are small but significant. The chief observable difference in molecular structure among the various $M(urea)_6^{3+}$ ions lies in the M-O bond length, which ranges from 1.88Å in the Al complex to 2.04Å in the Ti complex (perchlorate salt). Table II lists comparative values of the M-O bond length and distortions from octahedral geometry for these complexes.

In the crystal the complex ions form infinite linear chains along the three-fold axis (crystallographic c-axis). Successive ions are related by an inversion center and the separation between metal atoms is equal to one-half of the c-axial length. Each such chain is 'insulated' from its neighbors by six chains of anions. The ions forming each of the latter chains are related by a three-fold screw operation (hence the vertical separation between ions is equal to one-third of the c-axial repeat). The chains are very nearly linear, the ions being located very close to the axis of rotation. The entire structure is held together by a three-dimensional network of hydrogen bonds involving shared halide ions, but there are no direct hydrogen bonds between successive

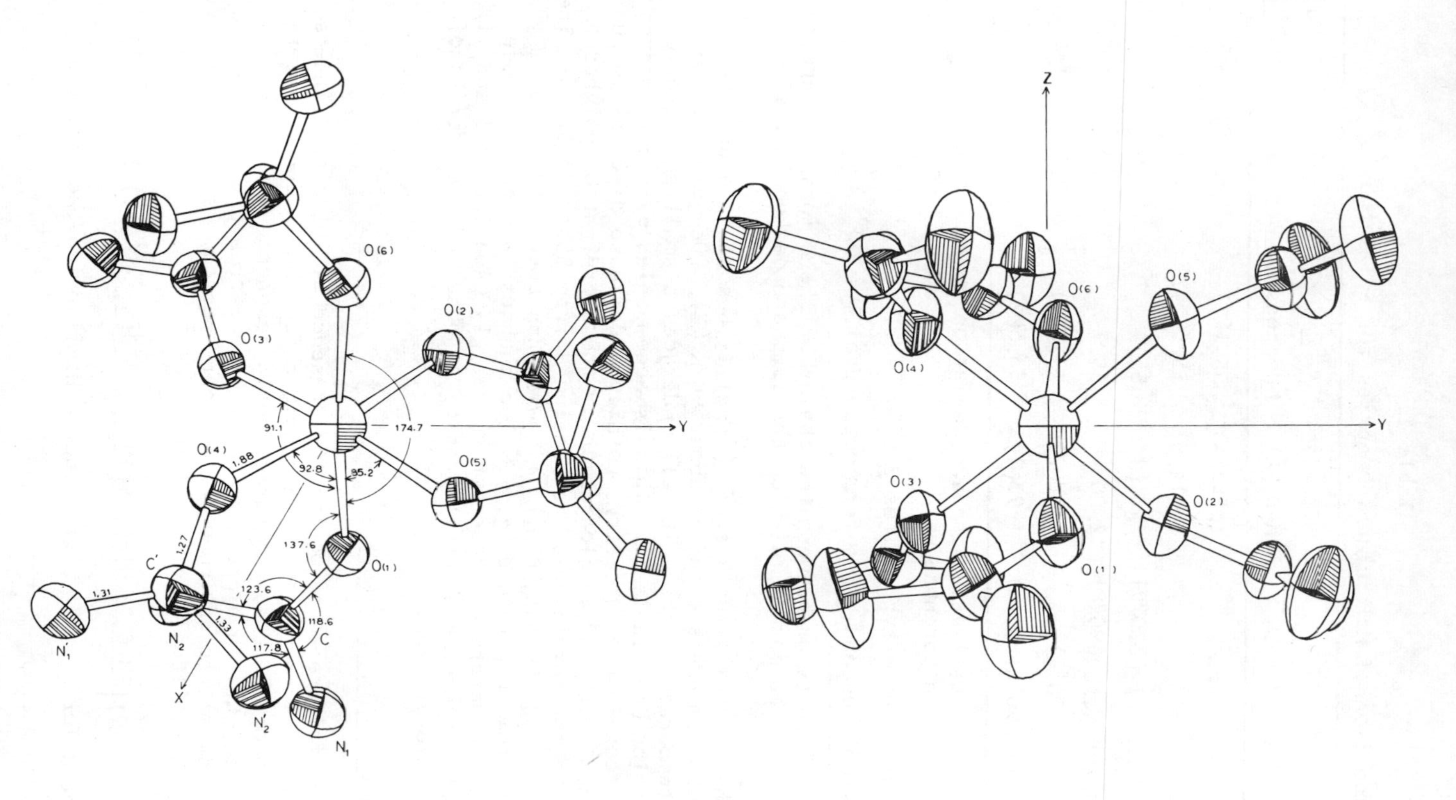

Figure 1. Two views of the structure of a representative $M(urea)_6^{3+}$ complex ion: $Al[CO(NH_2)_2]_6^{3+}$. Atoms represented by 50% probability thermal ellipsoids.

Table II

Comparison of Molecular Geometries for Salts $M(urea)_6X_3$[a]

Compound	M-O (Å)	θ (degrees)[b]	Δφ (degrees)[c]	Ref.
$Al(urea)_6I_3$	1.883(4)	55.1(3)	6.4	6
$Ti(urea)_6I_3$	2.014(5)	55.5(3)	5.5	3
$Ti(urea)_6(ClO_4)$ (300°K)	2.04(1)	55.4	5.0	4
$Ti(urea)_6(ClO_4)_3$ (90°K)	2.00	55.1	5.0	5
$V(urea)_6I_3$ (300°K)	1.98(2)	55.3(7)	7.0	7
$V(urea)_6I_3$ (90°K)	1.93(1)	53.5(2)	7.0	7

[a] Where known, estimated standard deviation in last significant digit given in parentheses.

[b] Octahedral geometry requires $\theta = 54.74^\circ$.

[c] Deviation plus and minus from octahedral $\varphi = 60^\circ$.

$M(urea)_6^{3+}$ units or between the chains of these ions. Figure 2 illustrates the manner in which the ions are packed in the crystal.

The crystal structure of $V(urea)_6I_3$ has been determined at both room temperature and at about 90°K (7). On going to the lower temperature, the unit cell undergoes a small anisotropic contraction (see Table I). This is accompanied by small changes in the coordination geometry, notably a decrease of 0.05Å in the V-O bond length and an elongation of the coordination polyhedron along the three-fold axis, decreasing θ from 55.3° to 53.5°. The structure of $Ti(urea)_6(ClO_4)_3$ likewise has been determined at high and low temperature (4,5). The changes in structure were parallel in nature to those for $V(urea)_6I_3$ (see Tables I and II), although matters were complicated by an observed disordering of the perchlorate ions.

Interactions in $Cr(urea)_6X_3$

The failure to recognize the presence of interionic interactions appears to have been responsible for the incorrect interpretation of the electron paramagnetic resonance (epr) spectrum of $Cr(urea)_6Cl_3$ in a previous paper (8). For a Cr^{3+} ion in an axially symmetric environment, the epr spectrum is usually fitted to a spin Hamiltonian of the form

$$\mathcal{H} = g_{||}\beta H_z S_z + g_{\perp}\beta(H_x S_x + H_y S_y) + D[S_z^2 - S(S+1)/3] \qquad (1)$$

evaluated for $S = 3/2$. From such a Hamiltonian one predicts three allowed ($\Delta M_s = \pm 1$) transitions occurring at fields $h\nu/g\beta$ and $(h\nu \pm 2\underline{D})/g\beta$ when the magnetic field is in the $\underline{z}$ direction and at approximately $h\nu/g\beta$ and $(h\nu \pm \underline{D})/g\beta$ when the magnetic field is anywhere in the $\underline{xy}$ plane. Figure 3 shows the epr spectrum of $Cr(urea)_6Cl_3$ powder as well as spectra of a single crystal of this compound at several orientations. Clearly the single crystal spectra are not consistent with the above Hamiltonian, yet when averaged over all possible orientations they give a powder spectrum which is similar in overall appearance to that which would be expected for Hamiltonian (1) (albeit poorly resolved). Interpreting the powder spectrum on this basis and assuming that the shoulders extend to the extrema ($dH/d\theta = 0$) associated with the parallel field position, one calculates values of $g = 1.98$ and $\underline{D} = 277 \times 10^{-4}$ cm^{-1} in agreement with those previously reported (8).

When $Cr(urea)_6^{3+}$ is present to the extent of a few mole-% in the $Al(urea)_6I_3$ lattice, the principal features of the epr spectrum, including not only the angular dependence of the resonance field positions of the allowed transitions but also that of the formally forbidden $\Delta M_s = \pm 2, \pm 3$ transitions which are predicted to have observable intensity when the magnetic field is approximately mid-way between the $\underline{z}$-axis and the $\underline{xy}$ plane, are adequately described by the Hamiltonian (1) with $g_{\parallel} = g_{\perp} = 1.9752$ and $\underline{D} = 483.4 \times 10^{-4}$ cm^{-1} (we also observe hyperfine coupling to the 9% abundant Cr^{53} isotope [$\underline{I} = 3/2$] with a nearly isotropic coupling constant $\Lambda \sim 18.7 \times 10^{-4}$ cm^{-1}). However, even here additional features, whose intensity increased with increased Cr^{3+} concentration, were observed (see Figure 4). We attribute the majority of these lines to pairwise interactions between Cr^{3+} ions occupying adjacent sites along the $\underline{c}$-axis of the host lattice.

The general spin Hamiltonian for a coupled pair of Cr^{3+} ions may be written (except for a usually small biquadratic exchange term, $j(\underline{S}_1 \cdot \underline{S}_2)^2$)

$$\mathcal{H} = g_{\parallel}\beta \underline{H}_z(\underline{S}_{1z}+\underline{S}_{2z}) + g_{\perp}\beta[\underline{H}_x(\underline{S}_{1x}+\underline{S}_{2x})+\underline{H}_y(\underline{S}_{1y}+\underline{S}_{2y})] + \underline{D}(\underline{S}_{1z}^2+\underline{S}_{2z}^2-\tfrac{5}{2}) + \underline{d}[2\underline{S}_{1z}\underline{S}_{2z}-(\underline{S}_{1x}\underline{S}_{2x}+\underline{S}_{1y}\underline{S}_{2y})] + \underline{J}\underline{S}_1\cdot\underline{S}_2 \qquad (2)$$

In all of our calculations on this system we have assumed that $\underline{g}$ and $\underline{D}$ have the same values in the interacting pair as in the isolated ion, although $\underline{D}$, especially, which is a measure of the distortion of the complex ion from octahedral geometry, might well be different for the pair. Employing the above Hamiltonian, calculations of the type described in the preceding paper predict a spectrum composed of eighteen predominant lines - three groups of six, each group roughly centered about one of the three allowed lines of the isolated ion spectrum (which fields form convenient zero'th order approximations for the perturbation calculations). Because the lines are narrower and we can measure their positions with greater certainty, we have concentrated our efforts on trying

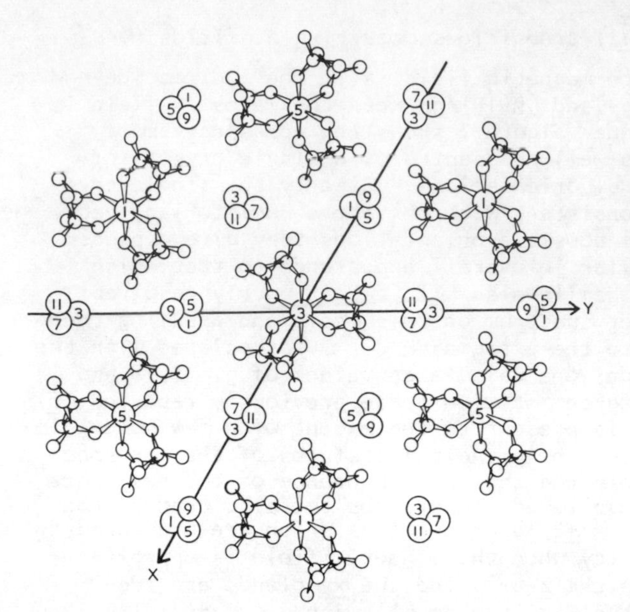

Figure 2. Packing in the $M(urea)_6X_3$ lattice. Numbers indicate the z-coordinates (in 1/12-ths of the c-axial length) of the metal and halogen atoms. Directly above and below each complex ion is another related by an inversion center to the one shown.

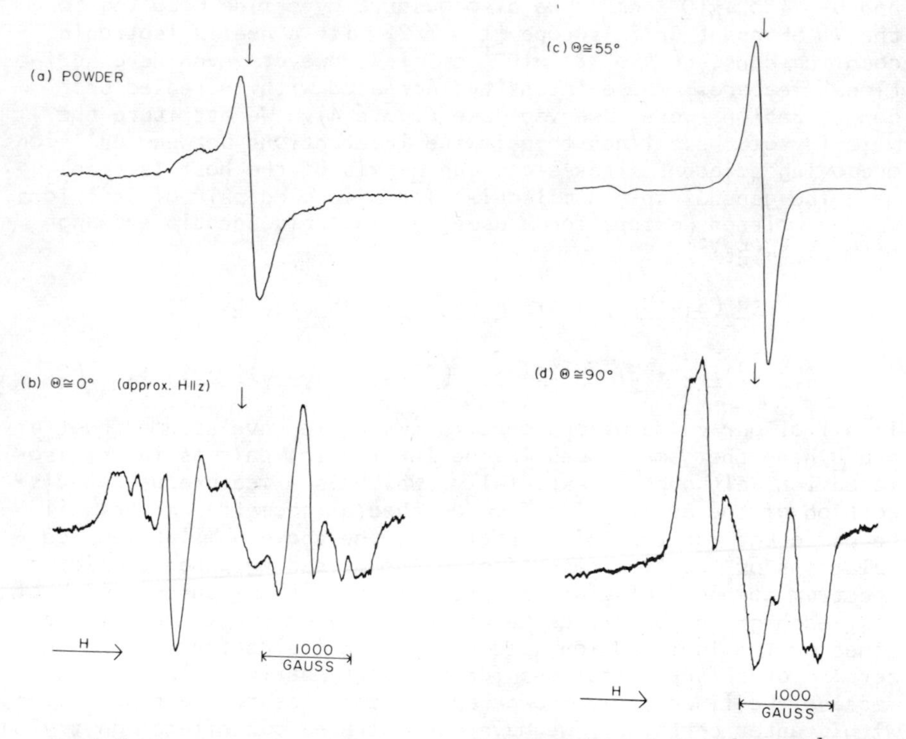

Figure 3. X-band epr spectrum of $Cr(urea)_6Cl_3$ powder and of single crystal at several orientations. For reference, arrow indicates position of dpph resonance.

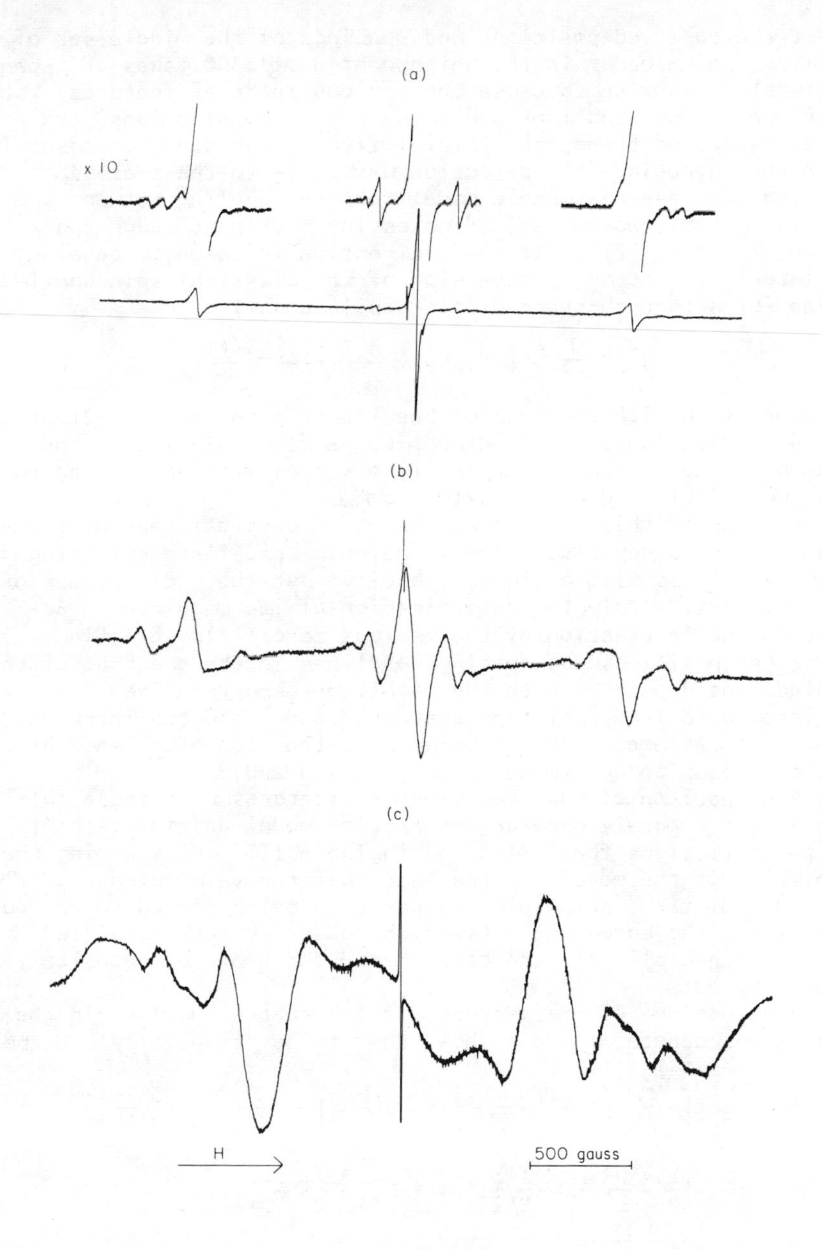

Figure 4. Concentration dependence of X-band epr spectrum of Cr-(urea)$_6^{3+}$: (a) ~ 1% Cr in Al(urea)$_6$I$_3$, (b) ~ 25% Cr in Al(urea)$_6$I$_3$, and (c) Cr(urea)$_6$I$_3$. For all spectra the magnetic field is directed along the z axis. Sharp spike near the center of each spectrum is due to dpph used as an internal standard.

to fit the observed positions and spacings of the middle set of six lines which occur in the neighborhood of 3300 gauss at χ-band (9.1 GHz). Likewise, because the various spectral features, both those due to isolated ions and those due to coupled ions, are better separated there, the initial fitting was donc for cases in which the magnetic field direction was close to the $\underline{z}$-axis.

The simplest reasonable model for the coupling between two ions is that of two magnetic dipoles interacting through space ($\underline{d} \neq 0$, $\underline{J} = 0$ in ($\underline{\underline{2}}$)). If the $\underline{z}$-direction is taken to be along the interionic vector $\underline{r}$, expansion of the classical spin Hamiltonian for a through-space dipole-dipole coupling

$$\mathcal{H} = g^2\beta^2 < \frac{1}{r^3} > [-\underline{S}_1\cdot\underline{S}_2 + \frac{3(\underline{S}_1\cdot\underline{r})(\underline{S}_2\cdot\underline{r})}{r^2}] \qquad (\underline{\underline{3}})$$

and comparison with the form of the interaction term in ($\underline{\underline{2}}$) allows the identification $\underline{d} = -g^2\beta^2<1/r^3>$ to be made. Taking as the value of r the crystallographically observed metal-metal separation, we calculate $\underline{d} = -50.75\times10^{-4}\ cm^{-1}$.

We note at this point that the above physical reasoning implies that $\underline{d}$ is negative. The intraionic zero field splitting parameter, $\underline{D}$, is also a signed quantity, but the ordinary epr experiment permits only the determination of its magnitude. Because of the interaction of the various zero field splitting terms, the positions and spacings of lines in the spectrum of the coupled ions depend on both the magnitudes and relative signs of the interionic interaction parameters $\underline{d}$ and $\underline{J}$ and the intraionic interaction parameter, $\underline{D}$. Knowledge of the sign of $\underline{d}$ should thus enable us to determine the signs of $\underline{D}$ and $\underline{J}$.

A comparison of observed spectral features with those calculated for a purely dipolar interaction model and for each of the possible signs for $\underline{D}$ is found in Table III. Considering the simplicity of the model and the fact that the value of $\underline{d}$ was calculated from first principles rather than being fitted to the observations, the agreement between the observed and calculated spectra (especially for the case when $\underline{d}$ and $\underline{D}$ are of opposite sign) is good.

Examination of the vectors for the states involved in the transitions denoted $\underline{1}$ and $\underline{6}$ shows them to be essentially (in the $|M_{S1},M_{S2}>$ basis)

$$1:\quad \frac{1}{\sqrt{2}}\{|-\tfrac{1}{2},-\tfrac{3}{2}> - |-\tfrac{3}{2},-\tfrac{1}{2}>\} \rightarrow \frac{1}{\sqrt{2}}\{|\tfrac{1}{2},-\tfrac{3}{2}> - |-\tfrac{3}{2},\tfrac{1}{2}>\}$$

$$6:\quad \frac{1}{\sqrt{2}}\{|\tfrac{3}{2},-\tfrac{1}{2}> - |-\tfrac{1}{2},\tfrac{3}{2}>\} \rightarrow \frac{1}{\sqrt{2}}\{|\tfrac{3}{2},\tfrac{1}{2}> - |\tfrac{1}{2},\tfrac{3}{2}>\}$$

Within each pair the effect of the isotropic exchange term in ($\underline{\underline{2}}$) is to shift the energy of each state by the same amount and in the same direction. The positions of and spacing between these two lines will thus be determined almost solely by the magnitude of the dipolar term in ($\underline{\underline{2}}$). That these two lines are fitted

Table III

Comparison of Observed and Calculated EPR Spectra for Coupled $Cr(urea)_6^{3+}$-$Cr(urea)_6^{3+}$ Pairs[a]

		Calculated Field[b] (gauss)		
Transition	Observed Field (gauss)	$\underline{D}$ = 483.4, $\underline{d}$ = -50.75, $\underline{J}$ = 0.0	$\underline{D}$ = -483.4, $\underline{d}$ = - 50.75, $\underline{J}$ = 0.0	$\underline{D}$ = -483.4, $\underline{d}$ = - 50.75, $\underline{J}$ = 45.11
1	3047 (2)	3047 (4)	3047 (4)	3047 (4)
2	3107 (10)	3117 (7)	3145 (9)	3107 (10)
3	3169 (1)	3196 (5)	3231 (3)	3168 (2)
4	3420 (1)	3392 (5)	3358 (3)	3420 (2)
5	3482 (10)	3472 (7)	3444 (9)	3482 (10)
6	3540 (2)	3542 (4)	3542 (4)	3542 (4)

[a] Spectra here limited to diagnostic group of six transitions centered around 3300 gauss (at 9.1 GHz) when magnetic field direction is parallel to $\underline{z}$-axis. Relative intensity in parentheses following field.

[b] For all calculations g = 1.9752. Remaining spin Hamiltonian parameters expressed in units of 10^{-4} cm^{-1}.

quite well and the remaining four lines less well by the purely dipolar interaction model indicates 1) that the magnitude of the dipolar coupling calculated from the crystal structure is appropriate and 2) that there is a significant amount of exchange coupling present.

If the other parameters of the spin Hamiltonian are held constant, the fields at which the six diagnostic transitions occur are each linear in the exchange parameter over a small range ($|\underline{J}| \leq |\underline{d}|$), facilitating the precise determination of $\underline{J}$. With the inclusion of exchange a fit between the observed and calculated spectra in the diagnostic region was obtained which was exact, within experimental uncertainty (see Table III). The parameters of best fit, which constitute a unique set, are g = 1.9752, $\underline{D}$ = -483.4×10^{-4} cm^{-1}, $\underline{d}$ = -50.75×10^{-4} cm^{-1} and $\underline{J}$ = 45.11×10^{-4} cm^{-1}. It is worth noting that while in the absence of exchange the best agreement between observed and calculated fields for the diagnostic transitions was obtained when the parameters $\underline{D}$ and $\underline{d}$ were of opposite sign, the behavior of the transition fields with the inclusion of exchange is consistent only with these parameters having the same sign. The positive value observed for $\underline{J}$ corresponds to an antiferromagnetic coupling.

Table III compares observed and calculated spectra in the diagnostic region for several combinations of spin Hamiltonian parameters, while Figure 5 illustrates the agreement between the calculation employing the parameters of best fit and the entire spectrum observed when the magnetic field is parallel to _z_. In Table IV observed transition fields at the Q-band frequency (35.1 GHz) and those calculated from the parameters of best fit (obtained at 9.1 GHz) are compared for several orientations. The computer program used for all spectral calculations in this paper was written by Mr. T. M. Lenhardt of the University of Illinois and is based on the frequency shift perturbation technique of reference (_9_).

If the crystals of $Cr:Al(urea)_6I_3$ are subject to temperature dependent changes in crystal and molecular structure of the type observed in the case of $V(urea)_6I_3$, a variation with temperature of most of the spin Hamiltonian parameters would be expected. A contraction of the lattice (at least in the _c_-direction) would reduce the metal-metal separation and have the effect of increasing both _d_ and _J_ in absolute value. The parameter _d_ will, of course, vary as $1/r^3$, while the rate of change of _J_ would depend on the mechanism(s) of the exchange, but in any case might reasonably be expected to be somewhat greater than that for _d_. A change in the coordination geometry about the Cr atom would result in a change in _D_, but it is difficult to make an _a priori_ prediction of the direction or magnitude of the change.

Experimentally, _D_, _d_, and _J_ all increase in magnitude with decreasing temperature. At about 90°K the parameters giving the best agreement between observed and calculated spectra (the fit was at least as good as that at room temperature) were g = 1.9752, $D = -514.8 \times 10^{-4}$ cm^{-1}, $d = -54.35 \times 10^{-4}$ cm^{-1}, and $J = 50.13 \times 10^{-4}$ cm^{-1}. The value of _d_, which was obtained from the spacing between diagnostic lines _1_ and _6_ (_vide supra_), corresponds to an apparent contraction of the lattice in the _c_-direction, or at least a reduction of the Cr-Cr pair distance, by about 2%. This is about twice the lattice contraction observed in the vanadium complex, but is not a physically unreasonable amount. The fractional increase in _J_ is about 1.6 times that observed for _d_; that is, _J_ appears to be varying approximately as $1/r^5$. While a pathway for direct exchange exists and the $1/r^5$ dependence does not seem unreasonable for this mechanism at the observed internuclear distance, a super-exchange mechanism is not conclusively ruled out.

When the magnetic field direction is appreciably away from the _z_-axis, most of the pair transitions which are calculated to have an observable intensity occur at fields such that they are masked by the much more intense isolated-ion lines, and the agreement between the calculated transition fields and those lines which are observed is not as good as for nearly parallel orientations (see Table IV). In addition, for the most dilute

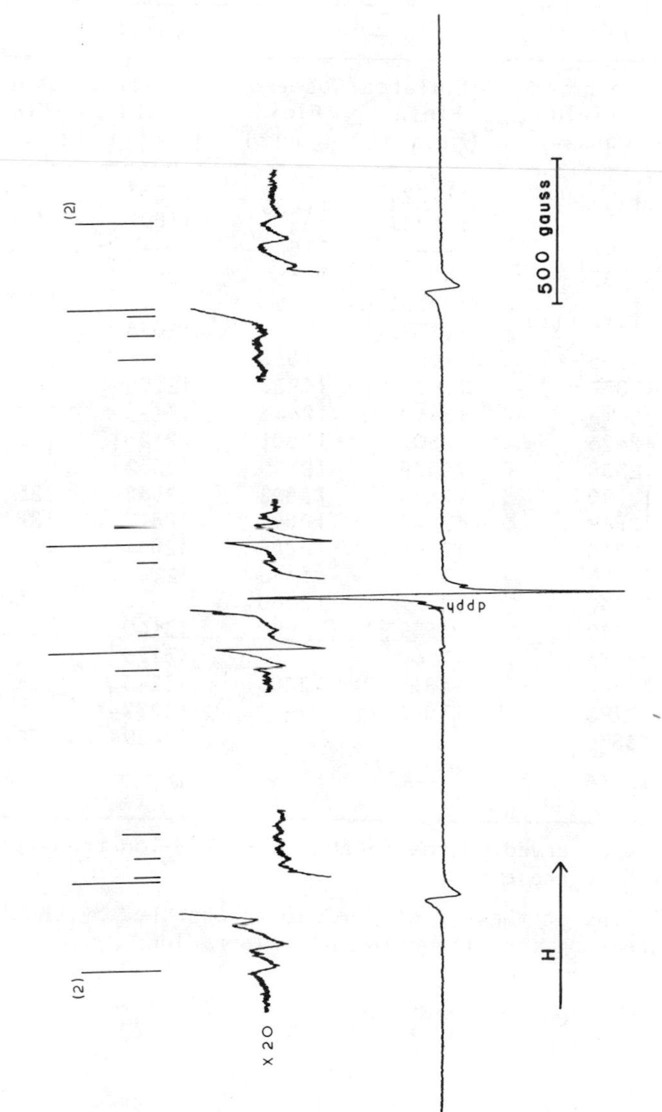

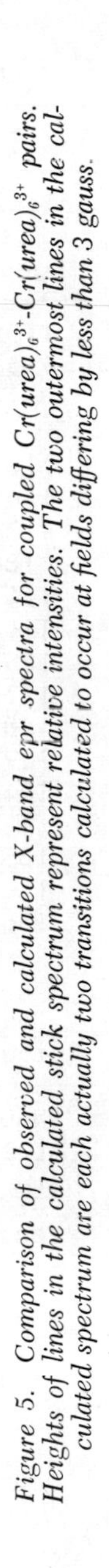

Figure 5. Comparison of observed and calculated X-band epr spectra for coupled $Cr(urea)_6^{3+}$-$Cr(urea)_6^{3+}$ pairs. Heights of lines in the calculated stick spectrum represent relative intensities. The two outermost lines in the calculated spectrum are each actually two transitions calculated to occur at fields differing by less than 3 gauss.

Table IV

Observed and Calculated Q-Band EPR Spectra at Several Orientations for Coupled $Cr(urea)_6^{3+}$-$Cr(urea)_6^{3+}$ Pairs

θ = 0° (H // z)

Calculated Field (gauss)	Observed Field (gauss)
11365 11368	11367
---	11447
---	11537
11679 11694	11692 (sh)[a]
11758	11765
11840	11837
12417	12415
12476	12476
12538	12538
12790	12790
12852	12849
12912	12910
13488	13485
---	13538
13570	13570 (?)
13635	13633
13649	13660 (sh)[a]
---	13793
---	13876
13960 13963	13958

θ = 15°

Calculated Field (gauss)	Observed Field (gauss)
11472 11483	11480
---	11500
---	11550
11776 11777	X[b]
11841	11838
11924	11930
12447	12446
12501	12501
12539	12540
12805	12802
12845	12842
12892	12888
13394	13395
13477	13460
13539 13540	X[b]
13815	13805
13846	---

θ = 30°

Calculated Field (gauss)	Observed Field (gauss)
11767	11765
11809	11803
---	11843
12008 12045 12073	X[b]
12155	---
---	12255
12525 12532	12524
12562	12563
12817	12815
12834 12840	12831
---	13000
13140	13120
13225 13245 13287	X[b]
13429	13420

[a] Shoulders are observed on the intense isolated-ion transitions at approximately these fields.

[b] Pair transitions at these fields would be obscured by the intense isolated-ion transition in the same region.

samples weak lines beyond those predicted by our model are observed in the region of the outermost lines associated with the isolated-ion, even at the parallel orientation (see Figures 4a and 5). These discrepancies, which we do not feel indicate any serious deficiency in the model, may arise from the terms which we have left out of our spin Hamiltonian. Another possible origin for the 'extra' lines would be the existence of two (or more) distinct $Cr(urea)_6^{3+}$-$Cr(urea)_6^{3+}$ sites, characterized by different $\underline{D}$ values. The effect of slightly different $\underline{D}$ values would be minimal on the diagnostic portion of the spectrum, but in the two outer regions would result in sets of lines displaced by $2\Delta\underline{D}/g\beta$. As the concentration of Cr^{3+} is increased, the intensity of the isolated-ion lines relative to the remaining features of the spectrum decreases, and all of the lines are broadened, reducing resolution (see Figure 4b). It is interesting to note that the 'extra' lines pointed out above have virtually disappeared in the spectrum shown in Figure 4b. The spectrum of pure $Cr(urea)_6I_3$ (Figure 4c) is different yet, although some general similarities between it and those of the diluted samples are evident. Undoubtedly the same interactions are at work here, but at the least we can expect it to be necessary to consider interactions between each Cr^{3+} ion and both of its axial nearest neighbors.

Additional Studies of Interactions in $M(urea)_6X_3$ Lattices

The various additional investigations of magnetic interactions in hexaurea-metal(III) halide lattices contemplated or currently underway in this laboratory are conveniently divided into three categories by the composition of the materials being studied.

The first group consists of systems involving a single paramagnetic species, either as a substitutional impurity in one of the aluminum salts or as the pure $M(urea)_6X_3$ salt. Here we wish to determine whether coupling between paramagnetic sites occurs and, if so, whether a mechanism of the type established for the chromium system appears appropriate. In the titanium system, the only other one for which experimental information is currently available, the lack of dependence of the epr spectrum on Ti^{3+} concentration suggests that interionic coupling is minimal. Analysis is complicated by the fact that, even for dilute samples, the spectrum of $Ti(urea)_6^{3+}$ is not that anticipated for a simple $\underline{S} = 1/2$ (d') ion and has not been adequately interpreted.

In the second group are systems in which one of the paramagnetic ions is doped into the lattice of a dissimilar paramagnetic ion. At the present our attention has centered on $Cr:Ti(urea)_6X_3$ and $Cr:V(urea)_6X_3$. Because of line broadening due to rapid relaxation, neither $Ti(urea)_6^{3+}$ nor $V(urea)_6^{3+}$ gives an observable epr spectrum at room temperature. For our purposes, sites occupied by $Cr(urea)_6^{3+}$ ions as guests in one of the other lattices may be classified according to the axial neighbors of a given Cr^{3+} ion. If the concentration of Cr^{3+} is low, most will have dissimilar ions in both neighboring axial positions, a few will have one similar and one dissimilar neighbor, while the number with two similar neighboring ions will be negligible. Depending on the strength of the interionic coupling, if any, spectral features due to ions in the first type of site potentially would provide information on the paramagnetic ion of the host lattice, since the $Cr(urea)_6^{3+}$ system is already well characterized. (Note, however, that differences in lattice dimensions may result in differing degrees of distortion of the $Cr(urea)_6^{3+}$ ion so that changes in the spectrum from the $Cr{:}Al(urea)_6X_3$ reference will probably be a combination of lattice and specific ion-coupling effects.) Preliminary epr spectra of samples of $Cr{:}Ti(urea)_6I_3$ are very similar to those obtained from $Cr{:}Al(urea)_6I_3$. The strongest features once again are consistent with the Hamiltonian (1). While there is no detectable change in g, D is larger in magnitude than observed in the aluminum lattice. There are several possible explanations of the failure to observe significant Ti^{3+}-Cr^{3+} coupling, but the most attractive would seem to be the assumption that the Ti^{3+} is relaxing so rapidly that the Cr^{3+} is not coupled to it, but sees only an averaged environment which would provide a very small paramagnetic shift. Similarly, the secondary features can be interpreted on the basis of coupled Cr^{3+}-Cr^{3+} pairs. The use of different isomorphous lattices coupled with measurements at low temperature provides a range of Cr-Cr separations which should help to further elucidate the precise nature of the coupling between the $Cr(urea)_6^{3+}$ ions and, in particular, the exchange mechanism.

As yet we have not prepared any crystals of the third type in which one of the $Al(urea)_6X_3$ salts is doped simultaneously with two dissimilar paramagnetic $M(urea)_6^{3+}$ ions. The primary aim here will be to study coupling in discrete $M(urea)_6^{3+}$-$M'(urea)_6^{3+}$ pairs. It is anticipated that a large part of any such investigations will have to be carried out at liquid helium temperature.

In summary, the isomorphous, isostructural $M(urea)_6X_3$ salts provide an opportunity for a variety of investigations of weak coupling between paramagnetic ions in the solid state. Epr investigations of these systems not only can provide information on the coupling itself, but indirectly, through the coupling, information about the individual ions involved can be obtained which would not normally be available.

Acknowledgements

This research was supported by the Petroleum Research Fund, the National Science Foundation, and the Advanced Research Projects Agency.

Literature Cited

1. Dingle, R., J. Chem. Phys. (1969) 50, 545.
2. Linek, A., Siskova, J., and Jensovsky, L., Acta Crystallogr. (1969) A25, 155S.
3. Davis, P. H. and Wood, J. S., Inorg. Chem. (1970) 9, 1111.
4. Figgis, B. N., Wadley, L. G. B., and Graham, J., Acta Crystallogr. (1972) B28, 187.
5. Figgis, B. N. and Wadley, L. G. B., Austral. J. Chem. (1972) 25, 2233.
6. Davis, P. H., Ph.D. Thesis, University of Illinois (1972), pp. 7-18.
7. Figgis, B. N. and Wadley, L. G. B., J. Chem. Soc. Dalton (1973), 2182.
8. Santangelo, M., Atti acad. sci. lettere ed arti Palermo, Part I (1957-58) 18, 123.
9. Belford, R. L., Davis, P. H., Belford, G. G., and Lenhardt, T. M., this volume.

7

Polymeric, Mixed-Valence Transition Metal Compounds

GILBERT M. BROWN, ROBERT W. CALLAHAN, EUGENE C. JOHNSON, THOMAS J. MEYER, and TOM RAY WEAVER

University of North Carolina, Chapel Hill, N.C. 27514

In our work we have prepared and characterized materials in which metal atoms or ions are held in close proximity by chemical linkages. One of our goals has been to establish the chemical and physical properties which arise from metal-metal interactions in such systems and to modify the properties in a controlled way by directed chemical synthesis. Three different kinds of systems have been studied which differ with regard to the nature of the metal-metal interaction: 1. Compounds in which there is strong, direct metal-metal bonding. 2. Cases where there are strong metal-metal interactions through bridging ligands. 3. Cases where there are weak metal-metal interactions through bridging ligands.

In compounds containing strong metal-metal bonds, the effect of the metal-metal bond(s) is to modify strongly the chemical and electronic properties of the compounds when compared to the component monomers.(1) One of the intriguing properties of such systems is that if certain bonding and/or structural features are present, they can exist in a variety of molecular oxidation states, e.g., $[(\pi\text{-}C_5H_5)Fe(CO)]_4^{2+/+/0/-}$ (2), $\{[(\pi\text{-}C_5H_5)Fe(CO)]_2(Ph_2P(CH_2)_3PPh_2)\}^{2+/+/0}$ (Ph is phenyl) (3). Ions like $[(\pi\text{-}C_5H_5)Fe(CO)]_4^+$ and $\{[(\pi\text{-}C_5H_5)Fe(CO)]_2(Ph_2P(CH_2)_3PPh_2)\}^+$ are formally mixed-valence cases, but evidence is now being obtained which indicates that the metal atoms are strongly coupled and that oxidation-reduction processes involve delocalized molecular orbitals.

Strong, Chemically-Significant Interactions Through a Bridging Ligand. μ-oxo-Bridged Complexes of Ruthenium (III).

We have prepared several μ-oxo-bridged complexes of ruthenium (III): $[(AA)XRu\text{-}O\text{-}RuX(AA)_2]^{2+}$ (AA is 2,2'-bipyridine (bipy) or 1,10-phenanthroline (phen); X is Cl or NO_2). The complexes have been characterized by elemental analysis (as hexafluro-phosphate salts), solution conductivity measurements, electro-chemical measurements, and spectrophotometric titrations using Cr^{2+} as reductant. Although x-ray crystallographic data is not yet available, in the complexes the X groups are almost certainly <u>cis</u> to the bridging oxide ion(4) and molecular models indicate that direct, through-space Ru-Ru interactions are probably not possible.

The properties of the μ-oxo-bridged dimers are unusual when compared to closely related bis(2,2'-bipyridine) complexes of ruthenium (II) and ruthenium (III). From electrochemical studies in acetonitrile, the system $[(bipy)_2ClRu\text{-}O\text{-}RuCl(bipy)_2]^{2+}$ also exists as mixed-valence +3 (Ru(III)-Ru(IV)) and +1 (Ru(II)-Ru(III)) ions. The +1 ion is chemically unstable on time scales longer than the cyclic voltammetry time scale (seconds). The electronic spectra of the +2 and +3 ions in acetonitrile include highly intense bands in the visible: for $[(bipy)_2ClRu\text{-}O\text{-}RuCl(bipy)_2]^{2+}$, λ_{max} 668nm(ε 17,000); for $[(bipy)_2ClRu\text{-}O\text{-}RuCl(bipy)_2]^{3+}$, λ_{max} 470nm (ε 17,000). Magnetic susceptibility data has been obtained on the salt $[(bipy)_2(NO_2)Ru\text{-}O\text{-}Ru(NO_2)(bipy)_2](PF_6)_2$ in the temperature range 77-275°K (5). An excellent fit of the data to the Bleaney-Bowers equation (6) has been obtained which indicates that the +2 ion has a singlet ground state with a low-lying triplet state (2J = -173 cm^{-1} with g = 2.48).

The unusual chemical and physical properties of the μ-oxo-bridged dimers appear to arise because of a strong, chemically meaningful interaction between the ruthenium ions. Many of the properties of the dimers can be explained using the qualitative molecular orbital model given in Figure 1. The model in Fig. 1 is slightly modified from the model used by Orgel(7) for the linear ion $Cl_5Ru\text{-}O\text{-}RuCl_5^{4-}$ in order to account for the lower symmetry and the possibility that the Ru-O-Ru linkage may be slightly bent.

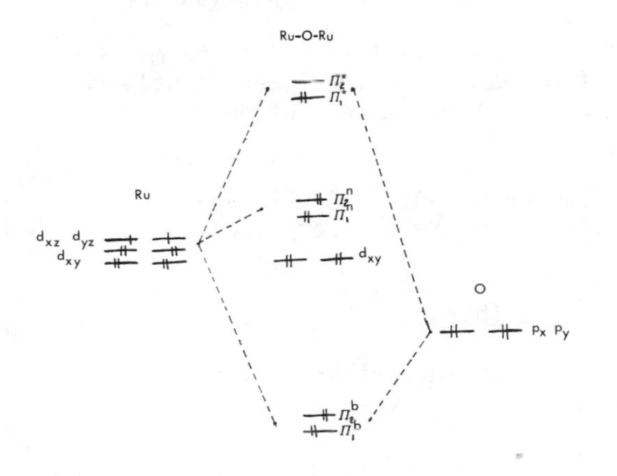

Figure 1. Qualitative molecular orbital scheme for $[(bipy)_2XRu\text{-}O\text{-}RuX(bipy)_2]^{2+}$. The z axis for each ruthenium ion is taken to be along the Ru–O axis.

Using the molecular orbital scheme in Fig. 1, the valence molecular orbitals are the slightly separated pair $\pi_1{}^*$ and $\pi_2{}^*$ which consist of antibonding combinations of ruthenium 4d and oxygen 2p orbitals. The $|2J|$ value measured in the magnetic study can be interpreted in terms of the energy separation between the singlet and triplet states $^1(\pi_1{}^{*2})$ and $^3(\pi_1{}^*\pi_2{}^*)$, although other interpretations can be given. From the molecular orbital model simple redox processes of the μ-oxo-bridged ions necessarily involve the gain or loss of electrons in net antibonding molecular orbitals which explains the relative stability of the +3 ion $[(bipy)_2ClRu\text{-}O\text{-}RuCl(bipy)_2]^{3+}$ and the somewhat surprising instability of the +1 ion $[(bipy)_2ClRu\text{-}O\text{-}RuCl(bipy)_2]^+$. By comparing reduction potential data for the μ-oxo-bridged dimers with data for related Ru(III)-Ru(II) couples, it can be estimated that the extent of destabilization of the π^* orbitals may be several kcal/mole, leading to an appreciable chemical modification of the system.

Systems, like the μ-oxo-bridged ions, in which there are strong metal-metal interactions across a bridging ligand should be of considerable interest in the future. In strongly coupled systems, as in systems containing direct metal-metal bonds, the strong coupling leads to significantly changed chemical and electronic properties when compared to the component monomers. Ultimately, such compounds may constitute a new class of materials having distinct and synthetically controllable properties of their own.

Weak Metal-Metal Interactions in Ligand-Bridged Ruthenium Complexes.

We have prepared a series of complexes in which ruthenium ions in different ligand environments are linked by organic bridging ligands. The complexes are of the type $[(NH_3)_5Ru(L)\text{-}RuCl(bipy)_2]^{3+}$ in which L is a dibasic, N-heterocyclic ligand, for example, pyrazine. From the results of spectrophotometric titrations (Br_2 in acetonitrile), and chemical and electrochemical isolation studies, the complexes also exist as the oxidized +4 and +5 ions.

From spectral and reduction potential data, in the mixed-valence +4 ions, the site of oxidation is localized largely on the $(NH_3)_5Ru$- end giving the oxidation state configuration $[(NH_3)_5Ru(III)(L)Ru(II)Cl(bipy)_2]^{4+}$. This is an expected result since the monomeric complexes $Ru(bipy)_2LCl^+$ are more difficult to oxidize by ~0.4 v than are the complexes $Ru(NH_3)_5L^{2+}$.

In the system

$$[(NH_3)_5RuN\langle\bigcirc\rangle\text{-}CH_2CH_2\text{-}\langle\bigcirc\rangle NRuCl(bipy)_2]^{5+/4+/3+},$$

where bis(4-pyridyl)ethane (BPA) is the bridging ligand, the electronic and redox properties of the +3 and +4 ions are essentially the superimposed properties of the monomeric ions $[Ru(NH_3)_5py]^{3+/2+}$ (py is pyridine) and $Ru(bipy)_2(py)Cl^{2+/+}$. No evidence has been obtained for Ru-Ru interactions, presumably, because the two π-systems of the bridging ligand are separated by a saturated ($-CH_2-CH_2-$) linkage.

Ru-Ru interactions can be "turned on" by using bridging ligands with extended, unbroken π-systems. For the dimers with the bridging ligands

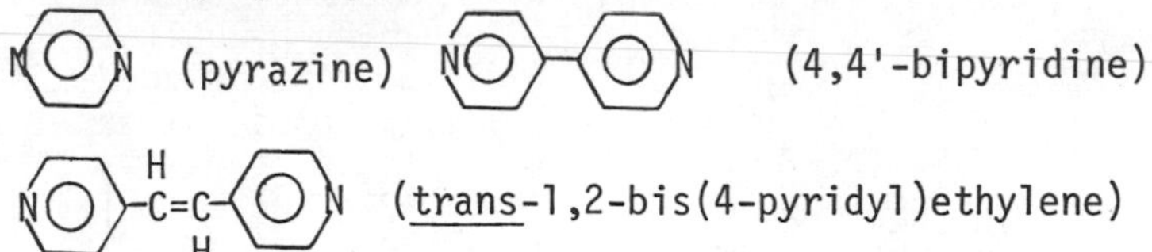

data in solution (acetonitrile) indicate that interactions do exist between metal centers. The interactions are weak but their presence is clearly seen in the spectral properties of the +3 and mixed-valence, +4 ions.

For the +3 ions, $[(NH_3)_5Ru(L)RuCl(bipy)_2]^{3+}$, qualitative band assignments can be made for the transitions $d \rightarrow \pi^*(L)$, $d \rightarrow \pi^*(bipy)$, and $\pi \rightarrow \pi^*(bipy)$. In comparing monomer and dimer spectra the only bands which are significantly shifted in the dimeric complexes are the $d \rightarrow \pi^*(L)$ charge transfer bands for which the transition moment necessarily lies along the Ru-Ru axis.

For the mixed-valence, +4 ions, Intervalence Transfer (IT) bands appear in the visible and near-infrared spectral regions. In IT absorption, light-induced electron transfer occurs between the metal centers,

$$[(NH_3)_5Ru(III)(L)Ru(II)Cl(bipy)_2]^{4+} \xrightarrow{h\nu} [(NH_3)_5Ru(II)(L)\text{-}Ru(III)Cl(bipy)_2]^{4+*}$$

giving the product ion in non-equilibrium vibrational and rotational states. The IT λ_{max} values for a series of complexes are given in Table I. Where it has been possible to test, we find that the IT bands have the solvent dependence and approximate band widths predicted by Hush. (7) The energies of the bands for the dimers in Table I are considerably higher than the energy of the IT band reported by Creutz and Taube (8) for the ion $[(NH_3)_5RuN\langle\bigcirc\rangle NRu(NH_3)_5]^{5+}$. The bands are expected to appear at higher energies for the unsymmetrical

Table I

Intervalence Transfer Bands in Acetonitrile

$$A_5Ru(III)(L)Ru(II)ClB_2^{4+} \xrightarrow{h\nu} [A_5Ru(II)(L)Ru(III)ClB_2^{4+}]^{*a}$$

Ion[a]	λ_{max} kK	λ_{max} nm	ε
$A_5RuN\bigcirc NRuClB_2^{4+}$	10.4	960	530
$A_5RuN\bigcirc NRu(NO_2)B_2^{4+}$	12.7	790	
$A_5RuN\bigcirc NRu(CH_3CN)B_2^{5+}$	13.3	750 (sh)	
$A_5RuN\bigcirc NRu(pyzRuA_5)B_2^{8+}$	13.9	720	920
$A_5RuN\bigcirc-\bigcirc NRuClB_2^{4+}$	14.4	695 (sh)	<300
$A_5RuN\bigcirc-CH{=}CH-\bigcirc NRuClB_2^{4+}$	14.7	680 (sh)	<300
$A_5RuN\bigcirc-CH_2-CH_2-\bigcirc NRuClB_2^{4+}$	—	—	—

[a]A is ammonia and B is 2,2'-bipyridine.

dimers since the product of light-induced electron transfer, $[(NH_3)_5Ru(II)(L)Ru(III)X(bipy)_2]^{4+}$, is a high energy oxidation state isomer in which the oxidation state configuration is reversed from the configuration of the ground state ion.

We have drawn two important conclusions from this work. The first is that if the bridging ligand includes an uninterrupted π-system, the effects of metal-metal interactions can be seen in the physical properties of the ions. Secondly, cooperative electronic interactions between metal centers, as viewed spectrally, and rates of intramolecular electron transfer, as estimated from the energies of IT bands(7), can be varied by a series of relatively simple chemical modifications.

The Effects of Weak Metal-Metal Interactions in Polymeric, Ligand-Bridged Compounds.

Several ligand-bridged, polymeric complexes of ruthenium(II) -- $[(bipy)_2ClRu(pyz)[Ru(bipy)_2pyz]_nRuCl(bipy)_2]^{+2n+2}$ (n=0, 1, 2, 3, 4; bipy is 2,2'-bipyridine; pyz is pyrazine) and $[(bipy)_2ClRu(L)[Ru(bipy)_2L]_nRuCl(bipy)_2]^{2n+2}$ (n=0, 1, 2; L is N⟨○⟩-⟨○⟩N or N⟨○⟩-CH_2-CH_2-⟨○⟩N) -- have been prepared and isolated as PF_6^- or ClO_4^- salts.(9) In the preparations a series of sequential, stepwise reactions are used based on the reactivity of the NO^+ and NO_2^- groups when bound to bis(2,2'-bipyridine)ruthenium(II).

We have investigated the oxidation state properties of the polymeric ruthenium complexes and of the series of 1,1'-polyferrocene compounds:

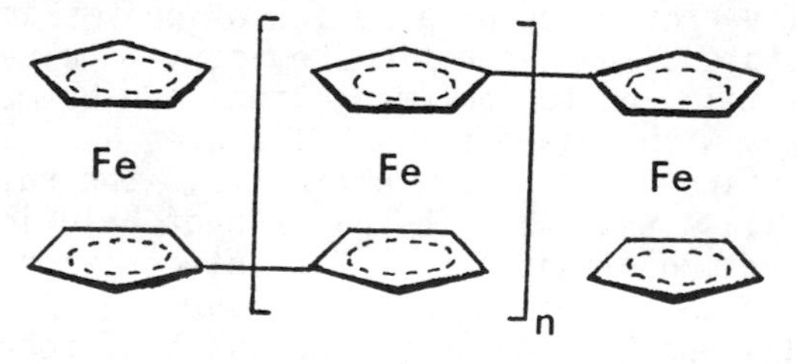

n = 0; biferrocene (Fc-Fc)

n = 1; 1,1'-terferrocene (Fc-Fc-Fc)

n = 2; 1,1'-quatreferrocene (Fc-Fc-Fc-Fc)

In the polyferrocenes, iron is present in the +2 formal oxidation state. Iron (III) is also an accessible oxidation state in the ferrocene coordination environment,and the mixed-valence biferrocenium ion, $[(C_5H_5)Fe(C_5H_4-C_5H_4)Fe(C_5H_5)]^+$, has been well characterized.(10)

From chemical and electrochemical studies, both the ruthenium(II) and ferrocene polymers undergo a series of distinct, chemically reversible, one-electron oxidations. By oxidizing the polymers at controlled potentials, or by using controlled amounts of a chemical oxidant, solutions containing mixed-valence or fully oxidized ions can be obtained, e.g., $[(bipy)_2ClRu(pyz)[Ru(bipy)_2pyz]_2RuCl(bipy)_2]^{10+/9+/8+/6+}$ and $[(C_5H_5)Fe(C_5H_4-C_5H_4Fe(C_5H_4-C_5H_4)Fe(C_5H_5)]^{3+/2+/+/0}$. The polymeric, mixed-valence ions are of interest when compared to related dimeric systems. Cooperative interactions between more than two metal ion sites may lead to molecular properties more normally associated with solid state materials.

For the ruthenium polymers, spectrophotometric titrations, using Ce(IV) in acidic aqueous solution, have shown that each ruthenium(II) site in the polymeric chains can be oxidized to ruthenium(III),

$$(Ru^{II})_n + n\ Ce(IV) \longrightarrow (Ru^{III})_n + n\ Ce(III)$$

A series of both fully oxidized and partly oxidized (mixed-valence) ions have been isolated as ClO_4^- salts. Recently, work has begun on the low temperature magnetic properties of the salts in collaboration with Professor W. E. Hatfield of the University of North Carolina. Work has also begun on the solid state electrical conductivity properties of the salts. It may prove possible, to some extent, to vary systematically the conductivity properties of a series of polymeric ions in the solid state. The synthetic chemistry involved is extremely versatile and the polymeric ions can be modified in a controlled way with regard to such features as: 1. The number of units in the polymeric chain. 2. The ratio of Ru(II) to Ru(III) sites. 3. The pattern of bridging ligands. 4. The non-bridging ligands. 5. The introduction of metal ions different from ruthenium.

For the 1,1'-polyferrocene compounds, electrochemical measurements have shown that each ferrocenyl group (-Fc) can be oxidized to ferrocenium ($-Fc^+$). For example, for 1,1'-terferrocene, there are three voltammetric waves at well-separated potentials. The voltammetric waves correspond to the electrode reactions:

	$E_{1/2}$,v
$(Fc\text{-}Fc\text{-}Fc)^{+} + e \longrightarrow (Fc\text{-}Fc\text{-}Fc)$	0.22
$(Fc\text{-}Fc\text{-}Fc)^{2+} + e \longrightarrow (Fc\text{-}Fc\text{-}Fc)^{+}$	0.44
$(Fc\text{-}Fc\text{-}Fc)^{3+} + e \longrightarrow (Fc\text{-}Fc\text{-}Fc)^{2+}$	0.82

The $E_{1/2}$ values are half-wave potentials vs. the saturated sodium chloride calomel electrode in 1:1 v/v dichloromethane-acetonitrile at 25±2°C.

Solutions containing mixed-valence ions such as $(Fc\text{-}Fc\text{-}Fc)^{+}$ and $(Fc\text{-}Fc\text{-}Fc)^{2+}$ can be prepared by controlled potential electrolysis. For the mixed-valence biferrocenium ion $(Fc\text{-}Fc)^{+}$, it has been concluded that discrete Fe(II) and Fe(III) sites exist and that electronic delocalization between the two sites is small.(10) Because of the similarities in spectral and redox properties between the biferrocenium ion and the mixed-valence polyferrocene ions, it appears that the mixed-valence polyferrocene ions also contain weakly interacting but discrete Fe(II) and Fe(III) sites.

For the 1,1'-polyferrocene compounds there are chemically different sites (Fc- and -Fc-) in the polymeric chains. Upon oxidation to the mixed-valence ions more than one oxidation state isomer can exist. The oxidation state isomers differ with regard to the site(s) of oxidation. For example, for the ion $(Fc\text{-}Fc\text{-}Fc)^{+}$ there are two energetically equivalent isomers -- $Fc^{+}\text{-}Fc\text{-}Fc$ and $Fc\text{-}Fc\text{-}Fc^{+}$ -- and one energetically nonequivalent isomer -- $Fc\text{-}Fc^{+}\text{-}Fc$.

From the effect of the ferrocenyl group as a substituent, it can be estimated that the isomers $Fc^{+}\text{-}Fc\text{-}Fc$ and $Fc\text{-}Fc^{+}\text{-}Fc$ are of similar energy. However, for other mixed-valence ions the difference in free energy between isomers can be significant. From $E_{1/2}$ data and the ferrocenyl substituent effect it is possible to estimate that for the reaction

$$Fc^{+}\text{-}Fc\text{-}Fc^{+} \longrightarrow Fc^{+}\text{-}Fc^{+}\text{-}Fc$$

$\Delta G \sim 2.8$ kcal/mole.

The properties of Intervalence Transfer (IT) bands are influenced by oxidation state isomerism. IT bands for several of the mixed-valence ions are given in Table II. For some of the ions the IT λ_{max} is shifted significantly from λ_{max} for $Fc^{+}\text{-}Fc$. The band shifts are expected when it is realized that the products of light-induced electron transfer are not symmetric for the mixed-valence polyferrocene ions. In some cases the products of light-induced electron transfer are energetically unfavorable oxidation state isomers, for example,

$$Fc^{+}\text{-}Fc\text{-}Fc^{+} \xrightarrow{h\nu} (Fc\overset{+}{-}Fc^{+}\text{-}Fc)^{*}$$

As described by Hush(7) the IT transition energy in such a case

Table II

Intervalence Transfer Bands for Mixed-Valence 1,1'-Polyferrocene Ions[a]

Transition	λ_{max} kK	λ_{max} nm	ε
Fc^{+}-Fc $\xrightarrow{h\nu}$ (Fc-Fc^{+})*	5260	1900	760
Fc-Fc^{+}-Fc $\xrightarrow{h\nu}$ (Fc^{+}-Fc-Fc)*	5020	1990	1560
Fc^{+}-Fc-Fc^{+} $\xrightarrow{h\nu}$ (Fc^{+}-Fc^{+}-Fc)*	5990	1670	1080
Fc^{+}-Fc-Fc^{+}-Fc $\xrightarrow{h\nu}$ (Fc^{+}-Fc^{+}-Fc-Fc)*	5590	1790	1720

[a]In 1:1 (v/v)CH_2Cl_2-CH_3CN, 0.1M in (n-Bu)$_4NPF_6$.

will include the difference in ground state energies between the isomers Fc^{+}-Fc-Fc^{+} and Fc^{+}-Fc^{+}-Fc in addition to the usual Franck-Condon energy barrier.

An equation has been derived by Hush which relates the energy of IT absorption to the rate of intramolecular electron transfer in mixed-valence compounds.(7)The equation allows estimates to be made for processes like:

$$Fc^{+}\text{-}Fc \longrightarrow Fc\text{-}Fc^{+} \qquad k \sim 3\times10^{10}\ sec^{-1}$$

$$Fc^{+}\text{-}Fc\text{-}Fc^{+} \longrightarrow Fc\text{-}Fc^{+}\text{-}Fc^{+} \qquad k \sim 1\times10^{10}\ sec^{-1}$$

The rate of electron hopping between ferrocenyl and ferricenium groups is slower in Fc^{+}-Fc-Fc^{+} than in Fc^{+}-Fc, mainly because of the energetically different chemical sites in Fc^{+}-Fc-Fc^{+}. The rate differences involved are small, but significant, since they indicate that in ligand-bridged complexes, rates of intramolecular electron transfer can be varied systematically by changing the chemical environments of the constituent ions.

Acknowledgements.

Acknowledgements are made to the Materials Research Center of the University of North Carolina under grant number GH33632 with DARPA and to the Army Research Office - Durham under grant number DA-ARO-D-31-124-73-G104 for support of this research.

Literature Cited

1. Meyer, T. J., Progr.Inorg. Chem., in press.
2. Ferguson, J. A. and Meyer, T. J., J. Amer. Chem. Soc., (1972), 94, 3409.
3. Ferguson, J. A., and Meyer, T. J., Chem. Commun., (1971), 1544.
4. Godwin, J. B. and Meyer, T. J., Inorg. Chem., (1971). 10, 471.
5. The magnetic data was obtained in collaboration with Professor W. E. Hatfield and his research group at the University of North Carolina.
6. Bleaney, B. and Bowers, K., Proc. Royal Soc. (London), (1952), 214A, 451.
7. Hush, N. S., Progr. Inorg. Chem., (1967), 8, 391.
8. Creutz, C. and Taube, H., J. Amer. Chem. Soc., (1973), 95, 1006.
9. Adeyemi, S. A.; Johnson, E. C.; Miller, F. J.; and Meyer, T. J.; Inorg. Chem. (1973), 12, 2371.
10. Cowan, D. O.; Le Vanda, C.; Park, J.; and Kaufman, F.; Accounts Chem. Res., (1973), 6, 1.

8

Exchange Interactions in Nickel(II), Copper(II), and Cobalt(II) Dimers Bridged by Small Anions

DAVID N. HENDRICKSON and D. MICHAEL DUGGAN

University of Illinois, Urbana, Ill. 61801

Introduction

In this paper some very recent work on "magnetic" exchange interactions in certain nickel(II) and copper(II) dimers will be summarized and some initial findings on analogous cobalt(II) and manganese(II) compounds will be presented. The compounds to be discussed include the following:

$$[M_2(tren)_2X_2](BPh_4)_2$$

where,

M = Ni(II), Cu(II), Co(II), Mn(II)
X = N_3^-, OCN^-, SCN^-, $SeCN^-$, CN^-(Cu only) and,
tren = 2,2',2"-triaminotriethylamine

One objective at the outset of this work was to keep the non-bridging ligand tren and the counterion tetraphenylborate constant while observing the effects on the exchange interaction of changing either the metal or the bridging ligand. The anticipation of obtaining similar dimer structures upon interchanging metals was not to be realized; however, interesting variations developed. It will be noticed that extended bridging groups have been selected to eliminate the consideration of direct metal-metal exchange interactions. In addition, the tetraphenylborate counterion should provide some degree of shielding between dimer units.

The phenomenon of magnetic exchange is, of course, electronic in origin. A particular distinction is made in our work between the parameterization of the observed phenomenon (as per some effective spin Hamiltonian) and the interpretation of the effective spin Hamiltonian parameters from the electronic wavefunctions of the system in question. This latter pursuit is the most important; at present it is only possible to provide qualitative or perhaps semi-quantitative interpretations of observed trends in parameters.

Nickel(II) Dimers

The four nickel systems are dimeric with two X^- anions end-to-end bridging such that the nickel atoms are octahedrally coordinated. Initially this was deduced from careful infrared and electronic absorption spectroscopy and X-ray powder pattern work. For example, the powder patterns of $[Ni_2(tren)_2NCO)_2]$-$(BPh_4)_2$ and $[Ni_2(tren)_2(oxalate)](BPh_4)_2$ are very similar and since the oxalate ($C_2O_4^{2-}$) ligand has been established (1) to bridge in a bis-bidentate fashion between two metal centers, the end-to-end di-bridging nature of the azide compound is indicated. As we examined the magnetic properties of this series of nickel dimers we became concerned with the details of the molecular structure.

The single-crystal X-ray structure of $[Ni_2(tren)_2$-$(NCO)_2](BPh_4)_2$ was determined (2). Discrete cationic $[Ni_2(tren)_2NCO)_2]^{2+}$ and anionic BPh_4^- units are found, where the metals are bridged by two cyanate groups bonding in an end-to-end mode. The structural characteristics of the cation are depicted in Figure 1. As a side-product of our exchange work we have thus established the first case of an authenticated oxygen-bonded cyanate metal complex. Each nickel atom in the cation is six coordinate with tren occupying four sites (Ni-N=2.047(7), 2.054(5), 2.095(7), and 2.130(7) Å) and the other two sites of the distorted octahedron are occupied by the N and O atoms of the bridging cyanate ions (Ni-N-2.018(7) and Ni-O=2.336(5)Å). The two cyanate bridges are essentially co-planar while the nickel atoms lie above and below the plane by ± 0.25Å. In respect to the exchange work there are three germane points. First, the Ni-Ni distance of 5.385(1)Å precludes a direct Ni-Ni exchange interaction. Second, packing diagrams show that the large BPh_4^- anions provide effective shielding between dimeric cations. Third, and most important to our discussion, it can be seen that each NCO^- in the cation is essentially bonding end-on at the nitrogen end ($\measuredangle$NiNC=155.0°) and almost at right angles at the oxygen end ($\measuredangle$NiOC = 117.1°).

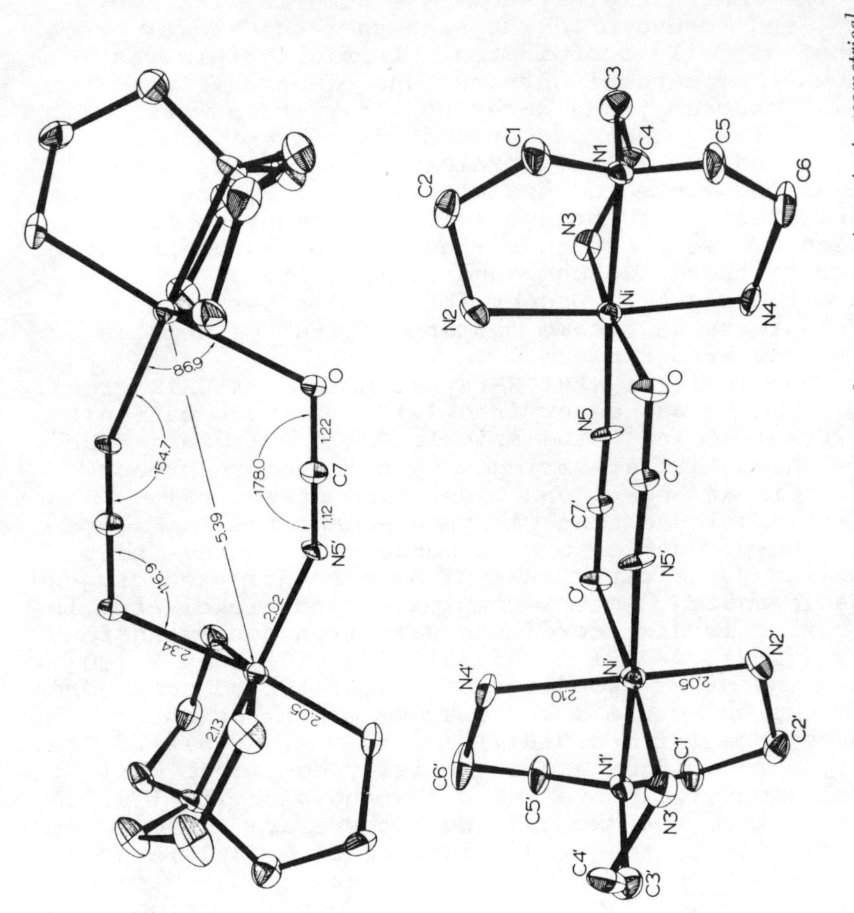

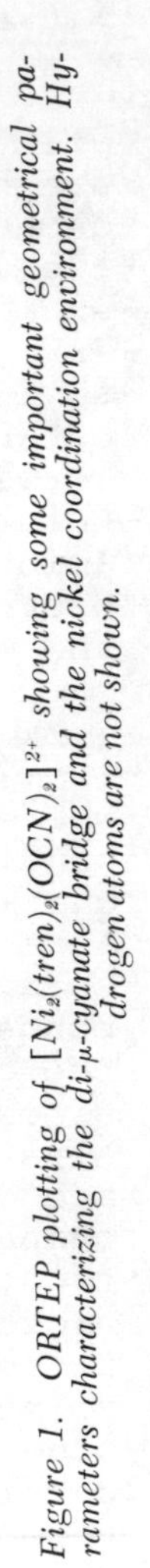

Figure 1. ORTEP plotting of $[Ni_2(tren)_2(OCN)_2]^{2+}$ *showing some important geometrical parameters characterizing the di-μ-cyanate bridge and the nickel coordination environment. Hydrogen atoms are not shown.*

Because there was available a structure of a related end-to-end bridged di-μ-thiocyanate nickel dimer, i.e., $[Ni_2(en)_4(NCS)_2]I_2$(3), and our work(4) on $[Ni_2(tren)_2(NCS)_2](BPh_4)_2$ had convinced us of their similarity of bridge structure, in our opinion it only remained to determine the structure of $[Ni_2(tren)_2(N_3)_2](BPh_4)_2$, which was solved very recently by Pierpont *et al*(5). The basic structural features of the $[Ni_2(tren)_2(N_3)_2]^{2+}$ cation are illustrated in Figure 2. The two end-to-end bridged azide groups are parallel and the nickel atoms are located ± 0.52Å from the azide plane. Each azide bridge is somewhat asymmetric with Ni-N-N angles of 135.3(7)° and 123.3(6)° and Ni-N distances of 2.069(8)Å and 2.195(7)Å. The coordination geometry at each nickel is again approximately octahedral.

We have measured the magnetic susceptibility of the four nickel dimers in the range of 4.2-283°K. In each case the (molar) paramagnetic susceptibility data were least-squares fit to an equation set out by Ginsberg et al(6):

$$\chi_M = (2Ng^2\beta^2/3k)\left[\frac{F_1}{T-4Z'J'F_1} + \frac{2F'}{1-4Z'J'F'}\right] + N\alpha$$

In this equation, N, β, and k have their usual meaning, Nα is the temperature-independent paramagnetism (taken as -200×10^{-6} cgs/mol of dimer), and F_1 and F' are complicated functions of temperature, single-ion zero-field splitting D, and the intradimer exchange parameter J. The effective interdimer exchange is $Z'J'$.

As we reported earlier(7), the compound $[Ni_2(tren)_2(N_3)_2](BPh_4)_2$ shows a relatively strong antiferromagnetic exchange interaction with a susceptibility maximum at ~100°K. In Figure 3 we have reproduced the susceptibility and effective moment curves for $[Ni_2(tren)_2(NCO)_2](BPh_4)_2$. In this case the susceptibility increases with decreasing temperature until a maximum is reached at 14°K, whereupon χ_m decreases rapidly to 4.2°K. The rapid decrease is indicative of a sample which is relatively free of paramagnetic impurities. The solid lines in Figure 3 are least-squares theoretical lines, fit to the above equation. Thus, qualitatively we see that, although the cyanate and azide nickel dimer structures are grossly similar, there is an appreciable attenuation in antiferromagnetic exchange interaction in going from the azide to the cyanate compound.

An even more striking change occurs when the nickel bridging groups are changed to either the

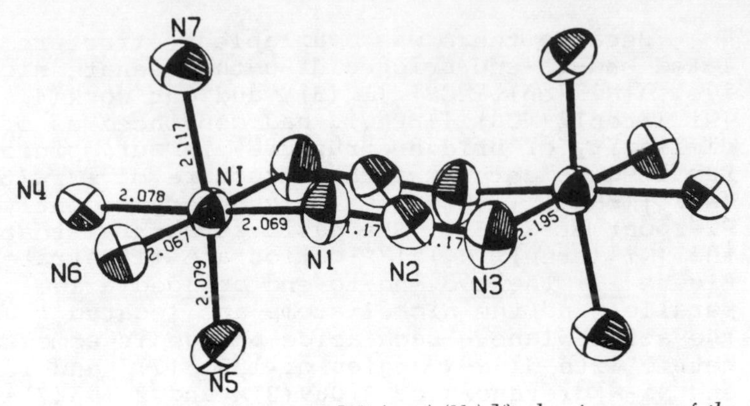

Figure 2. ORTEP plotting of $[Ni_2(tren)_2(N_3)_2]^{2+}$ showing some of the geometrical parameters; the dimer is located on a center of inversion, and carbon and hydrogen atoms are not shown.

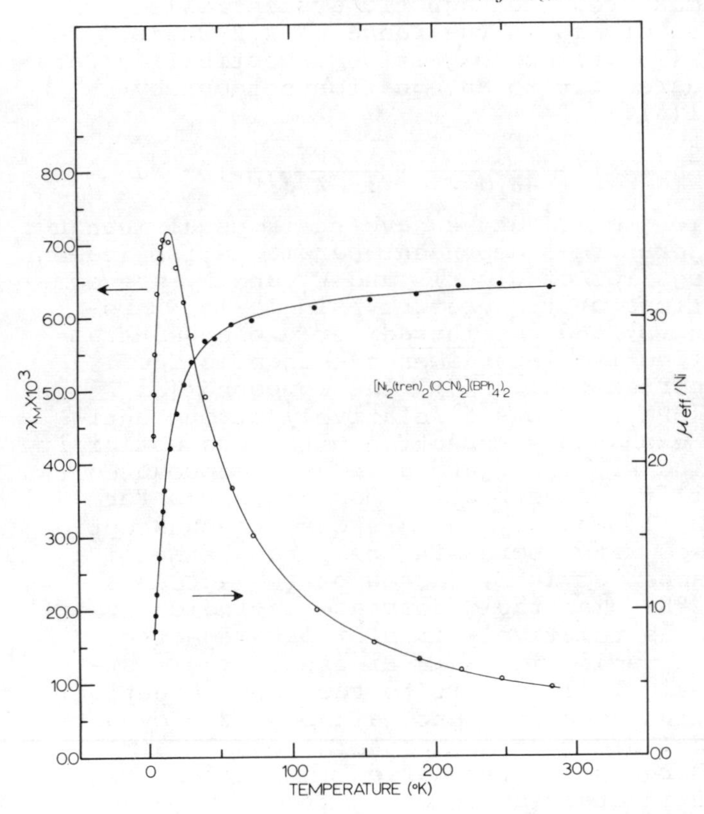

Figure 3. Experimental and calculated magnetic susceptibility data for $[Ni_2(tren)_2(NCO)_2](BPh_4)_2$. Lines are least-squares fit theoretical lines; for fitting parameters, see text.

thiocyanate (NCS^-) or selenocyanate ($NCSe^-$) groups. The effective moment curves for the nickel thiocyanate and selenocyanate dimers are shown in Figure 4. Both compounds have ferromagnetic interactions of approximately the same magnitude. For comparison purposes, the effective moment curve of one analogous, structurally characterized(8), cis-disubstituted tren nickel complex, i.e., $Ni(tren)(NCS)_2$, is also shown. In this last case there is a very small decrease in the moment at the lowest temperatures and this is undoubtedly due to single-ion zero-field splitting and perhaps a little intermolecular interaction. Parameterization of the susceptibility data for the four nickel dimers gives the following values:

Compound	J, cm^{-1}	g	D, cm^{-1}	Z′J′, cm^{-1}
N_3^-	-35	2.282	-10.1	0.71
NCO^-	-4.4	2.255	-0.45	-0.16
NCS^-	+2.4	3.347	-0.49	-0.07
$NCSe^-$	+1.6	2.181	-0.76	-0.02

It is important to mention that Ginsberg et al (6) reported a ferromagnetic exchange for $[Ni_2(en)_4(NCS)_2]I_2$ where they found J=+4.5 cm^{-1}.

With the magnetic and structural data in hand it is now appropriate to attempt to explain the change in sign and magnitude of the exchange parameter for these four nickel dimers. In our opinion it is the symmetry of the bridging units that is the most important factor in determining the sign and to a certain degree the magnitude of the exchange parameter J. Assistance in understanding the following explanation can be had by referring to Figure 5 where somewhat idealized (i.e., the azide system is represented as symmetric, because the figure was prepared before the structural details were known) plane projections of three of the nickel dimer bridging units are given. The geometry represented for the thiocyanate case is that drawn from the structure for $[Ni_2(en)_4(NCS)_2]I_2$(3), which we are convinced accurately approximates to that present in our compound. In the case of the most symmetric system, the azide dimer, the metals ions are bonding into the same bridge molecular orbitals. This leads to a pairing of electrons, that is a net antiferromagnetic exchange. Changing the bridge from azide to cyanate gives a bridging unit which is no longer symmetric and now the two nickel atoms bond into bridge orbitals that are not of necessity of equal construction with respect to the two metal centers. As a result there

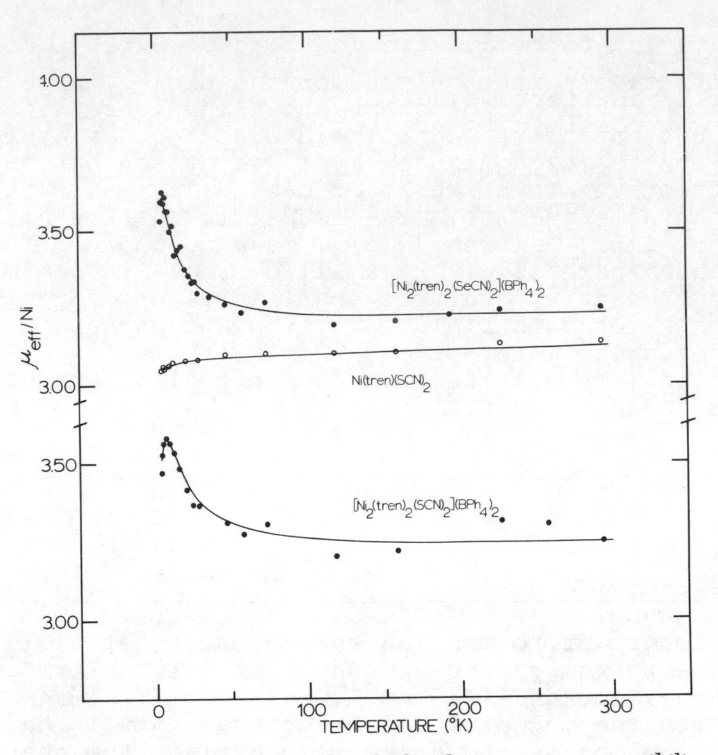

Figure 4. Experimental and calculated magnetic susceptibility data for $[Ni_2(tren)_2(NCSe)_2](BPh_4)_2$ (top, left), $Ni(tren)(NCS)_2$ (top, right), and $[Ni_2(tren)_2(NCS)_2](BPh_4)_2$ (bottom). The lines are least-squares fit theoretical lines.

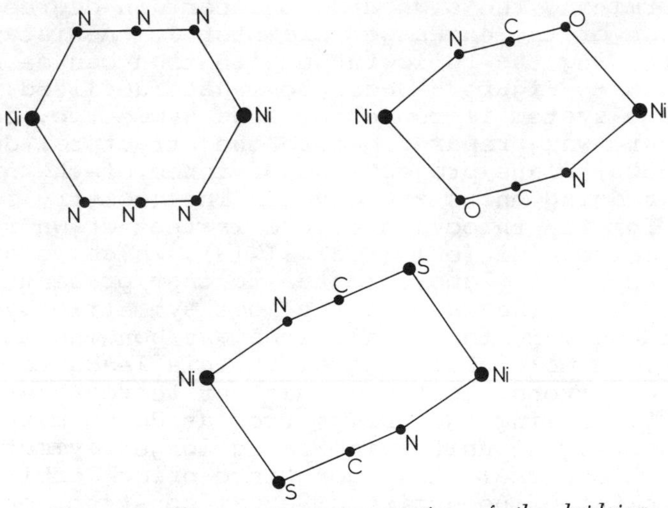

Figure 5. Diagramatic representation of the bridging structures of di-μ-azide, di-μ-cyanate, and di-μ-thiocyanate systems

are both antiferromagnetic and ferromagnetic pathways possible in the di-μ-cyanate system. The net exchange is weakly antiferromagnetic. It must be realized that it is not possible to predict from the X-ray structure that $[Ni_2(tren)_2(NCO)_2](BPh_4)_2$ would be weakly antiferromagnetic, however, it appears possible to rationalize the exchange in this compound relative to that in the azide compound.

If we continue the analysis, then, in the case of the thiocyanate- and selenocyanate-bridged dimers the transition from strong symmetry to strong antisymmetry has reached the point where the nickel-bridge bonding at the sulfur or selenium atom is closely approaching 90°. Because the bonding at the nitrogen atom is still close to 180°, ferromagnetic pathways develop and the net interaction can and does become ferromagnetic. We do not really know the details of the geometry of the selenocyanate bridging unit, but it is reasonable to assume that the geometry is closest to that for the thiocyanate case.

Before accepting this angularity or symmetry explanation of the observed exchange effects, there are three other considerations which must be discussed: 1. Is the exchange propagated through a σ-bonding or π-bonding pathway?, 2. Does the non-planarity of the bridging system affect the symmetry arguments? and, 3. What role might the Ni-X(X=O,N,S) distance play?

The third question in effect amounts to whether or not the Ni-X overlap integrals vary greatly for the distances of Ni-O=2.34, Ni-S=2.61, and Ni-N=2.02Å. These distances are a result of minimization of the energy of each system with respect to overlap, orbital energy, and nuclear repulsion, and as such the distances obtained are at least partially adjusted on the basis of overlap. The overlap integrals certainly do not vary as much as the distances would tend to indicate. It cannot be denied that, in comparison to the Ni-N bond in the azide, the longer Ni-O bond in the cyanate may contribute to a lessening of the exchange interaction. Certainly the change of sign of J down the series is not a reflection of differences in Ni-X distances.

The effect of differences in planarity, as guaged by the dihedral angle ϕ between the Ni-XYZ and XYZ-Ni′ planes, is difficult to judge. For the thiocyanate and cyanate systems ϕ is close to 0°, whereas, for the azide ϕ is 38.4°. These changes seem to parallel an expected increase in "allene-like" character of the bridge unit. In reference to possible σ-overlaps of the metal $d_{x^2-y^2}$ orbitals with π-symmetry bridge orbitals, it seems reasonable that for $0° < \phi < 90°$, molecular orbitals incorporating in a continuous bonding way both metal character and P_x and P_y type π-orbitals on the bridge would exist. The 38.4° dihedral angle of the azide dimer could, in part, lead to a larger antiferromagnetic exchange. An analysis of the dihedral angle effect is presented in greater detail for two nickel-azide compounds in a recent paper(5) and it is argued that the dihedral angle differences are not in the present case a major geometric factor.

Consideration of the first question revolves around whether the exchange interaction is of a first-order σ-exchange type or a higher-order "promoted" electron π-exchange. The unpaired "nickel d-electrons" are in $d_{x^2-y^2}$ and d_{z^2} type orbitals, which are sigma in construction. If there is overlap between these metal d-orbitals and the appropriate sigma bridge orbitals, then potentially viable σ-exchange pathways are present, and it is such a mechanism that we believe is present. In the promotion π-exchange mechanism one of the unpaired nickel electrons is promoted (configuration interaction with an excited state) into a π-type d-orbital (eg., d_{xz}) and this allows an overlap (ie., interaction) via the π-system of the bridge. The nature of say the d_{xz}-P_π(bridge) interaction would not be expected to change significantly as a function of Ni-bridge angle, and certainly not such as to affect the sign of the exchange interaction. Moreover, in a valence bond description (eg., sp^2 hybrids on the O and N atoms of OCN^- and sp hybridization at the carbon), the π-system of the bridge would involve an orthogonality at the carbon atom. To the extent that this description is accurate, the excited state π-exchange pathway is ineffective.

To account quantitatively for all factors contributing to the observed exchange interactions is at this time impossible in that to do so would require a thorough understanding of the bonding involved.

Copper(II) Dimers

We have found that the copper compounds prepared under identical conditions are not inner-sphere bridged dimers. The copper dimers are what we shall call "outer-sphere" dimers. The single-crystal X-ray structures of the cyanate and cyanide copper compounds have been determined. Figure 6 gives an ORTEP drawing of the copper-cyanate dimer, ie., $[Cu_2(tren)_2(NCO)_2]^{2+}$. The remarkable feature of this dimer is that the two halves of the cyanate dimer are bridged solely by means of two N–H···O hydrogen-bonding contacts between a cyanate oxygen and a tren nitrogen on the second copper center. The bridging hydrogen could not be seen, but its position can be calculated to be ∿0.2Å from the N-O vector. Each copper atom in the cation is trigonal bipyramidally coordinated with tren occupying four sites [Cu-N=2.076(5), 2.090(5), 2.119(5) and 2.083(5)Å] and with an axial carbon-bonded cyanide [Cu-CN=1.967(7) and C-N=1.127(9)Å]. The trigonal bipyramids are distorted from perfect three-fold symmetry.

Following the elucidation of the structural characteristics of these copper outer-sphere dimers, we have set about the task of constructing an extended series of such copper dimers in order to study what we consider to be a solid state analog of solution-state outer-sphere electron transfer as assisted by various anions. Przystas and Sutin (9) have studied several outer-sphere redox reactions as they are influenced in solution by added anions. Perhaps the hydrogen-bonding contact that is present in, for example, $[Cu_2(tren)_2(NCO)_2](BPh_4)_2$, is sufficient to establish an electron transfer between the copper centers. If this is the case and if we can measure the exchange parameter J, then we could study the effects of changing the bridge (ie., CN^- for OCN^-) on the exchange parameter J and therefore on the outer-sphere electron transfer rate.

We have verified that this same hydrogen-bonding association is found for a series of copper compounds. Figure 7 shows an ORTEP drawing of the outer-sphere dimeric cation in $[Cu_2(tren)_2(CN)_2](BPh_4)_2$ (10). Again we have essentially trigonal bipyramidal copper centers in an outer-sphere association. Inspection of the details of packing these outer-sphere dimers with BPh_4^- anions shows that the dimers are reasonably isolated by the very large counterions. As might be expected, exchange interactions are relatively weak for these outer-sphere systems. Figure 8 illustrates the susceptibility and effective moment curves for the copper-cyanide dimer. As can be seen there is no sign of an interaction until very low temperatures. However, it is quite clear that there is an interaction via the outer-sphere association in this compound. Fitting the susceptibility data to the usual exchange equation for a copper dimer gives $J=-1.8\ cm^{-1}$.

Because the cyanate, thiocyanate and selenocyanate anions are more extended, it was anticipated that J would be smaller than that for the cyanide. As such it became clear to us that we needed to use esr to

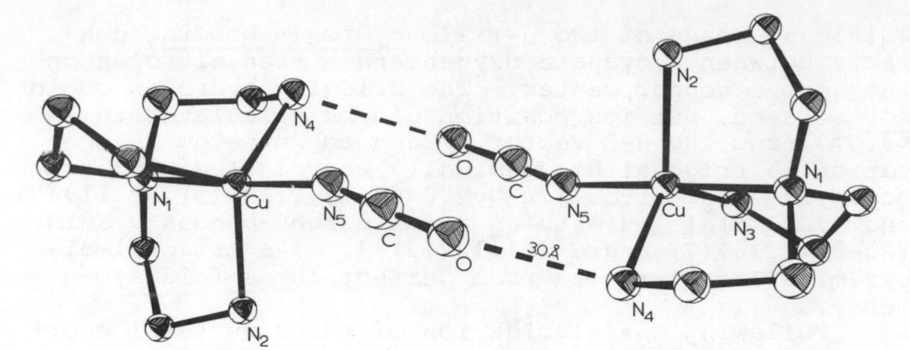

Figure 6. Molecular structure of the "dimer" cation in form I of $[Cu_2(tren)_2(NCO)_2]$-$(BPh_4)_2$ after refinement of atomic positions with isotropic thermal parameters. Hydrogen bonding contact is indicated by a dashed line.

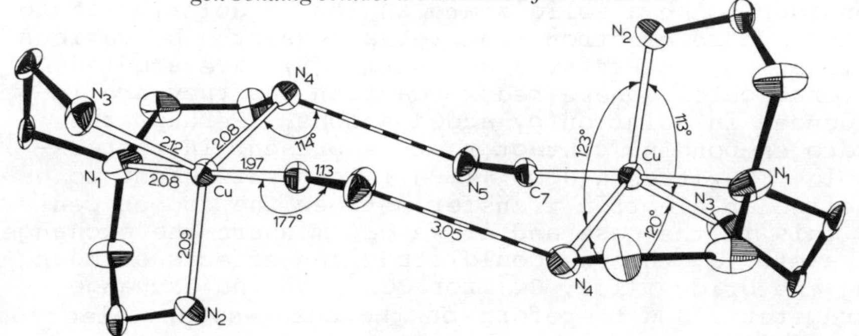

Figure 7. ORTEP plotting of $[Cu_2(tren)_2(CN)_2]^{2+}$. Hydrogen bonding contact is indicated by a dashed line.

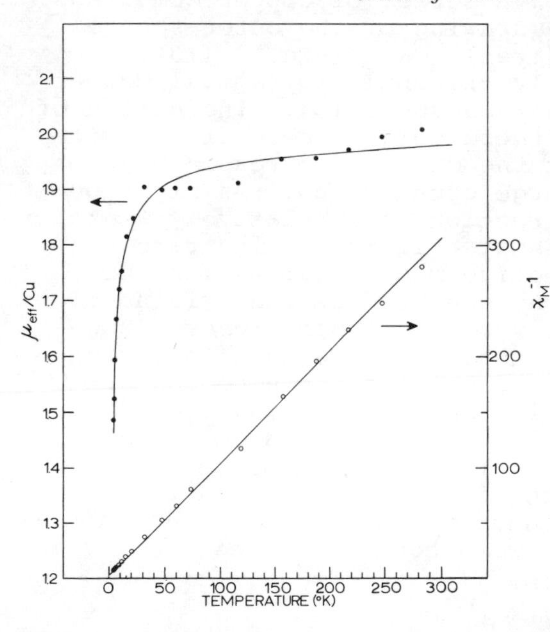

Figure 8. Experimental and calculated magnetic susceptibility data for $[Cu_2(tren)_2(CN)_2](BPh_4)_2$. Lines are least-squares fit theoretical lines.

determine J. Previous to this work, singlet-triplet esr transitions have been reported in only four cases: vanadyl tartrate dimer (11), nearest-neighbor exchange coupled Cu^{2+} ions doped into a potassium zinc sulfate lattice (12), for a Gd system (13), and for one discrete copper dimer (14).

Figure 9 shows an idealized esr spectrum for the case where J is of a magnitude that is comparable to the microwave energies ($\sim$0.3 cm^{-1} in X-band) used in an esr experiment. The zero-field splitting in our copper dimers is small in comparison with the exchange parameter J. By far the dominant feature in an X-band spectrum of such a system is the ΔM_s=1 transition at $\sim$3200 Gauss, represented in Figure 9 as a single featureless derivative. If the zero-field splitting D is of the proper magnitude, a ΔM_s=2 transition can be seen at half the field value of the ΔM_s=1 transition; the presence of this feature has been noted for several copper dimers. The two singlet-triplet transitions are indicated in Figure 9 as two transitions whose field positions are $\pm 2J/g\beta$ from the intense ΔM_s=1 transition. It is thus possible to determine J very accurately using esr, if the J value is in the esr range. As an aside it should be noted that the intensity of the singlet-triplet transition depends on $(A^2/4J^2)x$ $(M_{I_1}-M_{I_2})^2$ where A is the copper nuclear hyperfine interaction and the M_{I_i} values are nuclear spin projections for the two copper centers. Thus, there is a slight admixture of singlet and triplet functions as a result of electron-nuclear hyperfine.

A reproduction of the X-band spectrum of the outer-sphere copper cyanate dimer is given in Figure 10. In the 345°K spectrum (top tracing) the two singlet-triplet transitions are readily seen, one is close to the ΔM_s=2 transition. The two singlet-triplet transitions are equally spaced from the ΔM_s=1 transition, as expected, and their positions give a $|J|$ value of 0.09 cm^{-1}. When the temperature of $[Cu_2(tren)_2(NCO)_2](BPh_4)_2$ is lowered to 95°K, the two singlet-triplet features move to higher and lower field positions, respectively, and at 95°K the value of $|J|$ is calculated to be 0.16 cm^{-1}. We have, in fact, measured the temperature dependence of J for this compound and will report on this in a later paper (4).

Thus, with a combination of esr and variable-temperature magnetic susceptibility we have determined the J values for a series of outer-sphere copper dimers:

$[Cu_2(tren)_2(X)_2](BPh_4)_2$

	X	$J(cm^{-1})$
	N_3^-	---
	$SeCN^-$	---
	SCN^-	0.05 - 0.07
(I)	OCN^-	0.09 - 0.17
(II)	OCN^-	0.05 - 0.06
	CN^-	1.9
	Cl^-	3.2
	Br^-	3.5

For the azide and selenocyanate compounds we have not, as yet, been able to determine J; however, in both cases no signs of exchange interaction are seen in the susceptibility to 4.2°K nor are there any weak singlet-triplet esr transitions. Various physical data on the $SeCN^-$ compound indicate that it is isostructural with the cyanide and cyanate. The crystallography of the copper azide system is being pursued by Prof. Cort Pierpont. X-ray precession work on the SCN^-, Cl^-, and Br^- compounds indicates that they are also outer-sphere copper dimers. The thiocyanate compound shows singlet-triplet transitions in its X-band esr spectra, from which we calculated $|J|=0.05\ cm^{-1}$ at 345°K and $|J|=0.06\ cm^{-1}$ at 95°K. There are two crystalline forms of the copper cyanate compound; we have only determined the X-ray structure of the form labelled I. Both forms exhibit singlet-triplet transitions. It is perhaps relevant to point out that $[Ni_2(tren)_2(NCO)_2](BPh_4)_2$, even though it is composed of inner-sphere dimers, also has two different forms. Apparently, there are two ways of packing such dimers with the tetraphenylborate ion.

The halide-bridged outer-sphere dimers show the strongest interactions. The antiferromagnetic exchange interactions are large enough to easily discern with magnetic susceptibility to 4.2°K. However, it is interesting and at the same time puzzling that the bromide and chloride bridges support approximately the same magnitude of interaction if these interactions are propagated through hydrogen bonds. The X-ray crystal structure of the chloride compound is presently being worked on. Work (both experimental and theoretical) continues on these copper outer-sphere systems to understand the relationship between the "outer-sphere" exchange studies and the many reportings of "anion-assistance" of outer-sphere solution redox processes.

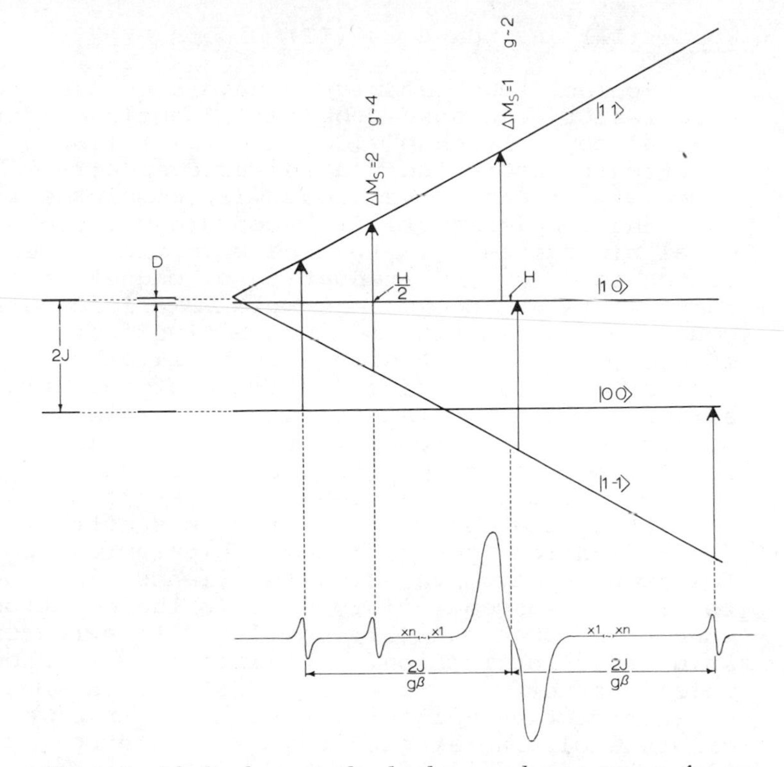

Figure 9. Idealized energy level scheme and esr spectrum for two S = 1/2 systems interacting to show both zero-field splitting (D) and exchange coupling (J). Allowed transitions ($\Delta M_s = 1$) are shown as large derivative feature in the spectrum whereas formally forbidden ($\Delta M_s = 2$ and singlet to triplet state) transitions are shown at an increased spectrometer gain (i.e., n *times greater).*

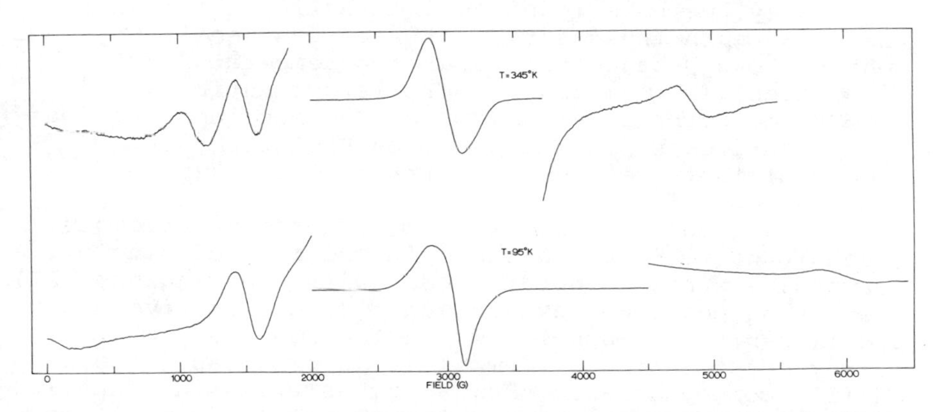

Figure 10. Temperature dependence of all visible features in the X-band (9 GHz) esr spectrum of $[Cu_2(tren)_2(NCO)_2](BPh_4)_2$, form I

Cobalt(II) and Manganese(II) Dimers

Schlenk tube techniques have been used to prepare a series of high-spin cobalt(II) complexes with the general composition of $[Co_2(tren)_2(X)_2](BPh_4)_2$. The electronic absorption and infrared spectra of these somewhat air-sensitive cobalt(II) complexes indicate that the complexes are five-coordinate, probably trigonal bipyramidal (TBP). The magnetic susceptibility curves indicate an attenuation of magnetism at low-temperatures. However, because zero-field interaction could be substantial in such a five-coordinate Co(II) species, it is not possible to easily extract the exchange parameter J from a fitting of the data. In agreement with this statement is our finding that the magnetism for five-coordinate $[Co(tren\text{-}Me_6)Cl]Cl$ also indicates a low-temperature attenuation.

Our attack on the cobalt(II) systems has been one of fitting the observed electronic spectra to ligand field equations for a trigonal bipyramidal complex, followed by attempts using the ligand field parameters to fit the susceptibility data to the equations for a monomeric TBP cobalt(II) complex with a reasonable spin-orbit interaction. Preliminary work shows that this does not work as well as fitting the magnetic data to an effective spin Hamiltonian for an S' = 3 Co(II) dimer, including single-ion zero field splitting. The ultimate check will be gauged by our success to fit the complicated low-temperature (4.2-90°K) X- and Q-band esr spectra we have measured for these compounds. Figure 11 shows three reasons (X-band at $\approx$ 15°K for A = N_3^-, B = OCN^- and C = SCN^-) why we are going to have considerable fun analyzing these systems. Empirically, we have found an interesting temperature dependence in these esr spectra as can see in Figure 12 which shows Q-band spectra at two temperatures for $[Co_2(tren)_2(NCO)_2](BPh_4)_2$. It is our anticipation and hope (perhaps using some of the machinery set out in an earlier symposium paper by Professor R.L. Belford) that these esr spectra hold the key to a determination of J.

The cobalt complexes have, after many attempts, been found to be unavailable in the form of single crystals. This is not the case with the manganese(II) compounds where we have obtained colorless crystals of $[Mn_2(tren)_2(X)_2](BPh_4)_2$, where X = NCO^- and NCS^-. X-ray crystal structure work is in progress. These Mn(II) compounds also show an attenuation in magnetism at very low temperatures. Large zero-field splittings in the esr spectra and infrared data point to the

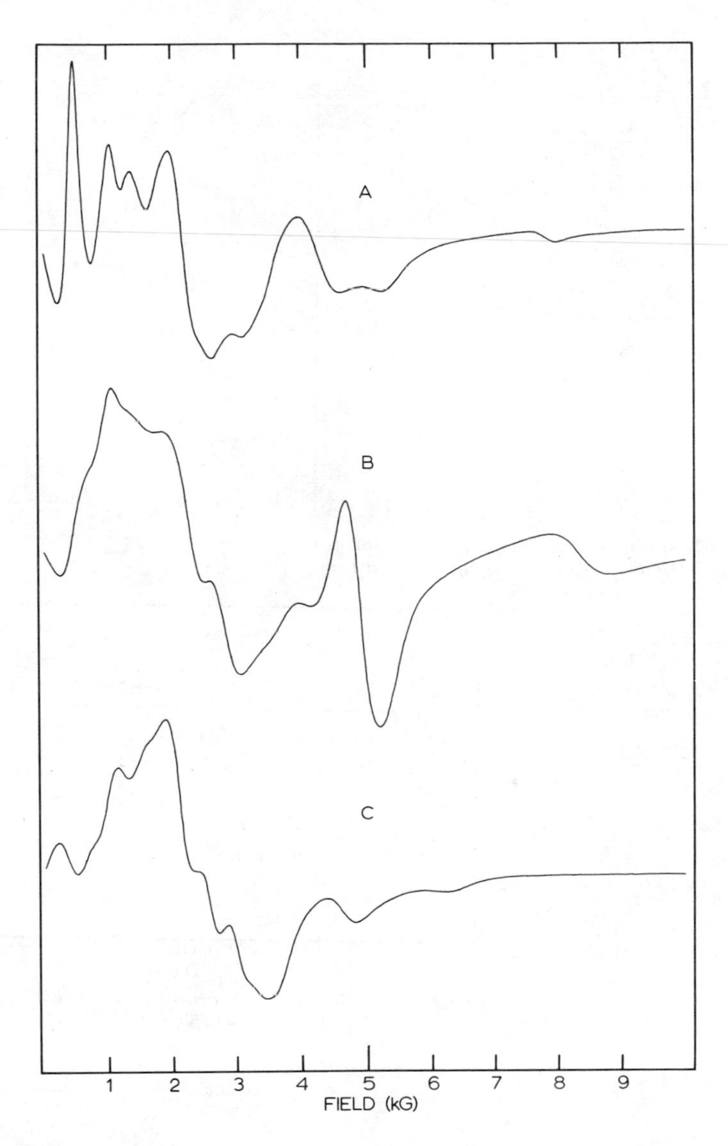

Figure 11. X-band (~15°K) esr spectra for three high-spin Co(II) compounds, $[Co_2(tren)_2X_2](BPh_4)_2$, where (A) X = N_3^-, (B) X = OCN^-, and (C) X = SCN^-.

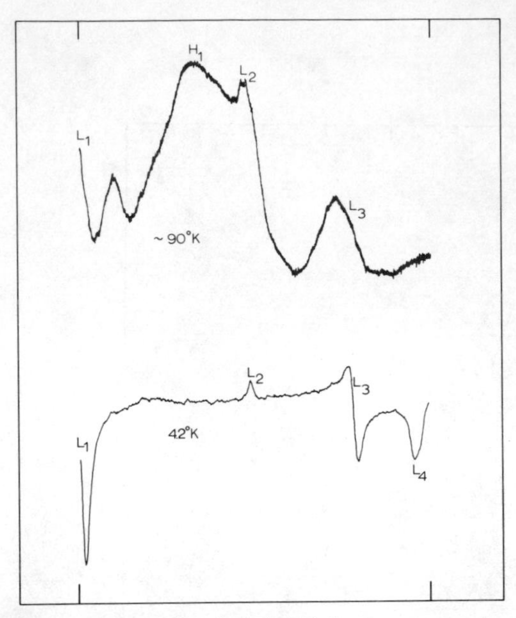

Figure 12. Q-band (35 GHz) esr spectra for $[Co_2(tren)_2(NCO)_2](BPh_4)_2$ at two different temperatures

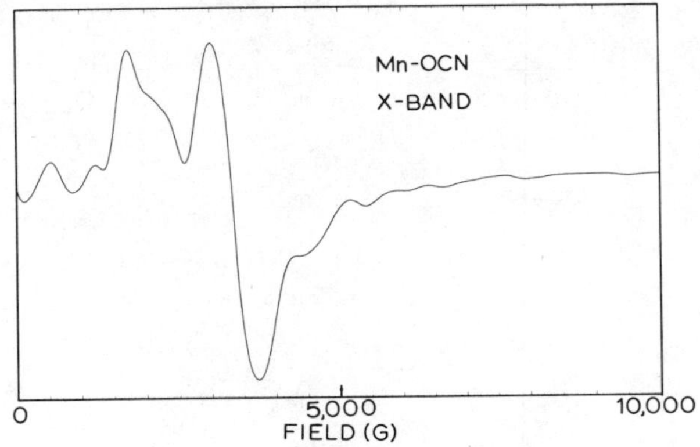

Figure 13. Room-temperature X-band esr spectrum of $[Mn_2(tren)_2(NCO)_2](BPh_4)_2$

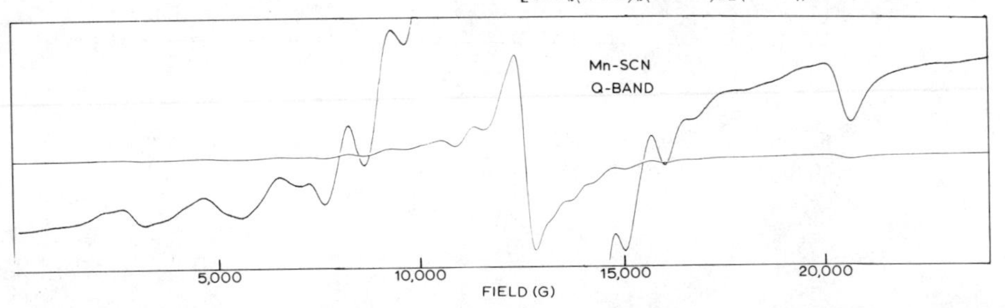

Figure 14. Room-temperature Q-band esr spectrum of $[Mn_2(tren)_2(NCS)_2](BPh_4)_2$

presence of outer-sphere dimers. Complicated and irregular X-band spectra such as that for $[Mn_2(tren)_2(NCO)_2](BPh_4)_2$ depicted in Figure 13 are clear indicators that we are dealing with dimers. There are a considerable number of features in this X-band spectrum, apparently encompassing the whole 10,000 Gauss range. The Q-band spectrum of the thiocyanate compound in Figure 14 is somewhat similar to the Q-band spectrum of the cyanate and as such it shows that there is an appreciable change in appearance upon changing microwave energies.

Literature Cited

1. Curtis, N.F., McCormick, I.R.N., and Waters, T.N., J. Chem. Soc., Dalton, (1973), 1537.
2. Duggan, D.M. and Hendrickson, D.N., Inorg. Chem., in press; **J. Chem. Soc., Chem. Commun., 411(1973).**
3. Shvelashvili, A.E., Porai-Koshits, M.A., and Antsyshkina, A.S., J. Struct. Chem. (USSR), (1969), 10, 552.
4. Duggan, D.M. and Hendrickson, D.N., Inorg. Chem., submitted for publication.
5. Pierpont, C.G., Hendrickson, D.N., Duggan, D.M., Wagner, F., and Barefield, E.K., Inorg. Chem., submitted for publication.
6. Ginsberg, A.P. Martin, R.L., Brookes, R.W., and Sherwood, R.C., Inorg. Chem., (1972), 11, 2884.
7. Duggan, D.M. and Hendrickson, D.N., Inorg. Chem., (1973), 12, 2422.
8. Cradwick, P.D. and Hall, D., Acta Cryst., (1970), B26, 1384.
9. Przystas, T.J. and Sutin, N., J. Amer. Chem. Soc., (1973), 95, 5545.
10. Duggan, D.M. and Hendrickson, D.N., Inorg. Chem., in press.
11. James, P.G. and Luckhurst, G.R., Mol. Phys., (1970), 18, 141.
12. Meredith, D.J. and Gill, J.C., Phys. Lett., (1967), 25A, 429.
13. Birgeneau, R.J., Hutchings, M.T., and Wolf, W.P., Phys. Rev. Letters, (1966), 17, 308.
14. Jeter, D.Y. and Hatfield, W.E., Inorg. Chim. Acta, (1972), 6, 440.

Acknowledgment. We are grateful for partial funding from National Institutes of Health grant HL13652.

9

Structural and Magnetic Properties of Chromium(III) Dimers

DEREK J. HODGSON

University of North Carolina, Chapel Hill, N.C. 27514

Introduction

A number of structural investigations in our laboratory and elsewhere have demonstrated that complexes of stoichiometry $[Cu(L)OH]_2^{2+}$, where L is a bidentate ligand, contain a dimeric unit in which two copper(II) centers are bridged by two hydroxo groups (1-5), and much of our recent research has been directed towards the correlation of the structural and magnetic properties of dimers of this type (6,7). We have recently extended these studies to chromium(III) complexes of the type $[Cr(L)_2OH]_2^{n+}$, where L is again a bidentate ligand, and in this paper I describe the results of our work in this area.

The structures with which we are concerned are molecules or ions of the type shown in figure 1, in which we have two chromium (III) centers which are bridged by two hydroxo groups in a planar array; the remaining coordination sites of the chromium octahedron are occupied by the bidentate ligands. For a symmetric bidentate ligand, there are two possible geometries for this dimer: if, in figure 1, atoms AN(1) and AN(10), BN(1) and BN(10), etc. form the chelate rings, the dimer lacks an inversion center but has approximately D_2 symmetry, but if, for example, the chelation at Cr(1) is changed to AN(1)-BN(10) and BN(1)-AN(10) while that at Cr(2) is unchanged the dimer has an inversion center and approximates C_{2h} symmetry. Each of these geometries is found. It should be noted, however, that in both of these geometries we maintain a planar bridging unit and octahedral coordination at the metal. Sinn (8) and Glick (9) have noted the influence of the geometry at copper on the magnetic interactions in copper dimers, but here we are able to keep the geometry at the metal center approximately constant.

The magnetic properties of these systems have been examined by my colleague, Professor W.E. Hatfield. The Van Vleck equation for exchange coupled Cr(III) ions (S = 3/2,3/2) can be written (10) as

$$\chi_m = \frac{Ng^2\beta^2}{kT}\left[\frac{2\ \exp(2J/kT) + 10\ \exp(6J/kT) + 28\ \exp(12J/kT)}{1+3\ \exp(2J/kT) + 5\ \exp(6J/kT) + 7\ \exp(12J/kT)}\right] \quad (1)$$

where J is the exchange coupling constant and $-2J$ represents the energy difference between the singlet ground state and the triplet first excited state. As a result of the presence of a manifold of relatively low-lying paramagnetic excited states, however, it has been suggested (11) that the Van Vleck expression should be modified by the inclusion of biquadratic exchange. This gives rise to the Hamiltonian

$$H_c = -2J(S_1 \cdot S_2) - j(S_1 \cdot S_2)^2 \quad (2)$$

and the expanded expression becomes

$$\chi_m = \frac{Ng^2\beta^2}{kT} \times \quad (3)$$

$$\frac{2\ \exp[(2J-6.5j)/kT] + 10\ \exp[(6J-13.5j)/kT] + 28\ \exp[(12J-9j)/kT]}{1.0+3\ \exp[(2J-6.5j)/kT] + 5\ \exp[(6J-13.5j)/kT] + 7\ \exp[(12J-9j)/kT]}$$

In this modified form of the Van Vleck equation, the energy separation between the singlet ground state and triplet first excited state, ΔE, is $-2J + 6.5j$.

Since it is our aim to correlate structural and magnetic properties, this discussion deals only with complexes whose structures have been precisely determined and does not include the large number of complexes (12-16) for which only magnetic data are available.

Glycinato Complex

The first structure which was determined was that of the glycinato complex $[Cr(gly)_2OH]_2$, which crystallizes in the monoclinic space group $P2_1/n$ with two dimers in a cell of dimensions a = 5.691(3), b= 16.920(9), c = 7.900(4) Å, and β = 79.90(3)° (17,18). With only two dimers in the cell, it is apparent that there must be an inversion center in the middle of the dimer, and that the molecule must be of the approximately C_{2h} type; an examination of the structure, which is shown in figure 2, verifies this conclusion. The Cr-Cr and O-O separations in the bridging unit are 2.974(2) and 2.575(6) Å, respectively, and the similarity of the two independent bridging Cr-O bond lengths of 1.966(4) and 1.968(4) Å demonstrates that the bridging in this unit is symmetric. The structural parameter of greatest interest is the value of the Cr-O-Cr bridging angle, ϕ, and in this case it is 98.2(2)° (18).

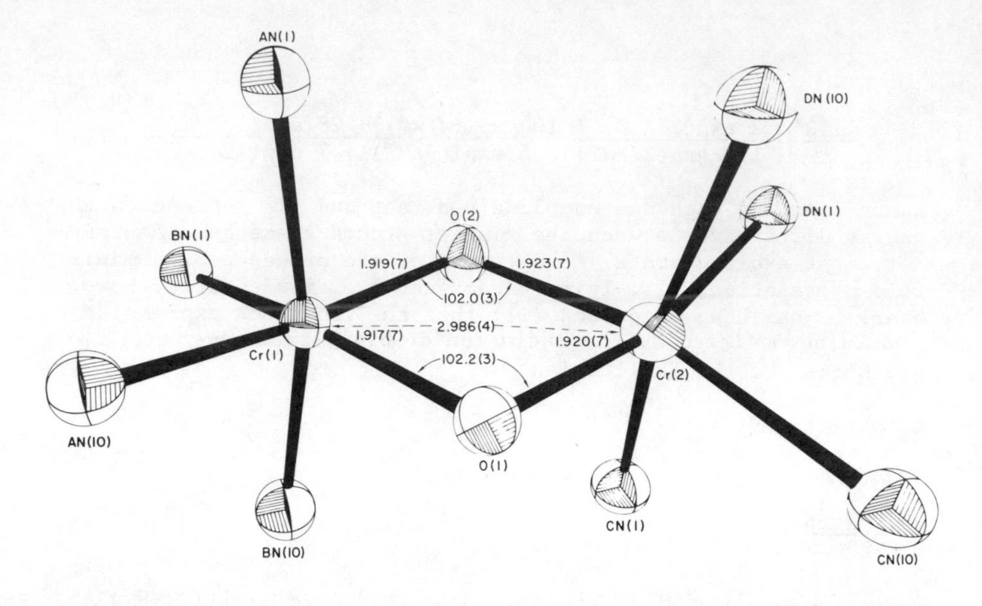

Figure 1. Coordination about chromium(III) centers in a typical dihydroxo-bridged dimer. O(1) and O(2) are the oxygen atoms of the hydroxo bridges. Chelate rings are formed by joining AN(1) to AN(10), BN(1) to BN(10), etc. Data are for the $[Cr(phen)_2OH]_2^{4+}$ cation in $[Cr(phen)_2OH]_2I_4 \cdot 4H_2O$.

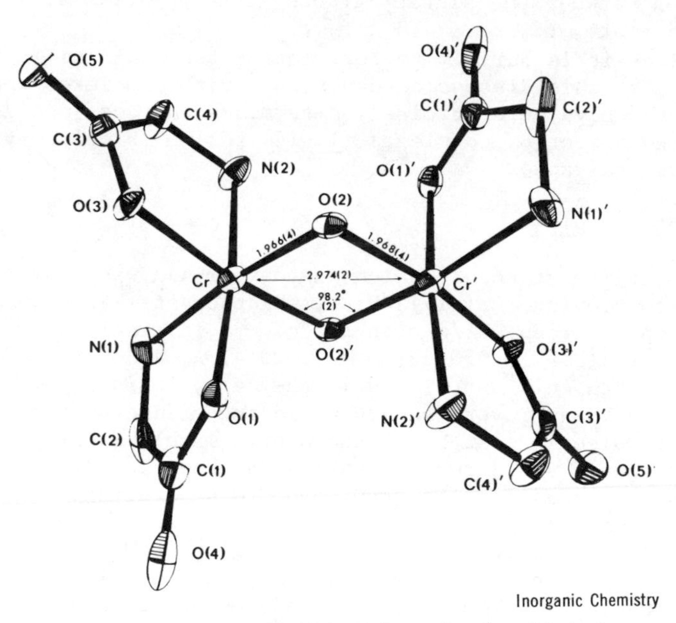

Figure 2. View of the $[Cr(gly)_2OH]_2$ molecule with hydrogen atoms omitted (18)

The low temperature magnetic susceptibility data for [Cr-(gly)$_2$OH]$_2$ are shown in figure 3, in which the dashed line represents the best least-squares fit to the unmodified form of the Van Vleck expression (equation (1)) while the solid line represents the best least-squares fit to equation (3). It is evident that, in this case, the observed susceptibility data are much more readily approximated by the solid line, *i.e.* the inclusion of biquadratic exchange is significant in this case. The magnetic susceptibility of [Cr(gly)$_2$OH]$_2$ maximizes near 20°K. The least-squares fitting process leads to values of $2J = -7.4$ cm^{-1} and $j = 0.04$ cm^{-1}, or $\Delta E = -10.0$ cm^{-1}. These values are in good agreement with the value of $2J$ predicted by Earnshaw and Lewis (12) on the basis of high temperature susceptibility data.

Phenanthroline Complexes

The second complex whose structure was determined was the 1,10-phenanthroline complex [Cr(phen)$_2$OH]$_2$Cl$_4$·6H$_2$O. This complex crystallizes in the triclinic space group $P\bar{1}$ with two dimers in a cell of dimensions a = 14.056(7), b = 11.296(6), c = 18.990(9) Å, α = 87.15(3), β = 107.63(2), and γ = 74.68(3)° (19). With two dimers in $P\bar{1}$, no crystallographic symmetry is imposed on the system, and this structure is an eighty atom problem, not counting the hydrogen atoms! The structure of the cation is shown in figure 4, and it is apparent that this ion closely approximates D_2 symmetry; there is no inversion center, but there are three approximate two-fold axes. If the dimer attempted to adopt the approximately C_{2h} geometry found in the glycinato complex, there would be very severe proton-proton interactions across the dimer, *e.g.* between the phenanthroline group labeled G2 and that labeled G3. Hence, for the bulky phenanthroline ligand, only the D_2 geometry is sterically feasible.

The coordination about the chromium(III) atoms is shown in figure 5. The Cr-Cr separation of 3.008(3) Å is a little larger than that in the glycinato complex, and this change is due to an increase of approximately 4.5° in the value of the bridging angle, ϕ. Thus, in this phenanthroline complex the average value of ϕ is 102.7°, while in the glycinato complex ϕ is 98.2° (vide supra).

The low temperature magnetic susceptibility of [Cr(phen)$_2$-OH]$_2$Cl$_4$·6H$_2$O exhibits a maximum near 110°K, and the best least-squares fit to equation (3) gives a value of ΔE of approximately -55 cm^{-1} (20). Single crystal epr examinations of this complex are currently nearing completion.

The magnetic data for this chloride salt of the phenanthroline complex are of considerable interest since, on the basis of high temperature measurements, Earnshaw and Lewis (12) have calculated that the corresponding iodide salt has a ΔE of approximately -14 cm^{-1}. Hence, if our contention that the magnetic properties of di-hydroxo-bridged dimers are principally determined by the geometry of the bridge were correct, it appeared that the geometry of the iodide salt must be considerably different from

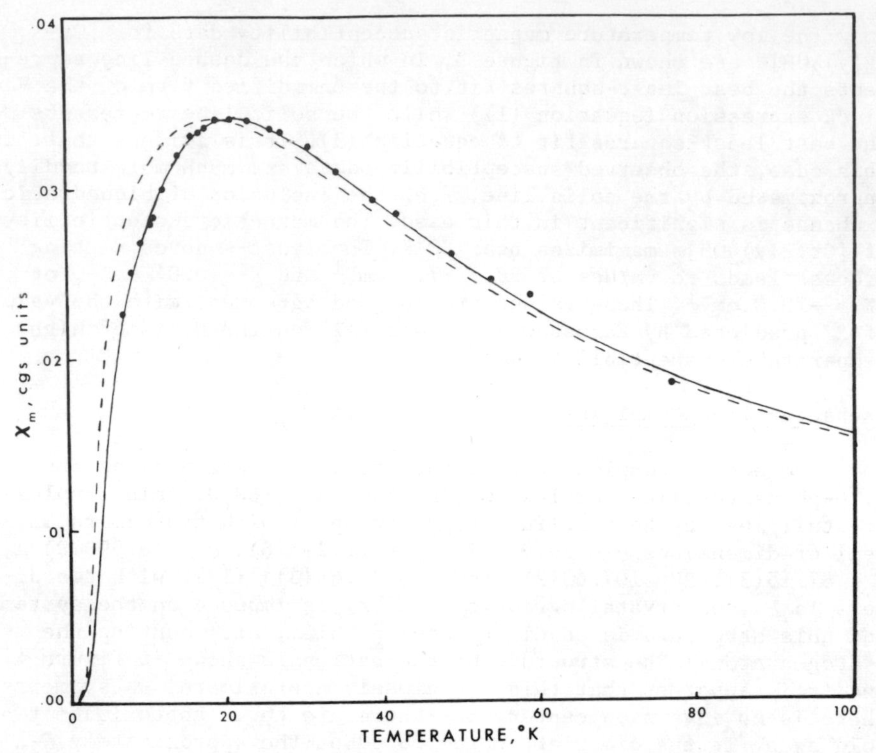

Figure 3. Temperature variation of the magnetic susceptibility of $[Cr(gly)_2OH]_2$. *Dashed line represents the best fit to equation (1); solid line represents the best fit to equation (3)* (see *text*) (18).

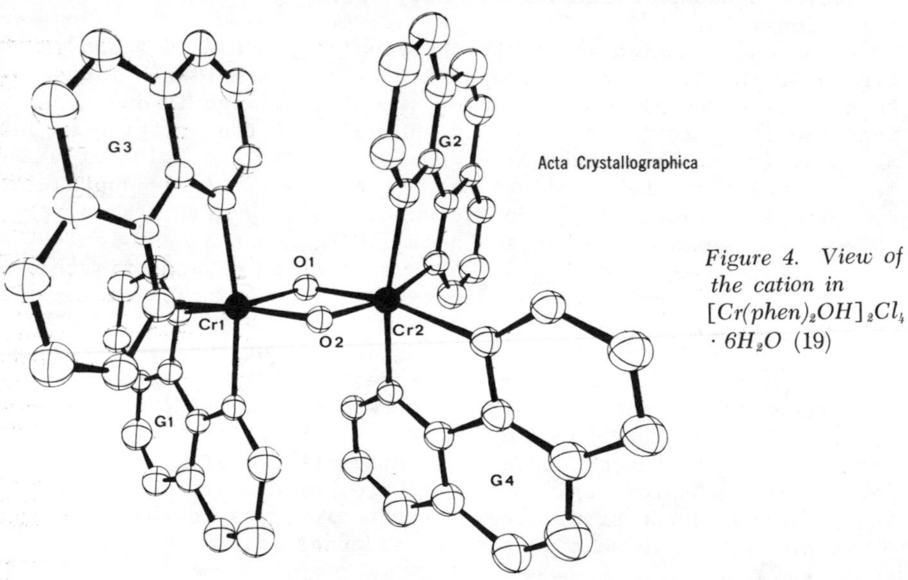

Figure 4. View of the cation in $[Cr(phen)_2OH]_2Cl_4 \cdot 6H_2O$ (19)

that of the chloride. Hence, we also examined the structure of the iodide salt, $[Cr(phen)_2OH]_2I_4 \cdot 4H_2O$.

The iodide salt also crystallizes in the triclinic space group $P1$ with two dimeric units in a cell of dimensions a = 11.464(12), b = 9.893(11), c = 22.757(25) Å, α = 90.06(2), β = 93.04(2), and γ = 82.82(2)° (21). The structure of the cation, which is shown in figure 6, is very similar to that found in the chloride salt, with the bulky phenanthroline ligands again forcing the dimer to adopt the roughly D_2 geometry. The inner coordination sphere for this dimer is shown in figure 1, and a comparison of this figure with figure 5 demonstrates that the $[Cr(phen)_2OH]_2^{4+}$ units in these two salts are structurally substantially similar (21). Thus, for example, the Cr-Cr separation and Cr-O-Cr bridging angle of 2.986(4) Å and 102.2(2)°, respectively, in the iodide salt probably do not differ significantly from the values (19) of 3.008(3) Å and 102.7(5)° in the chloride analog.

This structural result is, clearly, inconsistent with the magnetic properties reported by Earnshaw and Lewis (12), and so we have reexamined the magnetic susceptibility of the iodide salt (22). The low temperature susceptibility data are shown in figure 7, in which the solid line represents the best fit to the Van Vleck equation modified by the inclusion of biquadratic exchange. The magnetic susceptibility of $[Cr(phen)_2OH]_2I_4 \cdot 4H_2O$ is seen to maxmize near 110°K, and the least-squares fit to equation (3) yields $2J$ = -43.8 cm^{-1}, j = +1.5 cm^{-1}, and ΔE = -53.6 cm^{-1} (22). These values are very similar to those obtained (20) for the chloride but are substantially different from the value of ΔE = -14 cm^{-1} reported by Earnshaw and Lewis (12). Moreover, the similarity between these results and those for the chloride salt is consistent with the similarity of the two structures noted above. It is, however, noteworthy that while the magnetic properties of the corresponding nitrate salt, $[Cr(phen)_2OH]_2(NO_3)_4 \cdot 7H_2O$, with values of $2J$ = -42.2 cm^{-1}, j = 0.0 cm^{-1}, and ΔE = -42.2 cm^{-1} (23) are substantially similar to those of the chloide and iodide salt, the bromide salt, $[Cr(phen)_2OH]_2Br_4 \cdot 8H_2O$, apparently undergoes a weaker interaction with values of $2J$ = -29.2 cm^{-1}, j = 0.5 cm^{-1}, and ΔE = -32.5 cm^{-1} (23). It would appear, therefore, that the cation in the bromide salt may indeed be structurally different from that in the chloride and iodide cases, but no structural data are available to confirm or deny this hypothesis.

Oxalato Complex

The final structure of this type, which has recently been completed in our laboratories, is that of the oxalato complex $Na_4[Cr(OX)_2OH]_2 \cdot 6H_2O$. This material crystallizes in the monoclinic space group $P2_1/c$ with four dimers in a cell of dimensions a = 19.530(12), b = 9.860(7), c = 12.657(10) Å, and β = 106.93(4)°.

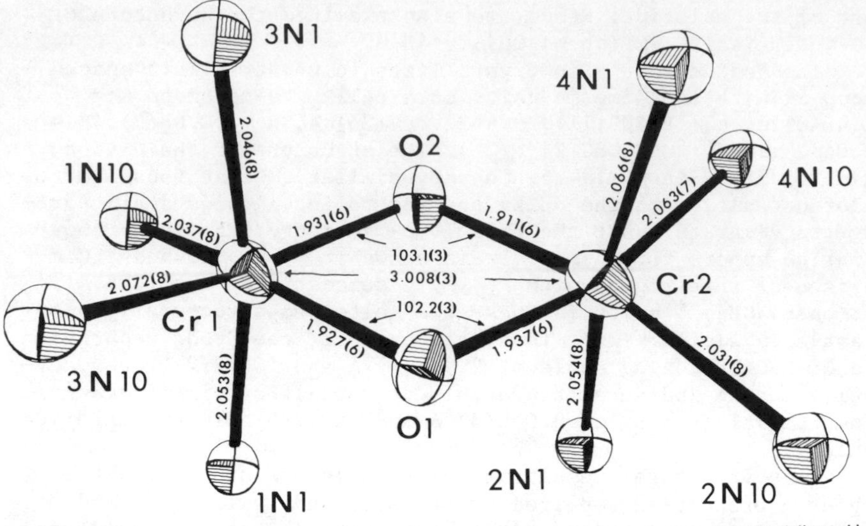

Figure 5. Coordination about the chromium(III) centers in $[Cr(phen)_2OH]_2Cl_4 \cdot 6H_2O$*. Chelate rings are formed by joining atoms 1N1 to 1N10, 2N1 to 2N10, etc. (19).*

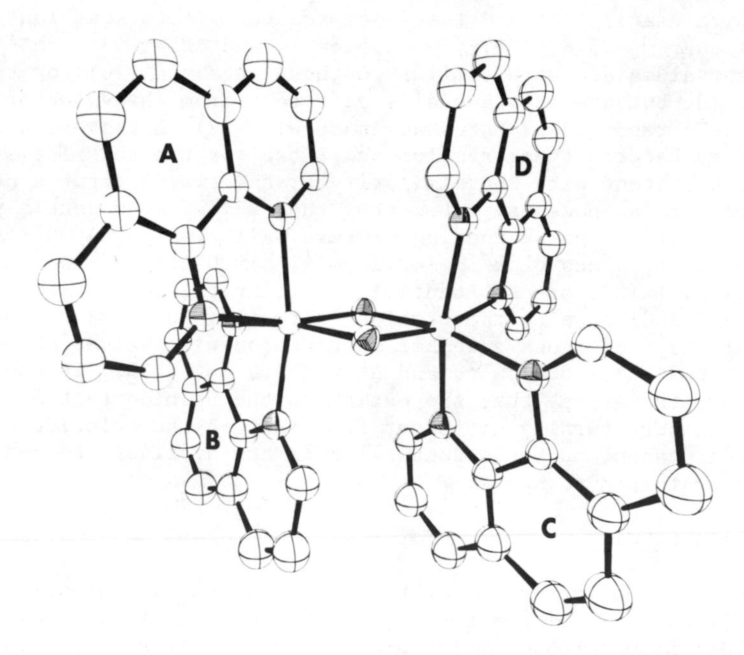

Figure 6. View of the cation in $[Cr(phen)_2OH]_2I_4 \cdot 4H_2O$

While no crystallographic symmetry is required for four dimers in this cell, it transpires that each dimer sits on a crystallographic inversion center so that there are, in effect, two separate independent "half-dimers" in the cell rather than one independent dimer. Hence, of course, the geometry of the anion must be of the C_{2h} type rather than the D_2 type. The structure of the anion is shown in figure 8, the bond lengths and angles given being the average of the values obtained for the two independent halves; the agreement between these two independent measurements is excellent (24). The values of the Cr-Cr separation of 3.000 Å and the Cr-O-Cr angle of 99.6(3)° are intermediate between those for the glycinato and phenanthroline complexes. Hence, while the only magnetic data available at present are the room temperature values (μ= 3.43 μB) obtained on the tetrahydrate (13), it is evident that the value of ΔE for this complex must lie between -10 and -53 cm^{-1} if there is a simple correlation between ΔE and ϕ.

Interpretation

The structural and magnetic data described above are correlated in figure 9, in which the singlet-triplet splitting ΔE is plotted against the bridging angle ϕ; a similar plot for the analogous copper(II) dimers $[Cu(L)OH]_2^{2+}$ is also included in figure 9. An examination of the chromium data suggests that the oxalato complex $Na_4[Cr(OX)_2OH]_2 \cdot 6H_2O$, which has a ϕ of 99.6° (vide supra), should have a ΔE of approximately -25 cm^{-1}. Figure 9 is noteworthy for two separate reasons: firstly, because for a given metal there is apparently an almost linear correlation between ΔE and ϕ, and secondly because the slope of the ΔE vs. ϕ plot for the chromium(III) complexes is considerably smaller than that for the copper(II) complexes. Each of these features is readily explained in terms of simple bonding theory.

The correlation between ΔE and ϕ. The correlation noted can be explained in terms of valence bond theory and the principles of super exchange(25). If the orbitals used by the bridging oxygen atoms are pure *p* orbitals, the bond angle is expected to be 90° and the ground state is predicted to be a triplet (*i.e.* $\Delta E > 0$); if the orbitals are purely *s* , the ground state is predicted to be a singlet (*i.e.* $\Delta E < 0$). Hence, since an increased value of the bridging angle implies greater *s* character in the bridging orbitals, we would expect a decrease in ΔE as the bridging angle is increased from 90°. For the six copper and three chromium cases which have been studied in detail, this trend is observed (2).

This correlation can also be expressed in terms of molecular orbital theory. The M-O-M-O ring is of approximate D_{2h} symmetry in these molecules, with the x-axis defined as the Cu-Cu direction and y-axis parallel to the O-O vector (26). Neglecting oxygen *s* orbitals, the eight σ-orbitals in this system transform in D_{2h}

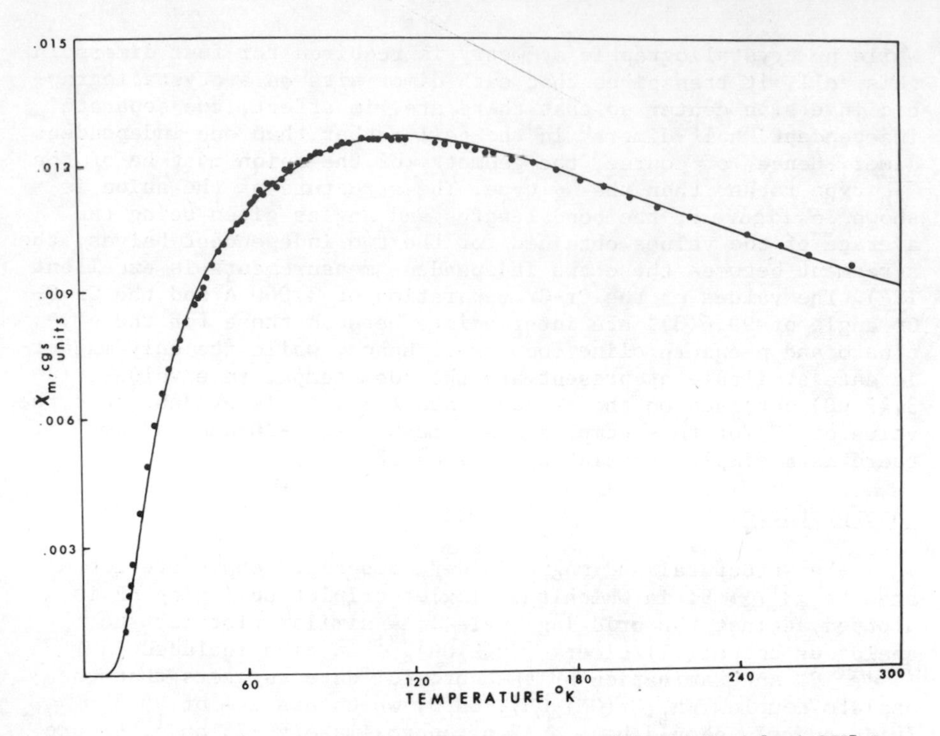

Figure 7. Temperature variation of the magnetic susceptibility of $[Cr(phen)_2OH]_2I_4 \cdot 4H_2O$*. Solid line represents the best fit to equation (3) (see text).*

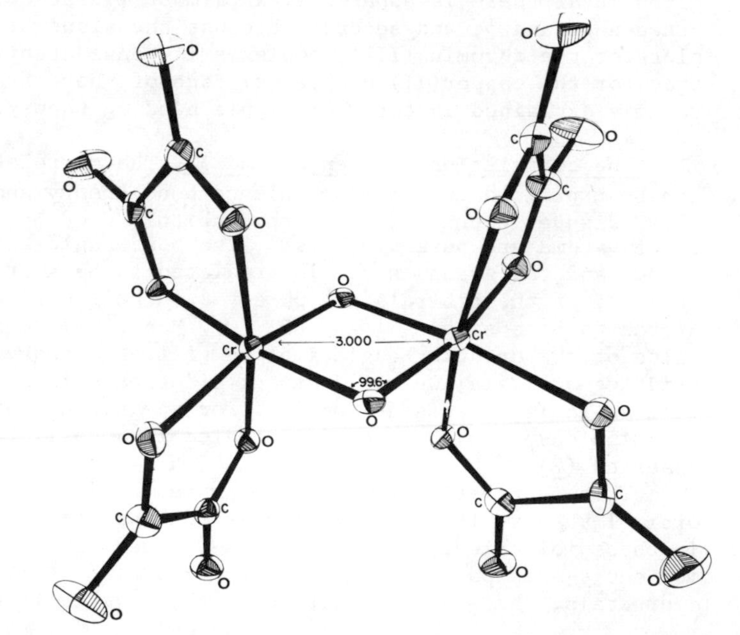

Figure 8. View of the anion in $Na_4[Cr(ox)_2OH]_2 \cdot 6H_2O$*. Data are the average values of the two crystallographically independent dimers.*

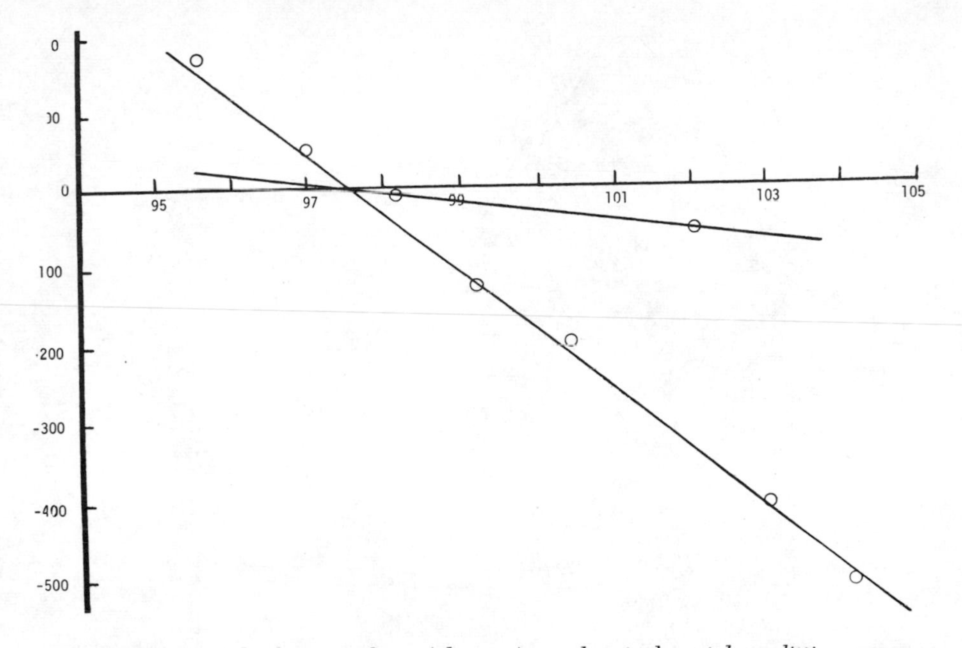

Figure 9. M–O–M bridging angle, φ (abscissa) vs. the singlet–triplet splitting energy, ΔE (ordinate) for dihydroxo-bridged complexes of Cu (steeper line) and Cr (flatter line)

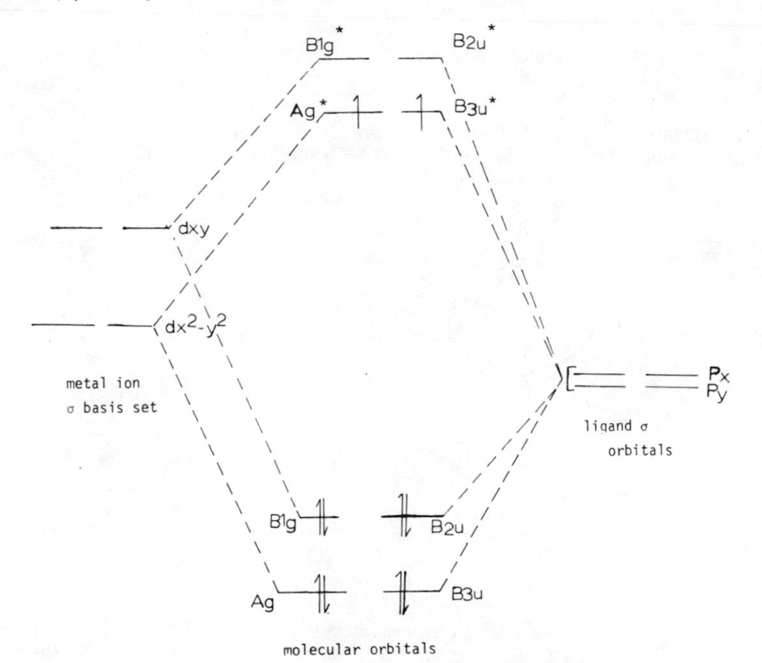

Figure 10. Molecular orbital diagram for the σ-orbitals in the Cr–O–Cr ring, assuming D_{2h} symmetry and a Cr–O–Cr angle of 90°

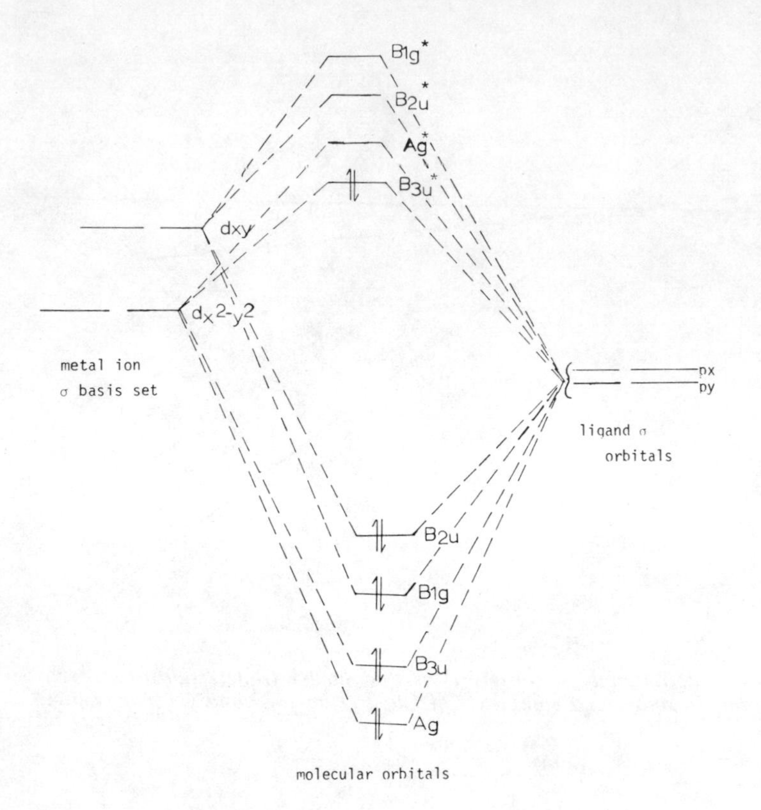

Figure 11. Molecular orbital diagram for the σ-orbitals in the Cr–O–Cr ring, assuming D_{2h} symmetry and a Cr–O–Cr angle considerably greater than 90°

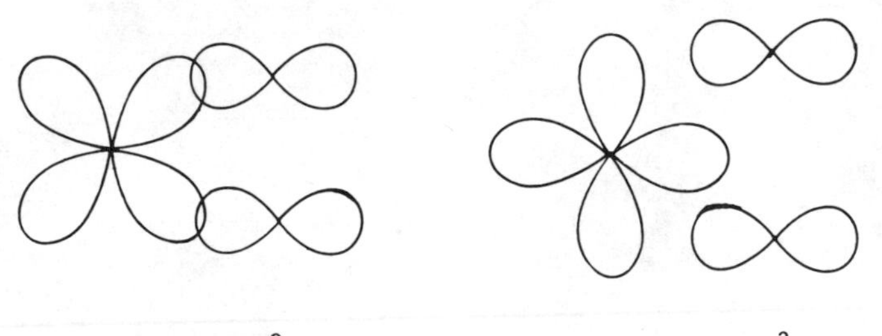

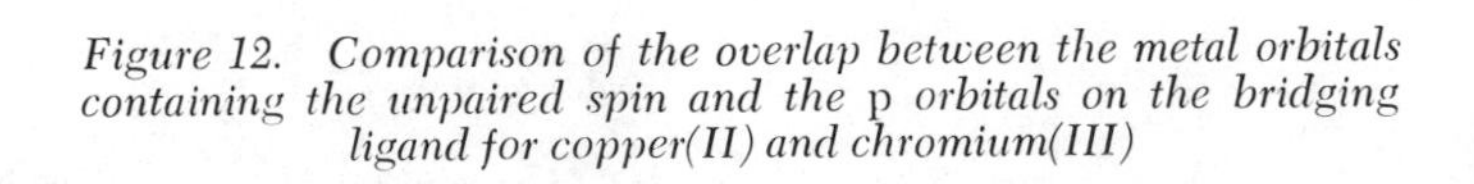

Figure 12. Comparison of the overlap between the metal orbitals containing the unpaired spin and the p orbitals on the bridging ligand for copper(II) and chromium(III)

symmetry as $2A_g + 2B_{1g} + 2B_{2u} + 2B_{3u}$, and there are bonding and anti-bonding combinations with all four of these symmetries. On the assumption that only oxygen *p* orbitals participate, the A_g and B_{3u} molecular orbitals would have identical energies when the Cr-O-Cr angle was 90°, as would the B_{1g} and B_{2u} orbitals. This situation is depicted in figure 10, which demonstrates that this ten electron system (for chromium) must give rise to a triplet ground state. As the Cr-O-Cr angle increases, however, the overlap of the A_g and B_{1g} combinations increases relative to that of the B_{3u} and B_{2u} combinations. Hence, the orbital degenercies in figure 10 are lifted, and at sufficiently large values of ϕ the molecular orbital diagram shown in figure 11 becomes operative. At these larger bridging angles, the splitting of the B_{3u}* and A_g* orbitals is sufficient to overcome the pairing energy, and the singlet state becomes the ground state. This molecular orbital view, therefore, is analogous to the valence bond approach above, and the experimental data may be interpreted on this basis; presumably, figure 11 becomes appropriate at ϕ values greater than approximately 97.6°, while at angles between 90° and 97.6° the splitting between the B_{3u}* and A_g* orbitals is less than the pairing energy, and so the triplet state remains lower in energy than the singlet (*i.e.* ΔE remains positive) (27). Unfortunately, for chromium(III) complexes of this general type there are no examples of positive ΔE values, but the presence of the triplet-ground state complexes in the copper(II) series lends strong support to this hypothesis. Moreover, of course, examination of the copper(II) line in figure 9 suggests that the ground state is the triplet if ϕ < 97.6°; hence, since for the chromium complexes the smallest value of ϕ yet obtained is 98.2°, none of the chromium complexes examined in detail is predicted to have a positive value of J (or ΔE).

The Slopes of the Cr and Cu Lines. In the copper(II) complexes the unpaired spin resides principally in atomic orbitals which point directly at the bridging oxygen ligands (the d_{xy} orbitals in figure 10, but more conventionally $d_{x^2-y^2}$ since the usual axial system is different from that forced upon us in D_{2h} symmetry), while in the chromium(III) complexes the unpaired spin is in t_{2g} orbitals which point between the bridging atom; this disparity is demonstrated pictorially in figure 12. Hence, since the overlap between the bridging orbitals and the metal orbitals containing the unpaired spin is much poorer for the chromium complexes than for the copper complexes, we predict that the magnitude of the spin-spin interaction should be greater for copper than for chromium. Hence, at a given value of ϕ, we predict that the magnitude of ΔE for copper is greater than that for chromium, *i.e.* that the slope of the line for copper is greater than that for chromium; this result, of course, is exactly what is seen in figure 9.

Acknowledgments

It is a pleasure to acknowledge the assistance which I have received from my present and former colleagues, especially Professor W.E. Hatfield, Dr. J.T. Veal, Dr. D.L. Lewis, Dr. D.Y. Jeter, Dr. R.F. Drake, Ms. E.D. Estes, Ms. D.E. Hartis, Mr. R.P. Scaringe, Mr. R.P. Eckberg, and Mr. K.T. McGregor.

Literature Cited

1. Lewis, D.L., Hatfield. W.E., and Hodgson, D.J., *Inorg. Chem.* (1974), 13, 147.
2. Lewis, D.L., Hatfield, W.E., and Hodgson, D.J., *Inorg. Chem.* (1972), 11, 2216.
3. Casey, A.T., Hoskins, B.F., and Whillans, F.D., *Chem. Commun.* (1970), 904.
4. Mitchell, T.P., Bernard, W.H., and Wasson, J.R., *Acta Crystallogr.* (1970), B26, 2096.
5. Majeste, R.J. and Meyers, E.A., *J. Phys. Chem.* (1970), 74, 3497.
6. McGregor, K.T., Watkins, N.T., Lewis, D.L., Drake, R.F., Hodgson, D.J., and Hatfield, W.E., *Inorg. Nucl. Chem. Letters* (1973), 9, 423.
7. Estes, E.D., Hatfield, W.E., and Hodgson, D.J., *Inorg. Chem.* (1974), in press.
8. Sinn, E. and Robinson, W.T., *Chem. Commun.* (1972), 359.
9. Glick, M.D. and Lintvedt, R.L., 167th. A.C.S. National Meeting, Los Angeles, California (1974).
10. Jezowska-Trzebiatowska, B. and Wojciedhowski, W., *Transition Metal Chemistry* (1970), 6, 1.
11. Ikeda, H., Kimura, I., and Uryu, N., *J. Chem. Phys.* (1968), 48, 4800.
12. Earnshaw, A. and Lewis, J., *J. Chem. Soc.* (1961), 396.
13. Morishita, T., Hori, K., Kyuno, E., and Tsuchiya, R., *Bull. Chem. Soc. Japan* (1965), 38, 1276.
14. Schugar, H.J., Rossman, G.R., and Gray, H.B., *J. Amer. Chem. Soc.* (1969), 91, 4564.
15. Kobayashi, H., Haseda, T., and Mori, M., *Bull. Chem. Soc. Japan* (1965), 38, 1455.
16. Jasiewicz, B., Rudolf, M.F., and Jezowska-Trzebiatowska, B., *Acta Physica Polonica* (1973), A44, 623.
17. Hodgson, D.J., Veal, J.T., Hatfield, W.E., Jeter, D.Y., and Hempel, J.C., *J. Coord. Chem.* (1972), 2, 1.
18. Veal, J.T., Hatfield, W.E., Jeter, D.Y., Hempel, J.C., and Hodgson, D.J., *Inorg. Chem.* (1973), 12, 342.
19. Veal, J.T., Hatfield, W.E., and Hodgson, D.J., *Acta Crystallogr.* (1973), B29, 12.
20. Eckberg, R.P., private communication.
21. Scaringe, R.P. and Hodgson, D.J., unpublished observations.

22. Eckberg, R.P., Scaringe, R.P., Hodgson, D.J., and Hatfield, W.E., unpublished observations.
23. Drake, R.F., Ph.D. Dissertation, University of North Carolina (1973).
24. Hodgson, D.J., and Scaringe, R.P., unpublished observations.
25. Goodenough, J.B., "Magnetism and The Chemical Bond", Interscience, New York, 1963.
26. Bertrand, J.A., and Kirkwood, C.E., *Inorg. Chim. Acta* (1972), 6, 248.
27. Hodgson, D.J., *Progr. Inorg. Chem.* (1974), in press.

10

Superexchange Interactions in Copper(II) Complexes

WILLIAM E. HATFIELD

University of North Carolina, Chapel Hill, N.C. 27514

This paper represents a brief survey of superexchange interactions (1) in copper(II) complexes with special emphasis on research at the University of North Carolina, which is principally concerned with structural, magnetic susceptibility, and EPR measurements. The systems to be discussed include (I) a series of hydroxo-bridged complexes of the general formula $[CuL(OH)]_2X_2 \cdot nH_2O$, where L is a bidentate amine such as 2,2'-bipyridine or an N-substituted 2-(2-aminoethyl)pyridine; (II) a series of chloro-bridged dimers including $[Co(en)_3]_2[Cu_2Cl_8]Cl_2 \cdot 2H_2O$, $[Cu_2(\text{guaninium})_2Cl_6]$, $[Cu_2(\alpha\text{-picoline})_4Cl_4]$, and $[Cu_2(\text{dimethylglyoxime})_2Cl_4]$; and (III) the compound $[Cu(\text{pyrazine})(NO_3)_2]_n$ and related chains.

Since most, if not all, of the copper(II) complexes to be considered have orbitally nondegenerate single-ion ground states the Hamiltonian appropriate for the problem is

$$H = -2J\sum_{i<j} \vec{S}_i \cdot \vec{S}_j \qquad (1)$$

For those cases in which antisymmetric exchange and anisotropic exchange become important the following terms may be added to (1):

$$\vec{D}_{ij} \cdot \vec{S}_i x \vec{S}_j + \vec{S}_i \cdot \underline{\underline{\Gamma}}_{ij} \cdot \vec{S}_j$$

where $\vec{D}_{ij}$ is the antisymmetric vector coupling constant and Γ_{ij} is the anisotropic coupling tensor. For orbital singlet single ions undergoing exchange, Moriya (2) has estimated that

$$D_{ij} \sim (\Delta g/g)J$$

and

$$\Gamma_{ij} \sim (\Delta g/g)^2 J$$

where $\Delta g = |g-2|$. For many copper compounds Δg is approximately

0.1, so $|D| \sim 5\ cm^{-1}$ for $J = 100\ cm^{-1}$. There is no limit on the magnitude of D and Γ for orbitally degenerate single ion states undergoing exchange, and a suggestion has been made that this may be an appropriate approach for the rationalization of the magnetic properties of the tetramers $[Cu_4OX_4L_4]$. (3) These problems will be discussed here.

I. Di-μ-Hydroxo-bridged Copper(II) Complexes

It has been known for some time that copper(II) forms complexes of the type $[CuLOH]_2X_2$, where L is a bidentate amine and X^- is an appropriate counterion. (4-8) These complex ions may be described as two planar or tetragonal pyramidal units sharing an edge which is defined by two bridging hydroxo oxygen atoms. The magnetic properties of the compounds give evidence for exchange interactions which differ widely. From a close examination of the structural and magnetic data for six of these compounds it has been possible to identify some of the factors which influence the exchange interactions, and the results of that study will be reviewed here. The chemical and structural features which will be examined include

1) the nature of the chelating amine.
2) the copper-oxygen (hydroxo) bond distance.
3) the nature of any out-of-plane coordination.
4) the geometry of the basal plane.
5) the single ion ground state.
6) hydrogen bonding by the hydroxo-bridge hydrogen atom.
7) the Cu-Cu separation.
8) the Cu-O-Cu bridge angle.

Magnetic parameters have been obtained from analyses of EPR spectra, and from the temperature variation of the magnetic susceptibility. The latter is characteristic of exchange coupled copper(II) pairs and the singlet-triplet splittings have been determined by fitting the data to the Van Vleck equation

$$\chi_m = \frac{g^2N\beta^2}{3kT} \{1 + \frac{1}{3} \exp(-2J/kT)\}^{-1} + N\alpha \qquad (2)$$

where the symbols have their usual meaning, Nα is the temperature independent paramagnetism, and the equation as written gives the susceptibility per copper ion. In some cases T has been replaced by (T-θ) to account for interdimer interactions.

A. Structural and Magnetic Data.

1. Di-μ-hydroxobis[N,N,N',N'-tetramethylethylenediamine-copper(II)] bromide. The first compound of this type to be characterized by an X-ray crystal structure determination was di-μ-hydroxobis[N,N,N',N'-tetramethylethylenediaminecopper(II)] bromide, $[Cu(tmen)OH]_2Br_2$.(9) The structure of the formula unit viewed along the b-axis is shown in Figure 1 along with some of

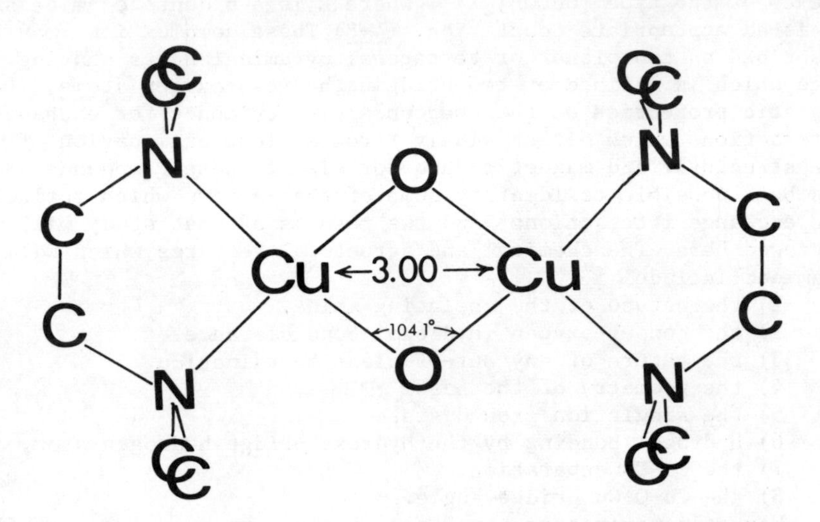

Figure 1. Structure of $[Cu(tmen)OH]_2^{2+}$ *viewed along the b axis (adapted from Ref. 9)*

the important structural parameters. The structural data are also collected for all compounds of this type in Table I. The Cu-N bond distance of 2.030 Å, the Cu-O bond distance of 1.902 Å, and the N-Cu-N angle of 86.7° are all normal for substituted ethylenediamine complexes. The coordination about the copper is square planar with the nitrogen atoms being 0.14 Å out of the plane of the Cu_2O_2 unit. The bromide ion is involved in hydrogen bonding with the hydroxo-bridge, since O-Br distance of 3.366 Å is comparable to the O-Br distances of 3.39 and 3.37 Å, respectively, in the hydrogen bonded systems $MnBr_2 \cdot 2H_2O$ and $CoBr_2 \cdot 2H_2O$. The infrared spectrum also indicates the presence of hydrogen bonding in that there is a strong OH stretching band at 3410 cm^{-1}, a value which is approximately 200 cm^{-1} lower than that of a free OH group. (10-12)

The magnetic susceptibility has been measured in the range 77-300°K both on a powdered sample and on a single crystal along the a,b, and c crystallographic axes. (13) The data were fitted to the Van Vleck equation (2) yielding the magnetic parameters $2J = -509\ cm^{-1}$, $g = 2.0$, and $N\alpha = \sim 150 \times 10^{-6}$ cgs units. Within the precision of the experimental measurement the crystal susceptibilities were isotropic.

2. Di-μ-hydroxobis[N,N,N',N'-tetraethylethylenediamine copper(II)] perchlorate. As a part of an investigation of the thermochromic properties of N-alkyl substituted ethylenediamine complexes of copper(II), Hatfield, Piper, and Klabunde(6) reported, in 1963, the temperature variation of the magnetic susceptibilities of the N,N,N',N'-tetraethylethylenediamine (teen) and N,N-diethyl-N'-methylethylenediamine complexes of the general formula $[Cu(diamine)OH]_2(ClO_4)_2$. The data for the latter compound were fitted to Equation (2) for the determination of the magnetic parameters, while 2J for the teen compound was determined from the expression $2J = -1.11\ T_{max}$ where T_{max} is the temperature at which the magnetic susceptibility attains the maximum value, and the constant has units of $cm^{-1}\ deg^{-1}$.

The structure of $[Cu(teen)OH]_2(ClO_4)_2$ has been completed only recently. (14) The structure consists of $[Cu(teen)OH]_2^{2+}$ units and discrete ClO_4^- anions. (While this geometry at the copper(II) ion is comparable to the situation described above(9) for $[Cu(tmen)OH]_2Br_2$, it is in marked contrast to the geometry of the structures of other compounds in this series with oxyanions, vide post.) The best least squares plane of the $N_2CuO_2CuN_2$ unit calls attention to the slight distortion in the molecule, the oxygen and nitrogen atoms are approximately 0.15 Å out of the plane. The Cu_2O_2 unit is planar owing to the inversion center. The structural features, which are given in Table I, are very similar to those determined for the tmen compound with the significant exception being the decrease in Cu-O-Cu angle from 104.1° in $[Cu(tmen)OH]_2Br_2$ to 103.0° in $[Cu(teen)OH]_2(ClO_4)_2$.

Apparently there is a hydrogen bonding interaction between

Table I

Structural and Magnetic Properties of $[Cu(diamine)OH]_2^{2+}$

	N-Cu-N	Cu-O-Cu	in-plane		out-of-plane	Cu-Cu, Å	2J, cm^{-1}	References
			Cu-O, Å	Cu-N, Å	Cu-O, Å			
$[Cu(tmen)OH]_2Br_2$	86.7(8)°	104.08(.17)°	1.902(3)	2.030(10)	-	3.000	-509	8,9,13
$[Cu(teen)OH]_2(ClO_4)_2$	87.8(2)°	103.0(2)°	1.899(4) 1.907(4)	2.013(5) 2.024(15)	-	2.978(2)	-410	5,14
β-$[Cu(DMAEP)OH]_2(ClO_4)_2$		100.4(1)°	1.900(3) 1.919(3)	2.003(3)* 2.066(3)	2.721(4)	2.935(1)	-201	15,17
$[Cu(EAEP)OH]_2(ClO_4)_2$	96.3(5)° 93.0(5)°	98.8(3)° 99.5(2)°	1.895(7) 1.913(7) 1.927(2) 1.930(8)	1.981(8)* 1.998(10) 2.001(8)* 2.053(10)	2.562(10) 2.618(9)	2.917(5)	-130	17,20,21
$[Cu(bipy)OH]_2SO_4 \cdot 5H_2O$	81°	97° 97°	1.92 1.94 1.95 1.95	1.99 2.00 2.00 2.02	2.21(SO_4^{2-}) 2.24(H_2O)	2.893(2)	+48	25,26,27, 28
$[Cu(bipy)OH]_2(NO_3)_2$	80.6(1)°	95.6(1)	1.920(2) 1.923(1)	1.998(2) 2.000(2)	2.379(2)	2.847	+172	29,30

*pyridine nitrogen

the hydroxo group and the perchlorate ion, since the average Cl-O bond distance for the three oxygens which are not involved is 1.389(5) Å, while the Cl-O bond distance for the oxygen which is probably hydrogen bonded is 1.434(5) Å.

3. β-Di-μ-hydroxobis[2-(2-dimethylaminoethyl)pyridinecopper-(II)]perchlorate. Two forms of the compound $[Cu(DMAEP)OH]_2(ClO_4)_2$ can be isolated. (15,16) The β-form is obtained, essentially uncontaminated with the α-form, by mixing equimolar quantities of copper(II) perchlorate hexahydrate and DMAEP in ethanol/ether while both forms may be found if methanol/ether is used. Uhlig and co-workers (17) had reported only one form of this compound in their study of the coordination chemistry of N-substituted 2-(2-aminoethyl)pyridine, and from a comparison of the magnetic properties of the two isomers, it is evident that they had the triclinic α-form. The α-form has, in addition to the two expected hydroxo bridges, two perchlorate bridges and it will not be considered further here. (16) The β-form has two hydroxo bridges (15) and, as shown in Figure 2, the copper ions are in a tetragonal pyramidal environment with oxygen atoms from perchlorate ions occupying the axial positions. The $N_2CuO_2CuN_2$ unit is essentially planar with the largest deviation from the best least squares plane being 0.09 Å. An unusual features obtains here in that the copper ions are not displaced out of the plane toward the axial ligand as has been observed in many tetragonal pyramidal copper(II) complexes. Presumably this absence of displacement is a reflection of the weak nature of the perchlorate coordination. The bond distances and angles are comparable to those observed in other similar complexes. The structural parameters pertinent to this discussion are listed in Table I where it may be seen that the Cu-pyridine nitrogen bond distance is somewhat shorter than the Cu-amine nitrogen bond. The Cu-O-Cu angle is 100.4(1).°

Hydrogen bonding involving the hydroxo bridging group is indicated by the oxygen-oxygen separation of 2.993 Å, which is less than twice the van der Waals radius of oxygen as given by Bondi (3.02 Å) (18) but slightly greater than the corresponding value given by Pauling (2.80 Å). (19) That the bonding is weak is evident from the strong, sharp O-H stretching band at 3580 cm^{-1}, a value which is only 40 cm^{-1} less than the value usually ascribed to free hydroxyl groups.

The temperature variation of the magnetic susceptibility of β-$[Cu(DMAEP)OH]_2(ClO_4)_2$ from 50-300°K is shown in Figure 3. (15) There is a very broad maximum in the magnetic susceptibility at about 175°K, and the data may be fitted to expression for exchange coupled pairs of copper(II) ions yielding $2J = -195\ cm^{-1}$ and $g = 2.00$, where the criterion for the best fit was the minimization of the function

$$A_{BF} = \sum_i [\{\chi(exptl)_i - \chi(calcd)_i\}T_i]^2$$

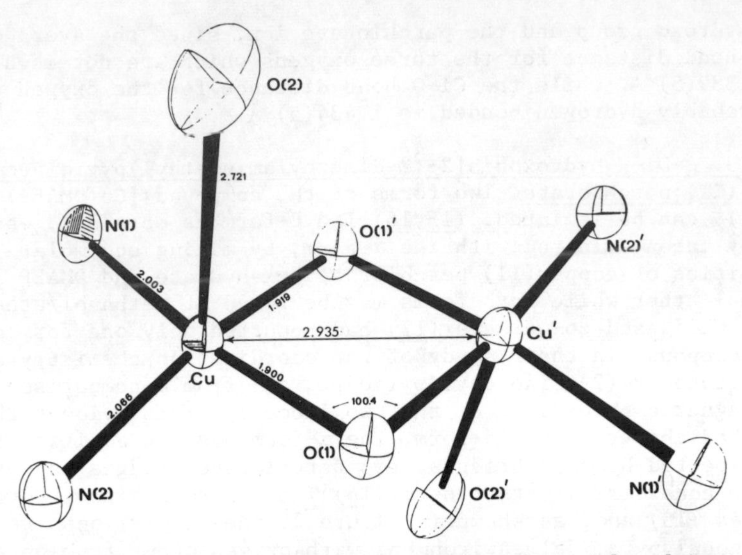

Figure 2. Structure of β-[Cu(DMAEP)OH](ClO$_4$)$_2$ (adapted from Ref. 15)

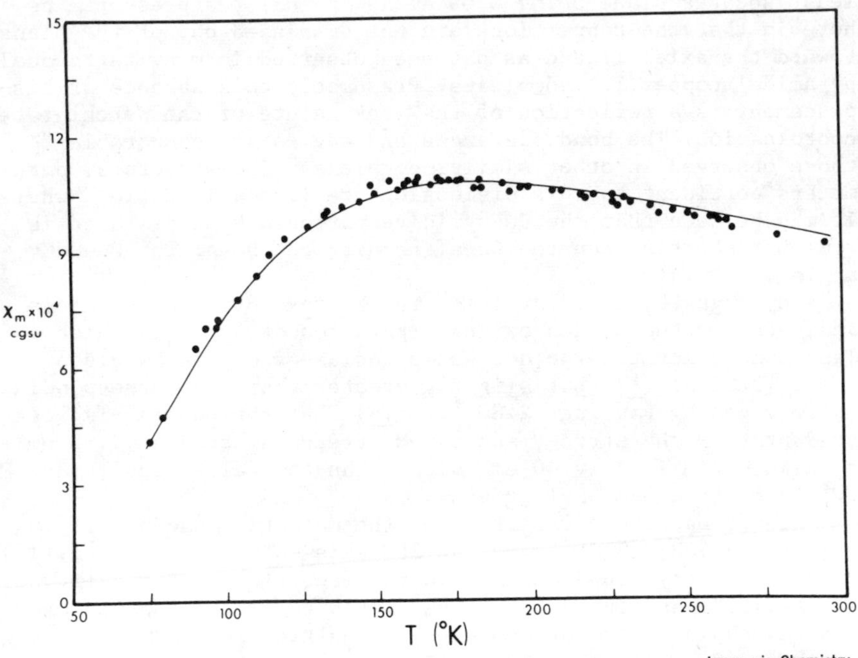

Figure 3. Susceptibility of β-[Cu(DMAEP)OH]$_2$(ClO$_4$)$_2$ per Cu atom as a function of temperature. Solid line represents values calculated from equation (2) with g = 2.03 and 2J = 201 cm^{-1} (15).

For this set of parameters $A_{BF} = 2.1x10^{-4}$. Since this best fit g-value is somewhat lower than the average g-value (2.03) found from the EPR spectrum taken at 77°K, a second calculation was made in which the g-value was held constant and only 2J was varied. The somewhat poorer fit ($A_{BF} = 5.7x10^{-4}$) gave 2J = -201 cm^{-1}. In view of these results a value of <g> = 2.03 is probably accurate within 2% and 2J = -200±10cm^{-1}.

<u>4. Du-μ-hydroxobis[2-(2-ethylaminoethyl)pyridinecopper(II)]-perchlorate</u>. The structure of the complex $[Cu(EAEP)OH]_2(ClO_4)_2$ and coordination geometry around copper is very similar to that described above for $[Cu(DMAEP)OH]_2(ClO_4)_2$. The pertinent structural details are tabulated in Table I. In this compound the copper ions are displaced approximately 0.12 Å from the best least-squares basal plane which is formed by the nitrogen atoms and the two bridging oxygen atoms. It is further noteworthy that the copper-oxygen(perchlorate) axial internuclear separations are significantly shorter in $[Cu(EAEP)OH]_2(ClO_4)_2$ than in the corresponding complex $[Cu(DMAEP)OH]_2(ClO_4)_2$ where the copper ions are not displaced from the basal plane. The two basal planes are nearly coplanar, with the angle between them being 1.4°. The average of the two Cu-O-Cu angles is 99.2(3)° and the Cu-Cu separation is 2.917(5) Å.

The structural data indicate that any hydrogen bonding involving the bridging hydroxo hydrogen atom is very weak. The short oxygen(bridge)-oxygen(perchlorate) separations are 2.89 and 2.95 Å, values which are less than twice the van der Waals radius of oxygen as given by Bondi but greater than that given by Pauling. The infrared spectra shows a strong, sharp band at 3580 cm^{-1}.

The magnetic susceptibility data (<u>21</u>) for $[Cu(EAEP)OH]_2(ClO_4)_2$ show a broad maximum at ~120°K and when the data are fitted to Equation (2) the parameters 2J = -130cm^{-1} and g = 2.04 result.

<u>5. Di-μ-hydroxobis[2,2'-bipyridinecopper(II)]sulfate pentahydrate</u>. The complexes of the general formula $[Cu(diamine)OH]_2X_2 \cdot nH_2O$ which are formed by 1,10-phenanthroline and 2,2'-bipyridine are of considerable interest since a variety of counter ions may be used and the magnetic properties depend on the identity of the counter ion. (<u>22</u>-<u>25</u>) The first compound of this type to be characterized fully was $[Cu(bipy)OH]_2SO_4 \cdot 5H_2O$ (<u>25</u>-<u>28</u>). The structural details necessary for this discussion are listed in Table I.

In this compound the copper ions are found in distorted tetragonal pyramidal environments with a water molecule coordinated to one copper and the sulfate ion coordinated to the other copper. The copper ions are displaced approximately 0.18 and 0.23 Å from the basal planes formed by the nitrogen atoms and the bridging oxygens. It is interesting to note that the greater displacement of 0.23 Å occurs in the portion of the molecule in which the sulfate ion is coordinated to copper and thatthat Cu-O (axial) bond distance is the shorter of the two. The dihedral angle between the basal planes is 7.9°. There is extensive hydrogen bonding with an O(hydroxo)-O(sulfate) separation of 2.77 Å.

The magnetic susceptibility of this compound has been measured as a function of temperature over the range 4.2-300°K, (26-28) and the χ^{-1} vs. T plot, Figure 4, clearly indicates a deviation at low temperature from the Curie-Weiss behavior exhibited at higher temperatures. The measured susceptibilities deviate in the manner expect for an exchange coupled pair of copper ions with a positive 2J value indicating a ferromagnetic interaction and a triplet ground state with a low lying singlet state. The best least square fit of the data yield 2J = +48(±10) cm^{-1} and g = 2.2.

The presence of the triplet state is confirmed by the EPR spectrum shown in Figure 5. The spectrum may be described using the spin Hamiltonian

$$H_{S'} = g_{||}\beta H_z S'_z + g_{\perp}\beta(S'_x H_x + S'_y H_y) + D(S'^{2}_z - 2/3)$$

where the resonance fields are

$$H_1(z) = (g_{||}\beta)^{-1}|h\nu - D|$$
$$H_2(z) = (g_{||}\beta)^{-1}(h\nu + D)$$
$$H_1(x,y) = (g_{\perp}\beta)^{-1}[h\nu(h\nu - D)]^{1/2}$$
$$H_2(x,y) = (g_{\perp}\beta)^{-1}[h\nu(h\nu + D)]^{1/2}$$
$$H_1(\mathrm{forb}) = (2g_{||}\beta)^{-1}h\nu$$
$$H_2(\mathrm{forb}) = (2g_{\perp}\beta)^{-1}(h^2\nu^2 - D^2)^{1/2}$$

Here the subscript 1 designates the low field transitions and the subscript 2 designates the high field transitions, while the $\Delta M_S = 2$ transitions are labeled forbidden.

6. Di-μ-hydroxobis[2,2'-bipyridinecopper(II)]nitrate. The structure of $[Cu(bipy)OH]_2(NO_3)_2$ is very similar to that of the sulfate salt with the difference being that nitrates are coordinated in the axial positions of the tetragonal pyramids. (29) The copper(II) ions are displaced from the basal planes toward the oxygen atom of the coordinated nitrate where the Cu-O (axial) bond is relatively short, being 2.379(2) Å. Structural details are listed in Table I.

The magnetic susceptibility has been measured as a function of temperature in the range 1.6-300°K and can be fit to Equation (2) yielding 2J = 172 cm^{-1} and g = 2.10. (30) The estimated

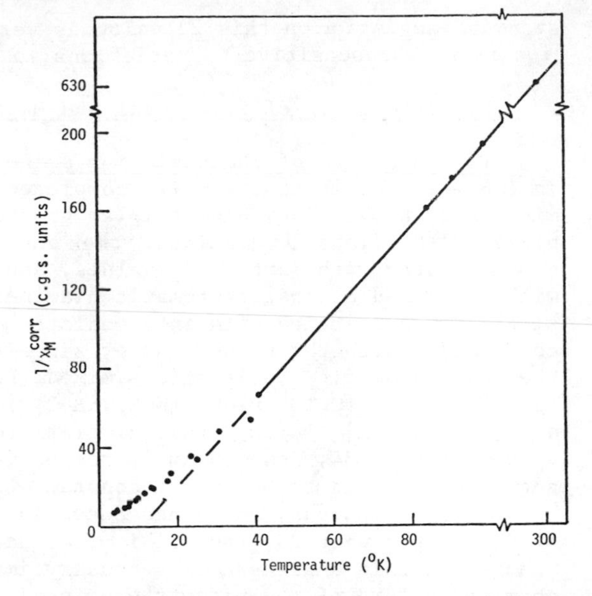

Inorganic Chemistry

Figure 4. Inverse of magnetic susceptibility per copper atom of $[(bipy)Cu(OH)_2Cu(bipy)]SO_4 \cdot 5H_2O$ *as a function of temperature. Solid line represents values calculated from the Van Vleck equation; dashed line is extrapolation of Curie-Weiss law. Observed values are black dots. Data above 80K from Ref.* 26 (28).

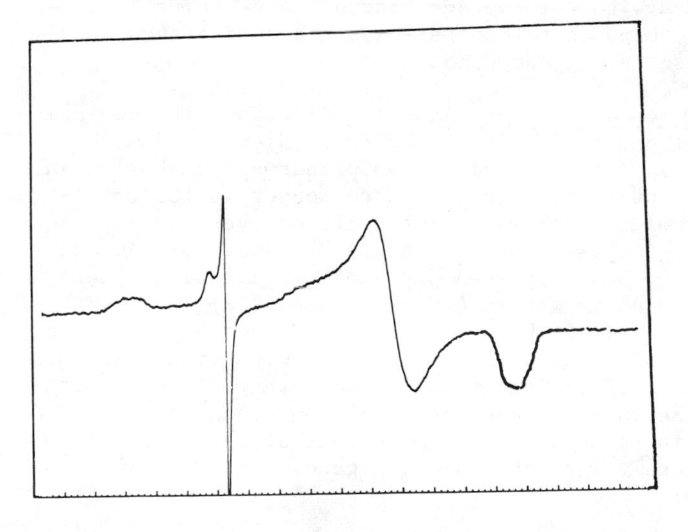

Figure 5. X-band EPR spectrum of $[Cu(bipy)OH]_2SO_4 \cdot 5H_2O$ *at 77°K, 0–10.0 kG*

standard deviation on this 2J value is very large since the equation is very insensitive to variations in large positive 2J values.

B. Correlation of Structural and Magnetic Properties.

1. The nature of the chelating amine. Inspection of the data in Table I reveals that the two complexes with the aromatic diamine, 2,2'-bipyridine, have triplet ground states, the two complexes with aliphatic diamines, tmen and teen, have singlet ground states with large |2J| values, and that the two complexes with the mixed aliphatic/aromatic diamines, DMAEP and EAEP, have singlet ground states with intermediate |2J| values. There is no correlation with the N-Cu-N angle, since these angles increase in the order aromatic < aliphatic < mixed. Furthermore, Casey (25) has shown that $[Cu(bipy)OH]_2(NCS)_2 \cdot H_2O$, $[Cu(bipy)OH]_2(NCSe)_2 \cdot H_2O$, and $[Cu(bipy)OH]_2Cl_2 \cdot 3H_2O$ have singlet ground states with 2J values of -6, -34, and -39 $cm.^{-1}$, respectively. Since the N-Cu-N angle is expected to be fairly constant for all of the 2,2'-bipyridine compounds, and since there is no apparent correlation of this angle with 2J, then it is reasonable to conclude that changes in this angle are of secondary importance. However, the chemical nature of the bidentate amine is probably important since as of yet there are no known overlaps of 2J values between the groups of compounds formed by the aromatic, mixed aromatic/aliphatic, and aliphatic diamines.

2. The copper-oxygen(hydroxo) bond distance. The copper-bridging oxygen bond distances are all comparable and fall in the range 1.9 to 1.95 Å with the average bond distance being 1.915 Å. For all practical purposes the copper-oxygen bond distance is constant in this series of compounds.

3. The nature of any out-of-plane coordination. The two complexes with the aliphatic diamines, $[Cu(tmen)OH]_2Br_2$ and $[Cu(teen)OH]_2(ClO_4)_2$, are formed by two planar units sharing an edge. The closest out-of-plane contacts to copper in $[Cu(tmen)OH]_2Br_2$ are to bromide ions which are centered over the five member chelate ring, these distances are 4.778 and 4.933 Å, and are considered to be too long even for semi-coordination. There are no out-of-plane atoms within 4.0 Å of copper(II) in $[Cu(teen)OH]_2(ClO_4)_2$. (31) The other four compounds are composed of tetragonal pyramidal units sharing an edge with apical ligands on opposite sides of the joined basal planes. There is an important trend here; the out-of-plane copper-oxygen bond distances decrease with an increase in the displacement of the copper ion from the basal plane toward the apical ligand, *viz*.,

compound	Cu-O, Å	displacement, Å
β-$[Cu(DMAEP)OH]_2(ClO_4)_2$	2.721(4)	none
$[Cu(EAEP)OH]_2(ClO_4)_2$	2.618(9)	0.11
	2.562(10)	0.13
$[Cu(bipy)OH]_2(NO_3)_2$	2.379(2)	0.16
$[Cu(bipy)OH]_2SO_4 \cdot 5H_2O$	2.21(SO_4^{2-})	0.23
	2.24(H_2O)	0.18

There is a general correlation of J with the copper-out-of-plane ligand distance. Except for $[Cu(bipy)OH]_2SO_4 \cdot 5H_2O$, the exchange coupling constant becomes more positive as the out-of-plane bond distance decreases.

4. The geometry of the basal planes. To a good approximation the bases of the tetragonal pyramids are all planar with deviations from the best least squares plane being on the order of 0.15 Å or less. The two basal planes [or the coordination planes in the case of $[Cu(tmen)OH]_2Br_2$ and $[Cu(teen)OH]_2(ClO_4)_2$]are frequently coplanar, with the largest deviation from coplanarity being a 7.9° dihedral angle between these planes in $[Cu(bipy)OH]_2$-$SO_4 \cdot 5H_2O$. Consequently, except for this latter compound, the geometry of the basal plane remains constant throughout the series.

5. The single ion ground states. It is very well established that the unpaired electron is in the σ* orbital for square planar copper(II) complexes. (32) Although spin-orbit coupling and the low symmetry crystal field components permit the mixing in of other states, to a good approximation the exchange mechanisms can be given in terms of this electronic configuration. Although the displacement of the copper(II) ion from the basal plane toward the axial ligand in the tetragonal pyramidal complexes complicates even further the nature of the single ion ground state, it will be assumed that the unpaired electron is in the $d_{x^2-y^2}$,σ*, orbital in these complexes. This assumption is admittedly less tenable for $[Cu(bipy)OH]_2SO_4 \cdot 5H_2O$ and $[Cu(bipy)OH]_2(NO_3)_2$ where the displacements are considerable and the axial bond distances are rather short. Precise descriptions of the ground states must await the completion of detailed EPR investigations which are presently underway. (33)

6. Hydrogen bonding by the hydroxo-bridge hydrogen atoms. The nature of the hydrogen bonding spans a wide range in these compounds. A suitable working hypothesis would suggest that the electron density on the bridging oxygen atom should increase with the strength of the hydrogen bond and that this variation in electron density should have a significant effect on the exchange

coupling. If either the O···X internuclear separation or $\Delta\nu(OH)$ (the deviation from the "free" hydroxyl stretching energy) are taken as gauges of the hydrogen bond, then it is clear from the data in Table II that there is no simple correlation between either of these parameters and the exchange coupling constant. Thus it is reasonable to conclude that hydrogen bonding is of secondary importance in determining the nature of the exchange coupling.

7. The copper-copper separation. There is an interesting correlation between the Cu-Cu separation and the singlet-triplet splitting. As the Cu-Cu separation increases from 2.847 Å in $[Cu(bipy)OH]_2(NO_3)_2$ to 3.000 Å in $[Cu(tmen)OH]_2Br_2$, the singlet-triplet splitting changes from +172 cm^{-1} to -509 cm^{-1}. If the data for $[Cu(bipy)OH]_2SO_4\cdot 5H_2O$ are omitted, since the basal planes in the compound are not coplanar, then the best line through the five available values of the copper-copper separation and 2J has a slope of -4545 cm^{-1}/Å and an intercept of 13,130 cm^{-1}.

8. The copper-oxygen-copper bridge angle. Since the copper-oxygen (bridge) bond distances are nearly constant at 1.915 Å, and since the Cu_2O_2 units are all nearly planar there is a similar striking correlation between the 2J values and the Cu-O-Cu bridge angle. This linear correlation (in the range $95.6° < \phi < 104.1°$) is illustrated in Figure 6, where the slope is -79.5 cm^{-1} deg^{-1} and the intercept is 7790 cm^{-1}.

C. Rationalization in Terms of a Molecular Orbital Model. The correlations which have been observed can be understood in terms of a simple molecular orbital model, which is most appropriate for the two joined-planar compounds, but which is still a good approximation for the joined tetragonal pyramidal compounds. (34) It is necessary to consider the two oxygen orbitals, p_x, p_y, and the two copper in-plane orbitals $d_{x^2-y^2}$, d_{xy}. If the following coordinate system in D_{2h} is adopted

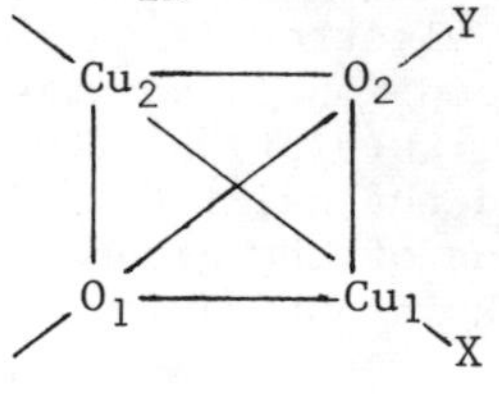

it can be seen that the orbitals transform as

$$A_g: \quad d_{x^2-y^2} + d_{x^2-y^2}$$

$$y_1 - y_2$$

Table II

Hydrogen Bonding Parameters for the Bridging Hydroxo Group.

Compound	ν(O-H), cm^{-1}	O····X, Å	X	Comments
$[Cu(tmen)OH]_2Br_2$	3410	3.366	Br^-	$MnBr_2 \cdot 2H_2O$(3.39 Å) $CoBr_2 \cdot 2H_2O$(3.37 Å)
$[Cu(teen)OH]_2(ClO_4)_2$	-	3.00	$OClO_3^-$	Bondi (3.02) Pauling (2.80)
β-$[Cu(DMAEP)]_2(ClO_4)_2$	3580	2.993	$OClO_3^-$	
$[Cu(EAEP)OH]_2(ClO_4)_2$	3580	2.89	$OClO_3^-$	
$[Cu(bipy)OH]_2SO_4 \cdot 5H_2O$	-	2.77	OSO_3^-	
$[Cu(bipy)OH]_2(NO_3)_2$	-	2.877	ONO_2^-	
"free" hydroxyl	3620			

B_{1g} : $d_{xy} + d_{xy}$

$x_1 - x_2$

B_{2u} : $d_{xy} - d_{xy}$

$y_1 + y_2$

B_{3u} : $d_{x^2-y^2} - d_{x^2-y^2}$

$x_1 + x_2$

where x_i and y_i designate the oxygen p_x and p_y orbitals. Symbolically these symmetry relationships are

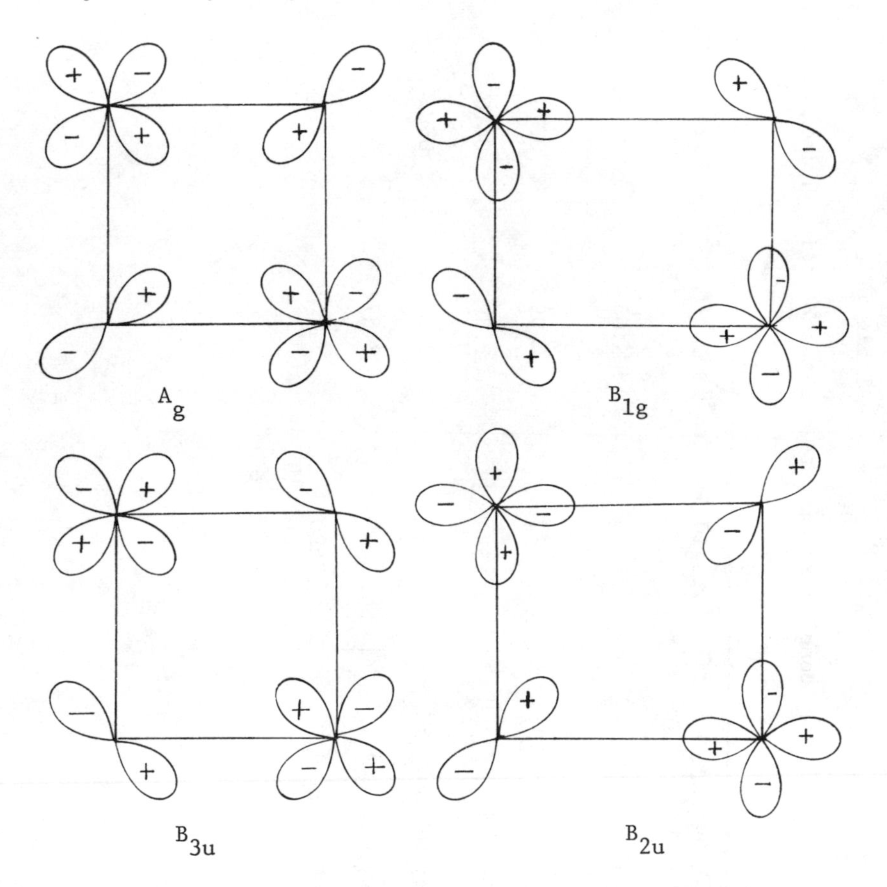

Inspection of these will show that the A_g and B_{3u} molecular orbitals will have identical energies as will the B_{1g} and B_{2u}

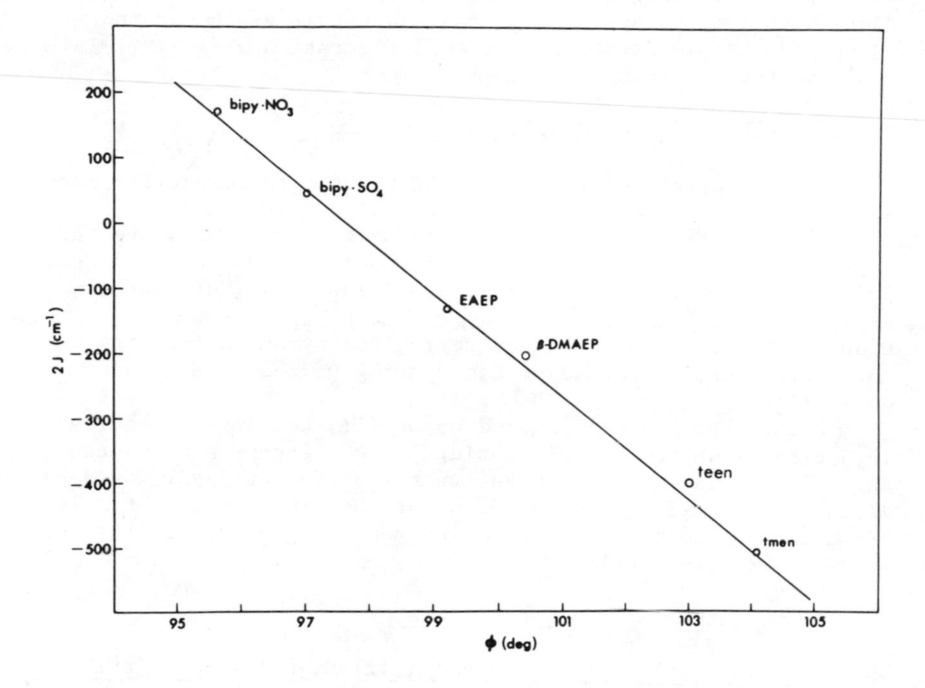

Figure 6. Correlation of singlet–triplet splittings with the Cu–O–Cu bridge angle

orbitals. With this information the molecular orbital diagram for this fourteen electron system may be constructed; it is given in Figure 7a. As the Cu-O-Cu angle increases the overlap of the A_g and B_{1g} combinations increase relative to that of the B_{3u} and B_{2u} combinations. At a sufficiently large enough angle it may be anticipated that the separation between the B_{2u}* and B_{1g}* molecular orbitals would exceed the pairing energy and the singlet state with configuration $(b_{2u}^*)^2$ would result as the ground state. The triplet state from the configuration $(b_{2u}^*)(b_{1g}^*)$ would be the low lying paramagnetic state.

II. Di-μ-Chloro-bridged Copper(II) Complexes

The properties of four di-μ-chloro-bridged copper(II) complexes will be described here. These include (Class I) $[Co(en)_3]_2[Cu_2Cl_8]Cl_2 \cdot 2H_2O$ and $[Cu(guaninium)Cl_3]_2$, where the copper(II) ions are in distorted trigonal bipyramids which share an equator-to-apex edge, (35-40) and (Class II) $[Cu(2\text{-methylpyridine})_2Cl_2]_2$ and $[Cu(dimethylglyoxime)Cl_2]_2$, where the coordination about copper is distorted tetragonal pyramidal and the dimeric structure is formed by the sharing of the base-to-apex edge. (41-44)

Unlike the hydroxo-bridged copper(II) complexes described in Section I where the Cu-O (bridge) bond lengths are constant, the chloro-bridged dimers have greatly different Cu-Cl (bridge) distances. It will be shown here that the singlet triplet splittings are dependent on the bridging bond lengths as well as the Cu-Cl-Cu bridging angle.

A. Structural and Magnetic Data.

1. Tris(ethylenediamine)cobalt(III) Di-μ-chlorobis[trichlorocuprate(II)] Dichloride Dihydrate. The unusual spectral properties (45) of the compound $Co(en)_3CuCl_5 \cdot H_2O$, as formulated by Kurnakow in 1898, led to an X-ray crystal structure examination (35,36) which revealed that the new and unusual $[Cu_2Cl_8]^{4-}$ ion was present.

The compound crystallizes in the orthorhombic space group Pbca with four molecules in a unit cell of dimensions a = 13.560(9), b = 14.569(9), and c = 17.885(12) Å. The bridging Cu-Cl distances are 2.325(5) and 2.703(5) Å, the Cu-Cu separation is 3.722(5) Å, and the angle at the bridge is 95.2(1)°.

The exchange interaction (37) in $[Cu(en)_3]_2[Cu_2Cl_8]Cl_2 \cdot 2H_2O$ has been precisely characterized by single crystal magnetic susceptibility measurements. (38) Data were collected in the temperature range 4.2-80°K on a crystal (5.2 x 2.9 x 2.5 mm) with the magnetic field applied along the crystallographic axes. The data are given in Figure 8, where it may be seen that the susceptibility maximizes at 12.9°K, and the usual g-factor anisotropy is observed ($\chi_c > \chi_a > \chi_b$). The data may be fitted to

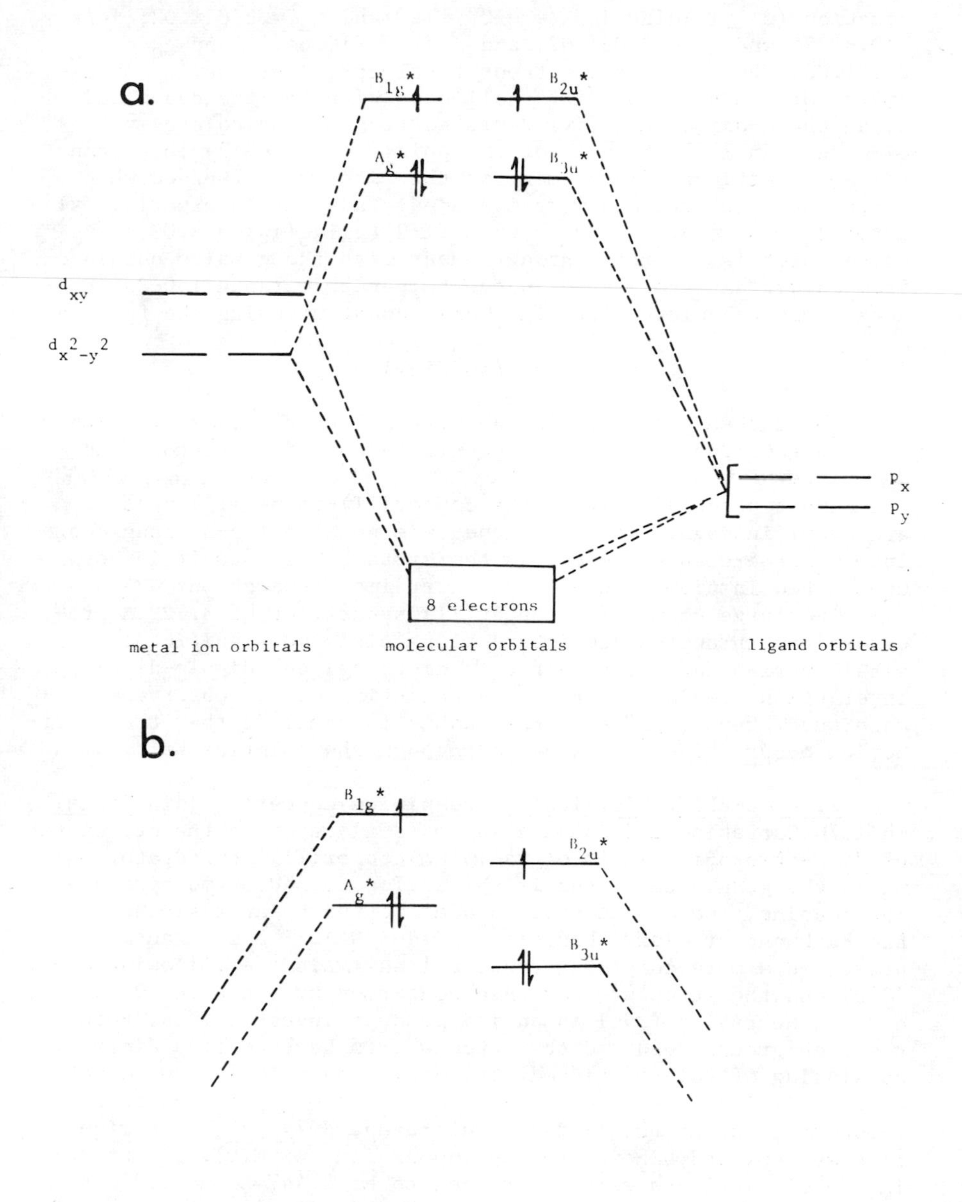

Figure 7. (a) Molecular orbital diagram for the in-plane orbitals in the Cu_2O_2 ring, D_{2h} symmetry, 90° Cu–O–Cu angle. (b) Antibonding molecular orbitals for Cu–O–Cu angle different from 90° (adapted from Ref. 34).

Equation (2) yielding $J_a/k = -10.7\pm.2°$ and $g_a = 2.07\pm.02$; $J_b/k = -10.8\pm.2°$ and $g_b = 2.03\pm.02$; and $J_c/k = -10.6\pm.2°$ and $g_c = 2.18\pm.02$. The data fit the theoretical curve very well. For example, of the 44 data points collected with the field applied along the c-axis, only five deviated from the calculated values by more than 1.5% while 34 of the points differed by less than 1%. No significant improvement of the fits were observed when interdimer interactions were included. Thus within experimental error $J_a = J_b = J_c = J$, and $\langle g \rangle = (1/3)(g_a+g_b+g_c) = 2.09$, a value which is in excellent agreement with the g value obtained from powder data collected in the temperature range 130-235°K. Here g was calculated from the Curie constant using the formula

$$g^2 = 3kC/N\beta^2S(S+1)$$

The exchange interaction observed along the c-axis, which is almost colinear with the copper(II)-copper(II) vector, and the exchange interaction observed along the a and b-axes, which are almost perpendicular to the copper(II)-copper(II) vector, are equal in magnitude. Also, there is no significant long range interdimer exchange present in the system. This result is not unexpected in view of the large interdimer separation.

The large copper(II)-copper(II) separation of 3.722 Å precludes any through-space interactions since no significant orbital overlap can occur over this distance, and dipole-dipole interactions could not produce a splitting of the observed magnitude. Hence, it seems reasonable to conclude that the interaction occurs via superexchange through the chloride bridges.

2. Di-μ-chlorobis[dichloro(guaninium)copper(II)]dihydrate. In 1970 Carrabine and Sundaralingam (39a) reported the structure of di-μ-chlorobis[dichloro(guaninium)copper(II)] dihydrate, where the guaninium ligand is the cation formed by monoprotonating guanine, one of the bases bonded to the sugar residues in the backbone of dioxyribonucleic acid (DNA). The same authors presented a more complete structural analysis the following year, (39b) and the structure was then confirmed by Declercq, Debbaudt, and Van Meerssche (39c) in an independent investigation. Both research groups reported the structure to be that of a dimer consisting of chloro-bridged, trigonal-bipyramidally coordinated copper(II) ions, as shown in Figure 9. The monoprotonation was shown to occur at the imidazole nitrogen, N(7), of the purine ring system, and binding to the copper ion, at N(9). The bridging Cu-Cl distances were determined to be 2.447 Å and 2.288 Å, with a Cu-Cl-Cu bridging angle of 98° and a Cu-Cu separation of 3.575 Å.

The temperature variation of the inverse susceptibility (calculated per copper(II) ion) of a powdered sample in the temperature region 1.6 to 255°K is represented by the black data points in Figure 10. (40) The maximum in the curve occuring at

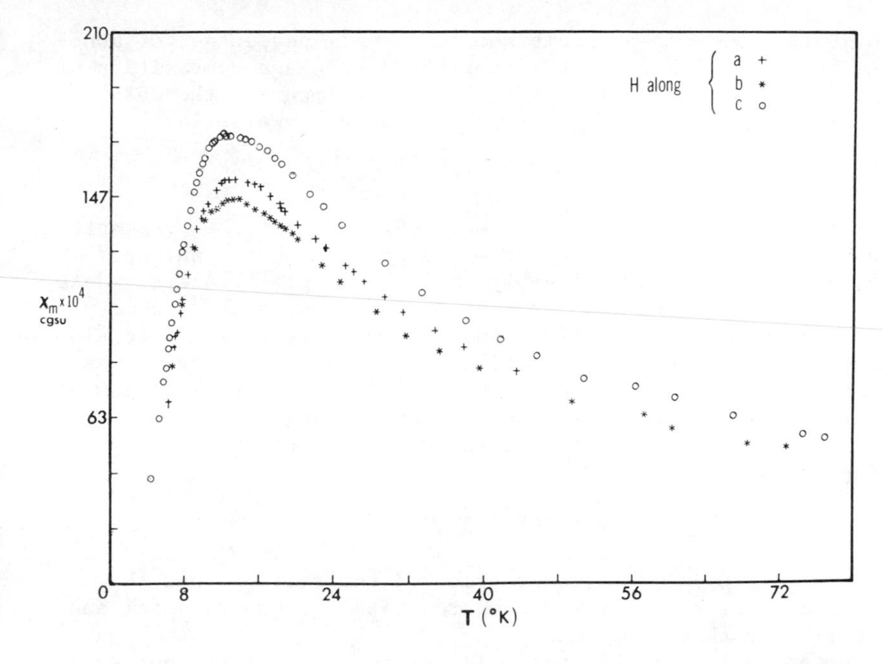

Figure 8. Single crystal magnetic susceptibility data for $[Co(en)_3][Cu_2Cl_8]$-$Cl_2 \cdot 2H_2O$ (38)

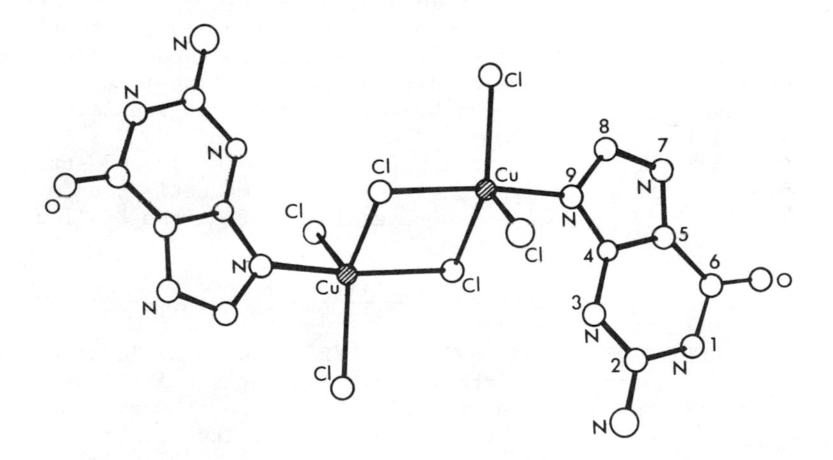

Figure 9. Structure of $[Cu(guaninium)Cl_3]_2$ *(adapted from Ref.* 39a)

approximately 15°K is probably due to a small percentage of monomeric impurity which did not affect the percentage composition of the elemental analysis. To correct for this impurity the data points from 1.6 to 11.2°K were fit to the Curie-Weiss law

$$\chi = C/(T - \theta)$$

Values of 7.86 x 10^{-3} and -3.22° were obtained for the constants C and θ, respectively. In this temperature region the susceptibility of the dimer is negligibly small (<u>vide post</u>). All the data points were then corrected for the contribution of the impurity to the observed susceptibility, and the corrected points are also plotted in Figure 10 as the un-filled circles. The impurity was estimated to be present to the extent of 1% in the following way: The susceptibility for an assumed monomer having a molecular weight equal to one-half that of the dimer was calculated from the expression

$$\chi = N\beta^2\mu^2/3kT$$

where $\mu = g\beta[S(S + 1)]^{1/2}$, at a selected temperature, and the calculated susceptibility was compared with the experimental susceptibility at that temperature.

The solid line in Figure 10 is the best fit of the corrected data to Equation (2), which yields the parameters 2J = -82.6±1.0 cm^{-1} and g = 2.12±0.02.

<u>3. Di-μ-chlorobis[chloro(dimethylglyoxime)copper(II)]</u>. The structure (<u>43</u>) of $[Cu(DMG)Cl_2]_2$ is shown schematically in Figure 11 with pertinant molecular dimensions indicated thereon. The copper atom is five-coordinated in a square-based pyramidal arrangement consisting of a nearly square planar arrangement of two nitrogen atoms and two tightly bound chlorine atoms with a chlorine atom from an adjacent unit in the apical position.

The magnetic data (<u>44</u>) can be described by the singlet-triplet equation (2) with 2J = 6.3 cm^{-1}, g = 2.06, and θ = -1.7°. Additional and convincing evidence for the triplet ground state is provided by the magnetization studies. The magnetization curves for S' = 1/2 and S' = 1 were evaluated from the Brillouin function (<u>46</u>)

$$B(X) = \frac{2S'+1}{2S'} \coth \frac{2S'+1}{2S'} X - \frac{1}{2S'} \coth \frac{X}{2S'}$$

where X = (H/T)(S'gβ/k) and S' is the effective spin. To account for <u>inter</u>dimer interactions the field H was set equal to the external field plus a molecular field, H_m, where it was assumed that H_m was proportional to the magnetization. Thus, $H_m = N_w N\beta\langle\mu\rangle$ where $\langle\mu\rangle = gS'B(X)$ and $N_w = (3k\theta')/[Ng^2\beta^2S'(S'+1)]$. (<u>47</u>) A best fit (least squares) of the experimental data was used to select a value of -1.2 for θ' thereby indicating an antiferro-

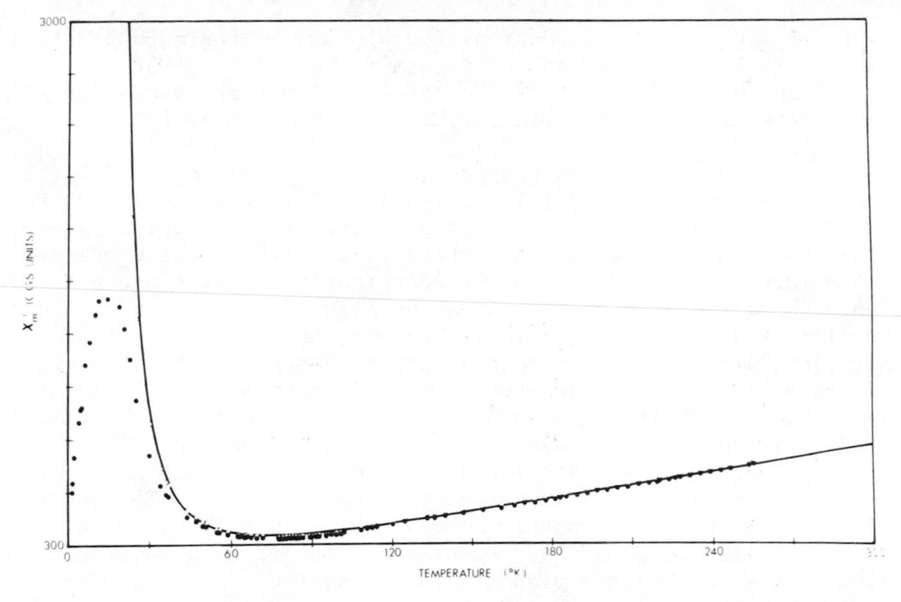

Figure 10. Temperature vs. *inverse susceptibility for the complex* [*(guaninium)-* $CuCl_3$]$_2$ · $2H_2O$ (38). *Experimental points,* ●; *experimental points corrected for monomeric impurity,* ○; *Van Vleck equation best fit,* ——.

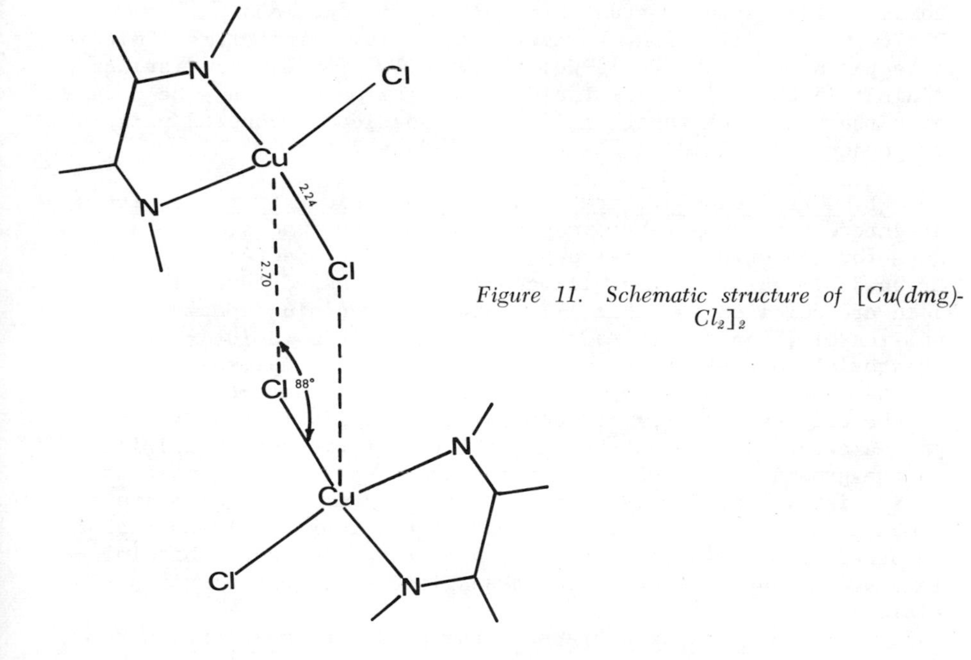

Figure 11. Schematic structure of [*Cu(dmg)-* Cl_2]$_2$

magnetic interdimer interaction. The θ' term in the magnetization studies is very similar to the θ term in Equation 2, but in view of the approximate nature of the theory, identical values for these interdimer interaction parameters are not expected.

4. Di-μ-chlorobis[chlorobis(2-methylpyridine)copper(II)]. In recent years many copper(II) complexes of the type CuL_2X_2, where X is chloride or bromide and L is pyridine or substituted pyridine, have been prepared and characterized. (48) These complexes are mainly polymeric, having six coordination about the copper ion with halide ligands from adjacent molecules occupying the out-of-plane coordination positions. However, the complexes of 2-methylpyridine were found to have properties somewhat different to those found for the analogous pyridine complexes, (49) and it was postulated that the methyl group in the 2-position provides steric hindrance to the usual octahedral coordination. Subsequently, Duckworth and Stephenson (41) determined that such was the case for dichlorobis(2-methylpyridine)copper(II). The coordination about copper in this complex is tetragonal pyramidal with the fifth position (out-of-plane) occupied by a chloride ligand from an adjacent planar moiety, and the sixth position is effectively blocked by the methyl groups of the pyridine ligands. The structural details are summarized in Table III.

As shown in Figure 12, the magnetic susceptibility (42) data for Cu(2-methylpyridine)$_2Cl_2$ obey the Curie-Weiss law, $\chi = C/(T-\theta)$, in the range 295°K to approximately 30°K. For the chloro-complex, the Curie constant C = 0.394, θ = 1°K. and $\mu_{eff} = 2.828C^{1/2}$ is 1.78 B.M. However, at the low temperature limit it is apparent that the Curie-Weiss law fails. There is a distinct minimum in the χ^{-1} versus T plot at approximately 7°K. The data obey the Van Vleck equation (2) for magnetically coupled pairs of copper ions yielding g = 2.15 and -2J = 7.4 cm^{-1}.

B. Correlation of Structural and Magnetic Properties. It is of interest to compare the magnetic parameters and structural data for the structurally- and magnetically-characterized chloro-bridged bimetallic copper(II) complexes discussed here. These data are compiled in Table III. Both the guaninium complex and the $[Cu_2Cl_8]^{4-}$ anion are made up of trigonal bipyramids sharing equatorial-to-apex edges, while the other two complexes listed in the table are square-based pyramids sharing base-to-apex edges. In the trigonal-bipyramidal complexes it is likely that the unpaired electrons are in the d_{z^2} orbitals of the copper(II) ions in the ground state, whereas in the square-pyramidal complexes it is likely that they are in the $d_{x^2-y^2}$ orbitals. A quantitative comparison of the exchange energies of the four complexes cannot be based on structural data because of the different orbitals involved in superexchange, but a qualitative comparison is possible.

The structural parameters of the guaninium complex and the

Table III. Magnetic and structural data for chloro-bridged copper(II) dimers.

COMPLEX	2J (cm^{-1})	STRUCTURE	Cu-Cl-Cu ANGLE (degrees)	Cu-Cl BOND in-plane (Å)	Cu-Cl BOND out-of-plane (Å)	REF.
$[(guaninium)CuCl_3]_2 \cdot 2H_2O$	-32.6	trigonal-bipyramidal	98	2.45	2.29	39,40
$Cu_2Cl_8]^{4-}$ anion	-14.6	trigonal-bipyramidal	95.2	2.70	2.33	35,36, 37,38
$[(2\text{-mepy})_2CuCl_2]_2$ [1]	-7.4	square-pyramidal	101.4	2.26	3.37	41,42
$[(DMG)CuCl_2]_2$ [2]	+6.3	square-pyramidal	88	2.24	2.70	43,44

[1] 2-mepy = 2-methylpyridine

[2] DMG = dimethylgloxime

$[Cu_2Cl_8]^{4-}$anion are compared by superimposition in Figure 13 with the solid line representing the guaninium complex. The smaller singlet-triplet splitting for the $[Cu_2Cl_8]^{4-}$anion in comparison to the guaninium complex accompanies an increase in the Cu-Cl-Cu bond angle from 95° to 98° and a decrease in the bridging bond lengths. Although it has been demonstrated that the bridging angle is important in determining the sign and magnitude of the splitting parameter 2J in the series of hydroxo-bridged copper(II) complexes it is unlikely this effect can be presented as the sole explanation in this comparison because the bridging bond lengths of these two chloro-bridged species are quite different, ranging between 2.3 and 2.7 Å, whereas they are nearly constant in the hydroxo-bridged species at 1.90 to 1.95 Å. (See Table I).

The complexes $[(2\text{-methylpyridine})_2CuCl_2]_2$ and $[(\text{dimethylglyoxime})CuCl_2]_2$ have different structures and presumably a different exchange coupling mechanism from that described above. In comparing these two square-pyramidal complexes it should be noted that there is a change in ground state multiplicity. The bridging bond length in $[(2\text{-mepy})_2CuCl_2]_2$ is 0.67 Å longer than the comparable bond in $[(DMG)CuCl_2]_2$ and the angle at the bridging chloride is 13.4° larger in the 2-methylpyridine complex than in the dimethylglyoxime complex. It is, therefore, not possible to attribute the change in the exchange coupling constant only to bridge angle changes.

Clearly a number of additional chloro-bridged copper(II) dimers of both structural forms must be studied before the bridge-angle effect on 2J can be separated from the bridging bond-length effect.

III. The Polymeric Compound $[Cu(\text{pyrazine})(NO_3)_2]_n$ and Related Chains.

The magnetic properties of the 1:1 copper(II) nitrate-pyrazine complex, $[Cu(C_4N_2H_4)(NO_3)_2]_n$, reflect an exchange coupling between the copper ions although as shown in Figure 14 the copper(II) ions are separated by 6.712 Å. (50) While exchange coupling across bidentate heterocyclic amine ligands had been suggested previously, (51) the preliminary investigation (52) provided the first demonstration of an antiferromagnetic interaction in a system which has been characterized by structural studies, magnetic measurements collected over a wide temperature range and EPR measurements.

The crystal structure of this compound reveals a chemical chain parallel to the a axis in the orthorhombic crystal. Copper atoms are bridged along this axis by the aromatic heterocyclic bidentate amine, pyrazine. Coordinated oxygen atoms from the nitrate ions complete the highly distorted octahedron around each copper atom.

The magnetic susceptibility data collected using a powdered

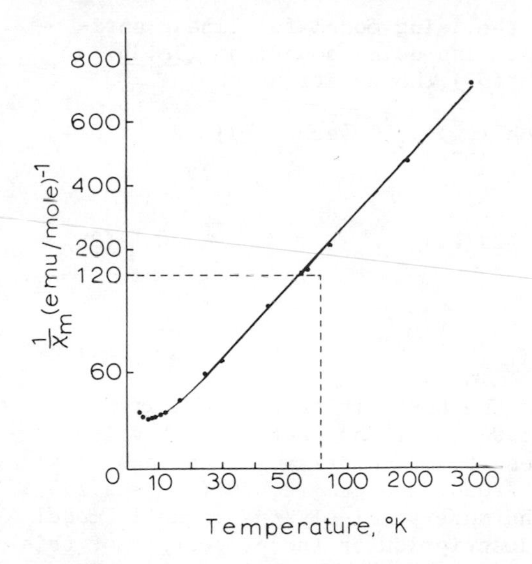

Figure 12. Temperature variation of the experimental inverse susceptibility of [Cu(2-methylpyridine)$_2$Cl$_2$]$_2$

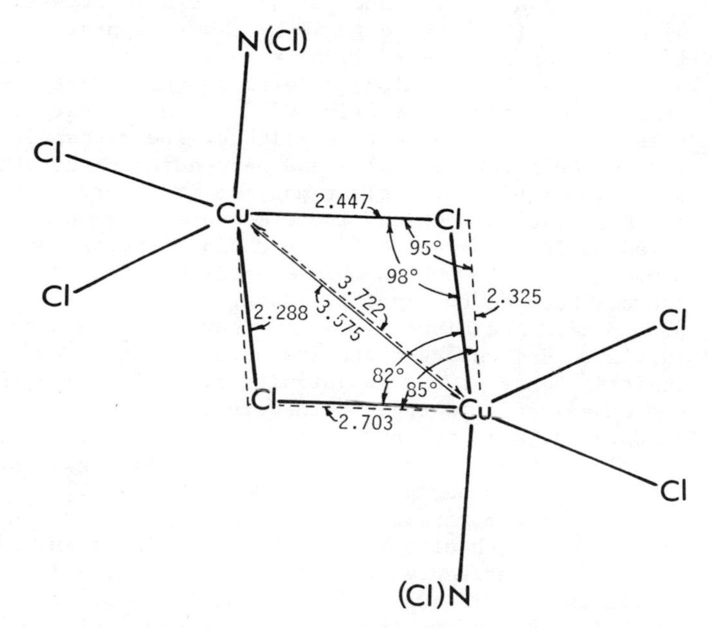

Figure 13. Comparison of the structural parameters for [(guaninium)CuCl$_3$]$_2$ · 2H$_2$O and [Cu$_2$Cl$_8$]$^{4-}$ with the solid line representing the guaninium complex (40)

sample were described (52) by the Ising model for linear antiferromagnetic interactions in chains using equations (3a,3b) which was developed by Fisher. (53) The equations are

$$\chi_{\perp} = \frac{Ng^2\beta^2}{8J} \left[\tanh\left(\frac{J}{kT}\right)+\left(\frac{J}{kT}\right) \operatorname{sech}^2\left(\frac{J}{kT}\right)\right] \quad (3a)$$

and

$$\chi_{||} = \frac{Ng^2\beta^2}{4kT} \exp(2J/kT) \quad (3b)$$

where

$$\langle\chi\rangle = \frac{1}{3}\chi_{||} + \frac{2}{3}\chi_{\perp} \quad (4)$$

The parameters which give the best fit to the experimental data are $\langle g\rangle = 2.22$ and $J = -6.04\ cm^{-1}$. The best fit $\langle g\rangle$ value is to be compared with the EPR results of Kokoszka and Reimann, (54) who reported $g_z = 2.295$, $g_x = 2.054$, and $g_y = 2.070$ ($\langle g\rangle = 2.133$).

In an attempt to determine more precisely the magnetic model which is appropriate for the description of the properties of this compound and to clarify the exchange pathway, measurements have been performed on single crystals of this material between 1.7° and 60°K. (55)

Reasonably large needle-shaped single crystals of $Cu(pyrazine)(NO_3)_2$ were grown by slow-evaporation of an aqueous 1:1 solution of copper(II) nitrate and pyrazine ($C_4H_4N_2$). Because of the crystal morphology several of the largest crystals were selected and carefully oriented under a microscope so that all a axes of the crystals were collinear. The susceptibilities from 1.7° to 60°K both parallel and perpendicular to the a axis are shown in Figure 15. A clear maximum is observed at approximately 6.8°K in each direction. These measurements reveal typical isolated antiferromagnetic linear chain behavior down to the lowest temperature achieved in this experiment. The Ising chain equations can not be used to fit these data.

Bonner and Fisher (56) have performed machine calculations on finite Heisenberg rings and have been able to estimate the limiting behavior for an infinite ring. Applying these results to the experimental data in both directions J/k equals -5.30 (±0.05) °K where J is as defined by the term $-2JS_i \cdot S_j$ in the Hamiltonian described by Bonner and Fisher. The g value giving the best fit to theory for the perpendicular direction was 2.10 (±0.01) and 2.03 (±0.01) for the parallel direction. As can be seen in Figure 15, the fit for both directions down to 1.7°K is in quite good agreement with experiment. Commonly "4+2" tetragonally distorted copper complexes exhibit two small g values ($g_{\perp}$) and one large g value ($g_{||}$). (57) If it is assumed that the smallest g value lies along the a axis, corresponding to the shortest bond distance, then one of the other small g values and the largest g value lie in the plane perpendicular to this axis. The susceptibility perpendicular

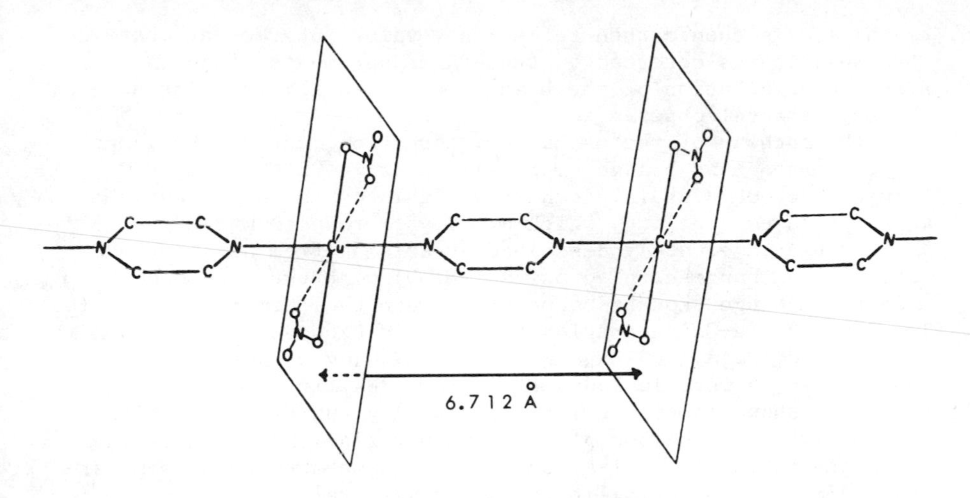

Journal of the American Chemical Society

Figure 14. Structure of $[Cu(C_4H_4N_2)(NO_3)_2]_n$ (52)

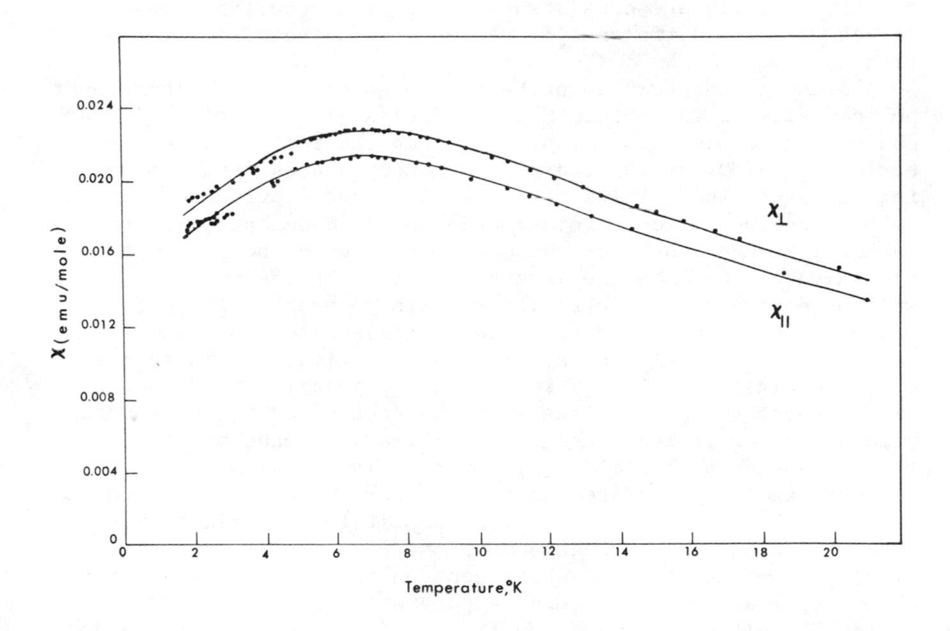

Journal of Chemical Physics

Figure 15. Best fit of the corrected magnetic susceptibility to the one-dimensional Heisenberg model with $J/k = -5.30°$ *for both directions and* $g_{||} = 2.03$ *and* $g_{\perp} = 2.10$ (55)

to the a axis should then reflect a g value intermediate between the two extremes detected by the EPR measurements if one assumes a random orientation of the b and c axes. This is consistent with the experimental observation.

The pathway of the exchange interaction must now be identified. Nitrate ions bridge copper(II) ions in $Cu(NO_3)_2 \cdot 2.5H_2O$ forming a crooked chain. Even though this chain is present the magnetic susceptibility, which displays a rounded maximum at 3.2 °K, has been adequately described as arising from antiferromagnetically exchange coupled pairs (58,59) with the predominant mode of exchange thought to occur between the chemical chains. (60) The shortest Cu-Cu separation (∿4.7 Å) is between copper atoms in the crooked chain. Much weaker antiferromagnetic interdimer exchange effects were included to improve the fit at low temperatures. As shown in Figure 16 a possibility for this same nitrate bridge exists in the b-c plane in $Cu(pyrazine)(NO_3)_2$ where the Cu-Cu separation is 5.1 Å. Because of these structural similarities dimeric susceptibility behavior was explored in Cu-(pyrazine)$(NO_3)_2$. When the dimer susceptibility equation was used to fit the data a value of $J/k = -5.4°K$ results when the g value was restricted to the range 2.00 to 2.20. However, the calculated susceptibilities are consistently lower than the experimental values with the difference increasing as the temperature decreases.

Although an exchange pathway through the nitrate ion cannot be completely ruled out in $Cu(pyrazine)(NO_3)_2$, the clear linear chain behavior in this compound as compared to the pair interaction in $Cu(NO_3)_2 \cdot 2.5H_2O$ argues strongly against this pathway for exchange. The results with the substituted pyridine complexes to be described below offers convincing evidence against this pathway, however, before that can be presented the postulated mechanism of the exchange interaction through the pyrazine bridge will be described. The pyrazine ring is perpendicular to the plane in which the bis(nitrato)copper(II) units lies, and the unpaired electron, in the single ion approximation, is in the $d_{x^2-y^2}$ orbital. Now it develops that the pyrazine ring is canted with respect to the xy plane such that the highest energy occupied molecular orbital, the B_{1g} π orbital, has the proper symmetry to overlap with the $d_{x^2-y^2}$ orbital. It is suggested here that the exchange is propagated in this manner. To test this mechanism complexes have been prepared with substituted pyrazines with the expectation that the exchange coupling would be affected by the electronic nature of the substituent on the pyrazine ring. Exchange by means of the alternate pathway through the nitrate "bridge" would be dependent on the steric properties of the substituent and only secondarily dependent on the electronic nature.

The experimental data which have been collected (61,62) are tabulated in Table IV. The similarity of the spectra and the g values can be taken as good evidence that the structures of the complexes are the same as that of the pyrazine complex. As can be

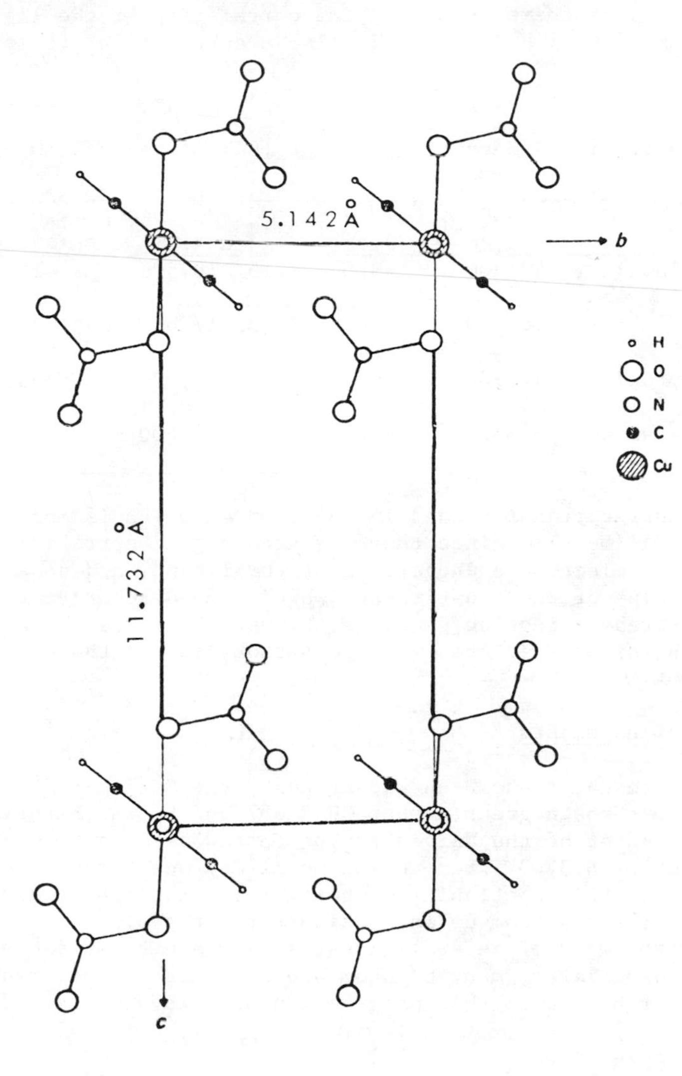

Journal of Chemical Physics

Figure 16. A view of Cu(pyrazine)(NO_3)$_2$ in the bc *plane which illustrates the weak nitrate bridge between copper atoms. For clarity only copper atoms lying at (011) have been included* (55).

seen in Table IV there is a striking correlation of the ligand π↔π* transition with 2J. There is also a correlation between 2J

Table IV

Spectral and Magnetic Data for $[Cu(R\text{-pyrazine})(NO_3)_2]_n$

LIGAND	ligand $\pi-\pi^*$, cm^{-1}	2J, cm^{-1}	<g>	complex d-d, cm^{-1}	complex $\pi\rightarrow\sigma^*$, cm^{-1}
quinoxaline	42,300	-9.0	2.15	18,500	29,000
pyrazine	38,460	-7.2	2.133	17,860	34,600
methyl pyrazine	37,900	-6.2	2.145	17,900	35,100
chloro pyrazine	33,700	-2.8	2.153	17,800	36,400

and the charge transfer band in the complex which is most likely π(ligand)→σ*(metal). Since there is such a good correlation of 2J with the electronic properties of the ligand, and none at all with the size of the substituent (which should separate the chains thereby affecting exchange through the nitrate ion), it seems conclusive that the exchange pathway is via the pyrazine bridge.

IV. Acknowledgements

This research has been supported by the National Science Foundation through grant number GP-22887 and by the Materials Research Center of the University of North Carolina through grant number GH-33632 from the National Science Foundation. I am grateful for this continuing support. This research effort has benefited greatly from collaboration with Professor D.J. Hodgson and from the work of several industrious graduate students and research associates, many of whom are named in the references. Their contribution to this program has been invaluable.

V. Literature Cited

1. For reviews see

a. R.L. Martin in New Pathways in Inorganic Chemistry, Edited by E.A.V. Ebsworth, A.G. Maddock, and A.G. Sharpe, Cambridge University Press, 1968.

b. E. Sinn, Coordn. Chem. Rev., 5, 313 (1970).

c. G.F. Kokoszka and G. Gordon in Transition Metal Chemistry Vol. 5, Edited by R.L. Carlin, Marcel Dekker, Inc., New York, 1969.

d. J.B. Goodenough, Magnetism and the Chemical Bond, Interscience Publishers, Inc., New York, 1963.

2. T. Moriya, Phys. Rev., 120, 91 (1960); T. Moriya in Magnetism, Treatise Modern Theory of Matter, 1, 85 (1963).
3. M.E. Lines, A.P. Ginsberg, R.L. Martin, and R.C. Sherwood, J. Chem. Phys., 57, 1 (1972).
4. P. Pfeiffer and H. Glaser, J. Prakt. Chem., 151, 134 (1938).
5. F.G. Mann and H.R. Watson, J. Chem. Soc., 2772 (1958).
6. W.E. Hatfield, T.S. Piper, and U. Klabunde, Inorg. Chem., 2, 629 (1963).
7. D.W. Meek and S.A. Ehrhardt, Inorg. Chem., 4, 584 (1965).
8. J.R. Wasson, T.P. Mitchell, and W.H. Bernard, J. Inorg. Nucl. Chem., 30, 2865 (1968).
9. T.P. Mitchell, W.H. Bernard, and J.R. Wasson, Acta Cryst., B26, 2096 (1970).
10. J.R. Ferraro and W.R. Walker, Inorg. Chem., 4, 1382 (1965).
11. W.R. McWhinnie, J. Inorg. Nucl. Chem., 27, 1063 (1965).
12. J.C.D. Brand and G. Eglinton, Applications of Spectroscopy to Organic Chemistry, Oldbourne Press, London (1965).
13. B.J. Cole and W.H. Brumage, J. Chem. Phys., 53, 4718 (1970).
14. E.D. Estes, W.E. Hatfield, and D.J. Hodgson, Inorg. Chem., in press.
15. D.L. Lewis, K.T. McGregor, W.E. Hatfield, and D.J. Hodgson, Inorg. Chem., 13, 1013 (1974).
16. D.L. Lewis, W.E. Hatfield, and D.J. Hodgson, Inorg. Chem., 13, 147 (1974).
17. P. Krähmer, M. Maaser, K. Staiger, and E. Uhlig, Z. Anorg. Allgem. Chem., 354, 242 (1967).
18. A. Bondi, J. Phys. Chem., 68, 441 (1964).
19. L. Pauling, The Nature of the Chemical Bond, 3rd ed., Cornell University Press, Ithaca, New York (1960).
20. D.L. Lewis, W.E. Hatfield, and D.J. Hodgson, Inorg. Chem., 11, 2216 (1972).
21. D.Y. Jeter, D.L. Lewis, J.C. Hempel, D.J. Hodgson, and W.E. Hatfield, Inorg. Chem., 11, 1958 (1972).
22. R.L. Gustafson and A.E. Martell, J. Amer. Chem. Soc., 81, 525, (1959).
23. D.D. Perrin and V.S. Sharma, J. Inorg. Nucl. Chem., 28, 1271, (1966).
24. C.M. Harris, E. Sinn, W.R. Walker, and P.R. Woolliams, Aust. J. Chem., 21, 631 (1968).
25. A.T. Casey, Aust. J. Chem., 25, 2311 (1972).
26. A.T. Casey, B.F. Hoskins, and F.D. Whillans, Chem. Commun., 904 (1970).
27. J.A. Barnes, W.E. Hatfield, and D.J. Hodgson, Chem. Commun., 1593 (1970).
28. J.A. Barnes, D.J. Hodgson, and W.E. Hatfield, Inorg. Chem., 11, 144 (1972).
29. R.J. Majeste and E.A. Meyers, J. Phys. Chem., 74, 3497 (1970).
30. K.T. McGregor, N.T. Watkins, D.L. Lewis, R.F. Drake, D.J. Hodgson, and W.E. Hatfield, Inorg. Nucl. Chem. Letters, 9, 423 (1973).

31. E.D. Estes, private communication.
32. See, for example, C. Chow, K. Chang, and R.D. Willett, J. Chem. Phys., 59, 2629 (1973).
33. K.T. McGregor and W.E. Hatfield, to be published.
34. D.J. Hodgson, Progress in Inorganic Chemistry, Edited by S.J. Lippard, Wiley-Interscience, New York, in press.
35. D.J. Hodgson, P.K. Hale, J.A. Barnes, and W.E. Hatfield, Chem. Comm., 786 (1970).
36. D.J. Hodgson, P.K. Hale, and W.E. Hatfield, Inorg. Chem., 10, 1061 (1971).
37. J.A. Barnes, D.J. Hodgson, and W.E. Hatfield, Chem. Phys. Letters, 7, 374 (1970).
38. K.T. McGregor, D.B. Losee, D.J. Hodgson, and W.E. Hatfield, Inorg. Chem., 13, 756 (1974).
39. a. J.A. Carrabine and M. Sundaralingam, J. Amer. Chem. Soc., 92, 369 (1970);b. M. Sundaralingam and J.A. Carrabine, J. Molec. Biol., 61, 287 (1971); c. J.P. Declercq, M. Debbaudt, and M. Van Meerssche, Bull. Soc. Chim. Belges, 80, 527 (1971).
40. R.F. Drake, V.H. Crawford, N.W. Laney, and W.E. Hatfield, Inorg. Chem., 13, 1246 (1974).
41. V.F. Duckworth and N.C. Stephenson, Acta Crystallogr., Sect. B, 25, 1795 (1969).
42. D.Y. Jeter, D.J. Hodgson, and W.E. Hatfield, Inorg. Chim. Acta, 5, 257 (1971).
43. D.H. Svedung, Acta Chem. Scand., 23, 2865 (1969).
44. N.T. Watkins, E.E. Dixon, V.H. Crawford, K.T. McGregor, and W.E. Hatfield, Chem. Commun., 133 (1973).
45. W.E. Hatfield and T.S. Piper, unpublished observations.
46. J.S. Smart, Effective Field Theories of Magnetism, W.B. Saunders Co., Philadelphia, 1966.
47. For example, see J.A. Bertrand, A.P. Ginsberg, R.I. Kaplan, G.E. Kirkwood, R.L. Martin, and R.C. Sherwood, Inorg. Chem., 10, 240 (1971).
48. W.E. Hatfield and R. Whyman, Transition Metal Chemistry, Edited by R.L. Carbin, Marcel Dekker, Inc., New York, Vol. 5, p. 53ff (1969).
49. D.P. Graddon, R. Schulz, E.C. Watton, and D.G. Weeden, Nature, 198, 1299 (1963).
50. A. Santoro, A.D. Mighell, and C.W. Reimann, Acta Cryst., B26, 979 (1970).
51. See for example, D.E. Billing, A.E. Underhill, D.M. Adams, and D.M. Morris, J. Chem. Soc. (A), 902 (1966); M.J.M. Campbell, R. Grzeskowiak, and F.B. Taylor, ibid., 19 (1970).
52. J.F. Villa and W.E. Hatfield, J. Amer. Chem. Soc., 93, 4081 (1971).
53. M.E. Fisher, J. Math Phys., 4, 124 (1963).
54. G.F. Kokoszka and C.W. Reimann, J. Inorg. Nucl. Chem., 32, 3229 (1970).
55. D.B. Losee, H.W. Richardson, and W.E. Hatfield, J. Chem. Phys., 59, 3600 (1973).

56. J. Bonner and M. Fisher, Phys. Rev. Sect. A, 135, 640 (1964).
57. E. König Magnetic Properties of Transition Metal Compounds., Springer-Verlag, Berlin, 1966.
58. L. Berger, S. Friedberg, and J. Schriempf, Phys. Rev., 132, 1057 (1963).
59. B. Myers, L. Berger, and S. Friedberg, J. Appl. Phys., 40, 1149 (1969).
60. J. Bonner, S. Friedberg, H. Kobayashi, and B. Myers, Proc. 12th International Conf. on Low Temp. Physics, Kyoto (1970).
61. H.W. Richardson and W.E. Hatfield, to be published.
62. H.W. Richardson, W.E. Hatfield, H.J. Stoklosa, and J.R. Wasson, Inorg. Chem., 12, 2051 (1973).

11

Electronic and Magnetic Properties of Linear Chain Complexes Derived from Biscyclopentadienyl Titanium(III) and of the Infinite RMX_3 Linear Chain Complexes

D. SEKUTOWSKI, R. JUNGST, and G. D. STUCKY

Materials Research Laboratory, University of Illinois, Urbana, Ill. 61801

Introduction

The focus of this paper will be on the results of experimental studies of the structural and electronic properties of two types of linear chain systems containing metal atoms:

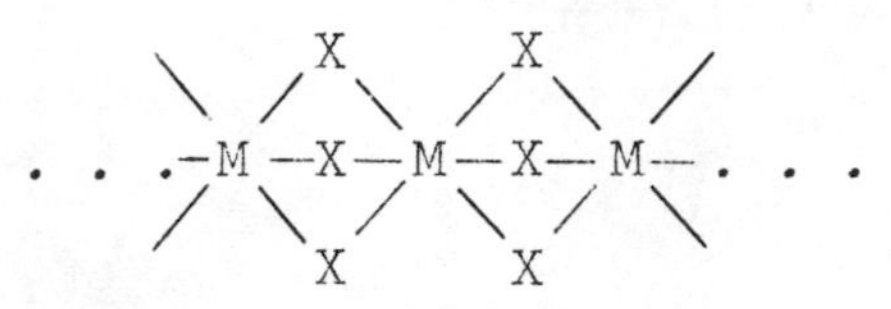

and

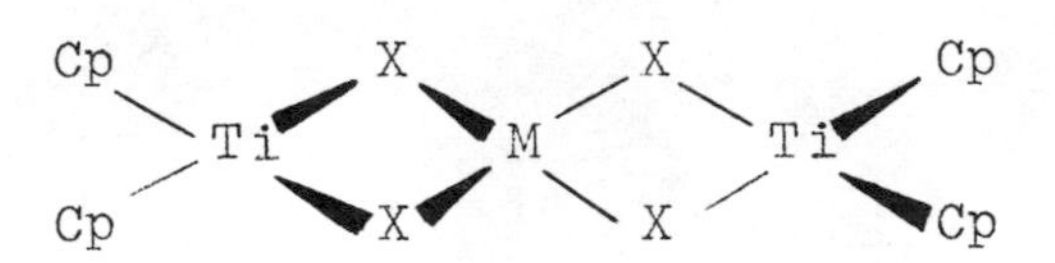

The results we have obtained for the biscyclopentadienyl complexes, which are recent and of a substantially more preliminary nature, will be described first. A part of our earlier work on the infinite linear tri-bridged chain systems will then be described partly by way of introduction to the following two papers by Holt and McPherson on their studies of these materials.

Biscyclopentadienyl Titanium Derivatives

The reduction chemistry of Ti(IV) halides was first extensively studied as part of the development of Ziegler-Natta catalysts in the late 1950's. It was recognized that an important feature of titanium(III) chemistry is the ability of Ti(III) to act as a Lewis acid, and one can obtain a variety of complexes depending upon the solvent, reducing agent and anion. For example, the reduction of Cp_2TiCl_2 with Zn proceeds according to:

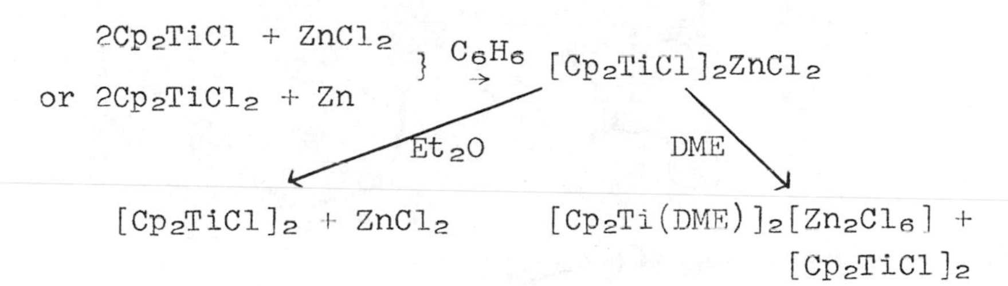

DME = dimethoxyethane

The 2:1 complex was first reported by Salzmann in 1968 (1). The 1:1 complex has not been previously reported and was obtained by addition of dimethoxyethane to a solution of the 2:1 complex.

The structural chemistry of these mixed metal Ti(III) compounds is interesting. The molecular structure of the 1:1 zinc complex as prepared with dimethoxyethane was recently determined by us and is shown in figure 1. It contains $[Cp_2Ti(DME)]^+$ cations and the first example of the $Zn_2Cl_6^{2-}$ anion, which is isoelectronic with Ga_2Cl_6. It will be important to our later discussion to have some idea of what one might expect for the intermolecular magnetic interactions of biscyclopentadienyl d^1 systems. We have examined the magnetic susceptibility of $[Cp_2Ti(DME)]_2[Zn_2Cl_6]\cdot C_6H_6$ to $4.5^\circ K$ and found that the intermolecular exchange is less than 1 cm^{-1}.

The molecular structure of the dibenzene solvate of the 2:1 compound was determined by us and independently by Vonk (2). Our coordinates are shown in figure 2. The paramagnetic molecules are separated from each other in the lattice by benzene molecules. The Zn-Cl-Ti angle is 90° within experimental error, while the Cl-Ti-Cl angle is $81.9(1)^\circ$. The Ti-Zn-Ti angle is $173.2(1)^\circ$ and the Ti-Ti distance is 6.820(5)Å.

An obvious question of interest is whether or not there is any evidence of exchange coupling between the two d^1 metal atoms which are separated by the diamagnetic zinc atom. The rotation of the two Cp_2Ti groups through 90° with respect to each other suggests that any exchange pathway through the zinc atom which involves zinc orbitals which are orthogonal with respect to a 90° rotation about the long orthogonal axis of the

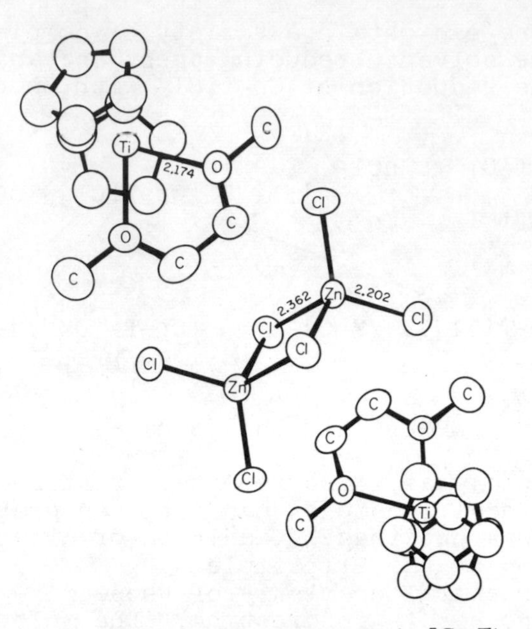

Figure 1. Molecular structure of $[Cp_2Ti(DME)]_2[Zn_2Cl_6] \cdot C_6H_6$. Benzene molecule omitted.

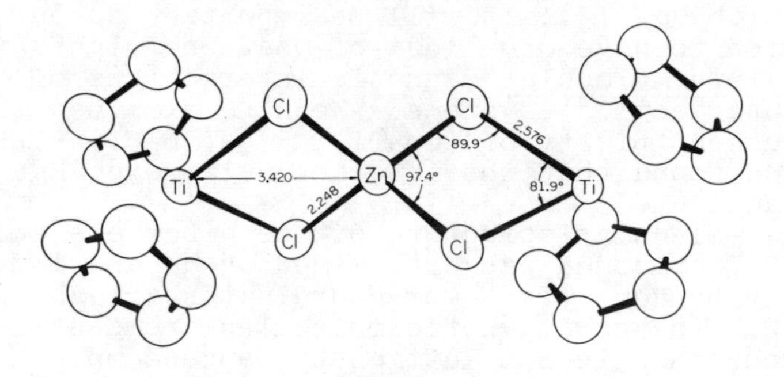

Figure 2. Molecular structure of $[Cp_2TiCl]_2ZnCl_2 \cdot 2C_6H_6$. Solvent molecules not shown.

molecule must be ferromagnetic. Thus, exchange through p_x, p_y or sp^3 hybrid orbitals on the zinc atom would result in ferromagnetic coupling.

To our knowledge, there are no known examples of linear or nearly linear trinuclear metallic systems in which 1,3 magnetic exchange has been detected through a diamagnetic metal atom. Sinn (3) has reported on the magnetic interactions of a number of trinuclear systems of the type shown below in which the approximate arrangement of the metal atoms is an isosceles triangle

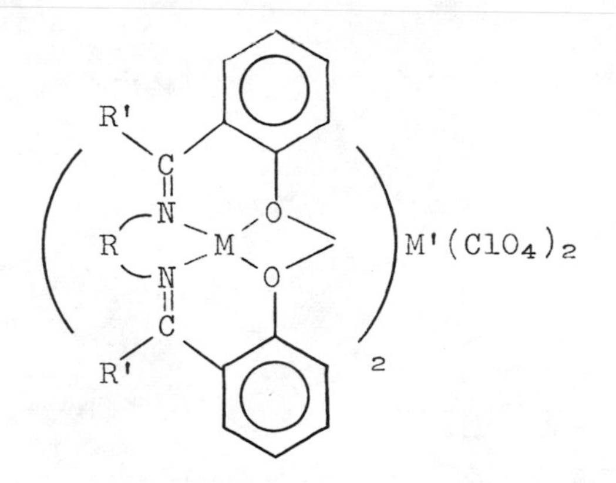

where M = Cu and M' is a variety of metals including Mg. No 1,3 magnetic coupling was detected, although it should be pointed out that the magnetic data were measured only to liquid nitrogen temperature and the absence of magnetic exchange was based on the magnitude of the Weiss constant, θ. Studies of the temperature dependence of the magnetic susceptibility of $Ni_3(acac)_6$ have shown that an antiferromagnetic exchange between the terminal nickel atoms via the paramagnetic nickel(II) central atom is necessary to fit the experimental data (4).

The 2:1 titanium-zinc complexes do in fact exhibit antiferromagnetic behavior as shown by the magnetic susceptibility data (figure 3) for the compound

$$[Cp_2TiBr]_2ZnBr_2 \cdot 2C_6H_6$$

These data were fit by the Van Vleck expression for a singlet-triplet system:

$$\chi_m = \frac{2N\beta^2g^2}{3k(T-\theta)} [1 + 1/3 \exp(-2J/kT)]^{-1} + N\alpha$$

The data were measured at a magnetic field of 10 K gauss with a Zeeman energy of ~ 1 cm^{-1}. In Table 1

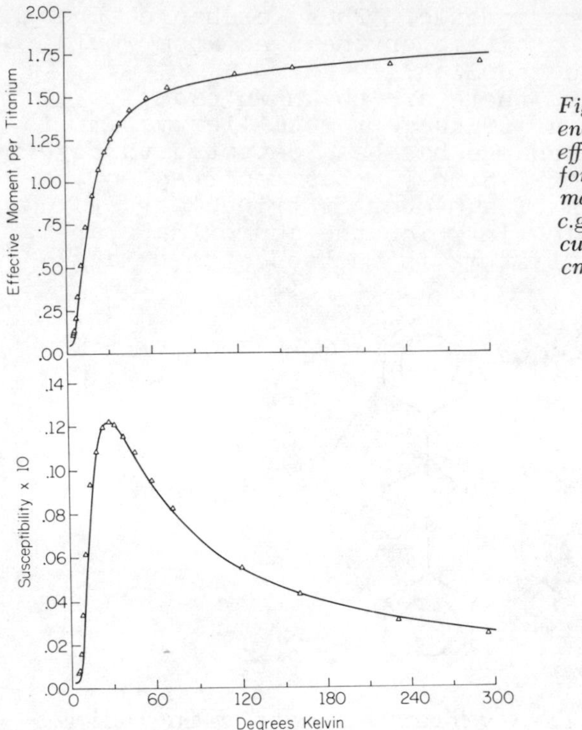

Figure 3. Temperature dependence of molar susceptibility and effective moment per titanium (BM) for $[Cp_2TiBr]_2ZnBr_2 \cdot 2C_6H_6$. Diamagnetic correction of -483×10^{-6} c.g.s. was applied. Theoretical curves calculated using: $J = -15.66$ cm^{-1}, $g = 1.94$, $\theta = -1.37°K$, $N\alpha = 260 \times 10^{-6}$ c.g.s.

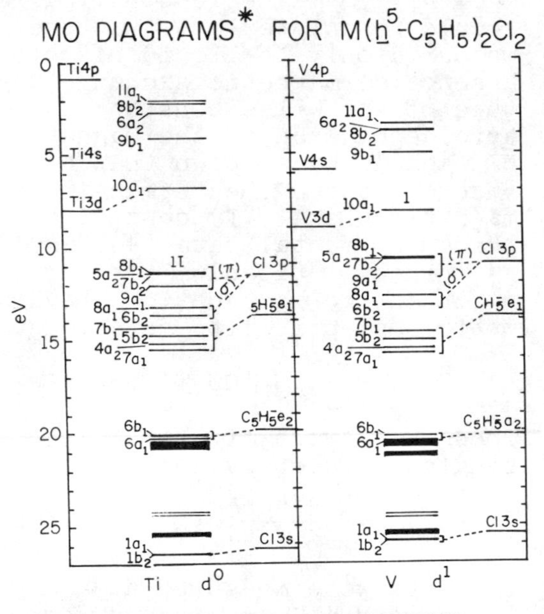

Figure 4. Molecular orbital diagrams for d^0 *and* d^1 *biscyclopentadienyl systems*

are the results of a similar analysis for the chloride and for the dimeric species; $[Cp_2TiX]_2$, X = Cl, Br as reported by Coutts, Wailes and Martin (5). Martin (6)

Table 1

	J(cm^{-1})	
$[Cp_2TiCl]_2ZnCl_2 \cdot 2C_6H_6$	-8.93	J_{Br}/J_{Cl} = 1.75
$[Cp_2TiBr]_2ZnBr_2 \cdot 2C_6H_6$	-15.66	
$[Cp_2TiCl]_2$[a]	-78	J_{Br}/J_{Cl} = 1.60
$[Cp_2TiBr]_2$[a]	-125	

a. From reference 5.

has pointed out that for a superexchange mechanism, the transfer integral and hence the exchange coupling should become larger as the electronegativity of the anion decreases and this appears to be the case for the trimetallic systems as well. However, as indicated below, this similarity may very well be fortuitous and, in fact, there is evidence that the exchange mechanisms in the binuclear and trinuclear systems are quite different.

Fortunately, there is some experimental and theorectical information available concerning the electronic distribution about the metal atom in biscyclopentadienyl d^1 systems which leads to a plausible mechanism for the observed antiferromagnetic coupling. Dahl and Petersen (7),(8) have recently demonstrated by single crystal esr experiments and photoelectron spectroscopy that the unpaired electron in the d^1 system, Cp_2VCl_2, is primarily in a molecular orbital of a_1 symmetry which is 3-5 ev below the lowest unoccupied molecular orbital (figure 4). The unpaired electron density is primarily in a metal orbital which is perpendicular to the twofold symmetry axis of Cp_2VCl_2 and in the plane of the VCl_2 group. A smaller amount of unpaired density is in a d orbital which lies along the bisector of the Cl-V-Cl group (Table 2). The same ordering of levels was obtained theoretically for Cp_2TiCl_2, and it seems likely that a similar electronic distribution about the titanium atom will prevail in $[Cp_2TiX]_2ZnX_2$ (X = Cl, Br).

In figure 5, two antiferromagnetic exchange pathways are shown which are consistent with the symmetry of the molecule and with the theoretical and experimental results of Dahl and Petersen. Both are of symmetry b_2 in the point group D_{2d}. A d orbital can also

Table 2

% Orbital Character (a_1 symmetry)

	M		Cl		Cp		
	d_{z^2}	$d_{x^2-y^2}$	p_x	p_z	a_2	e_1	e_2
Cp_2TiCl_2	69.5	4.9	0.9	16.1	0.3	0.4	6.1 (Humo)
Cp_2VCl_2	67.6	3.3	0.8	22.3	0.2	0.3	4.4 (Homo)

L. Dahl and J. Petersen
Private Communication

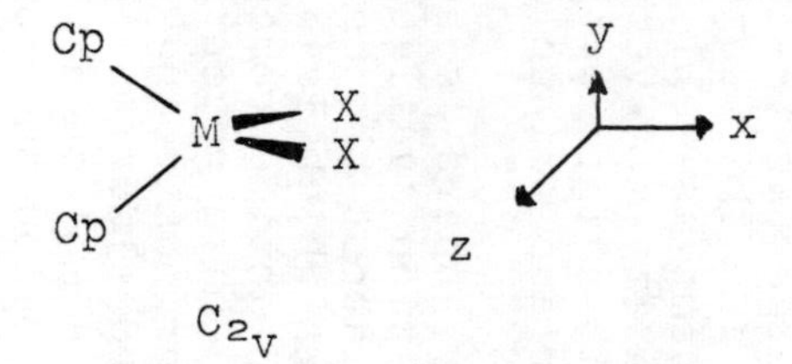

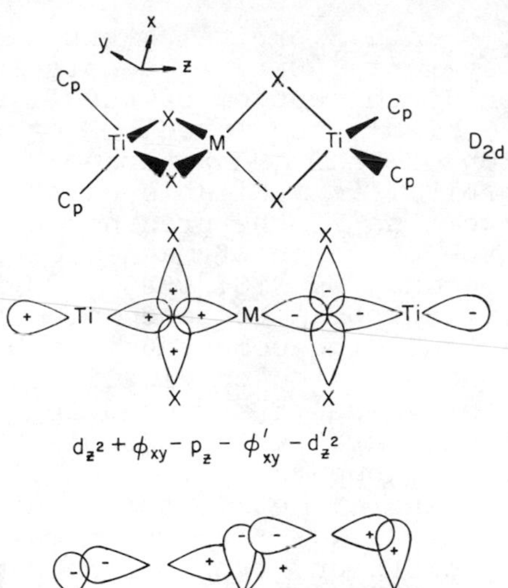

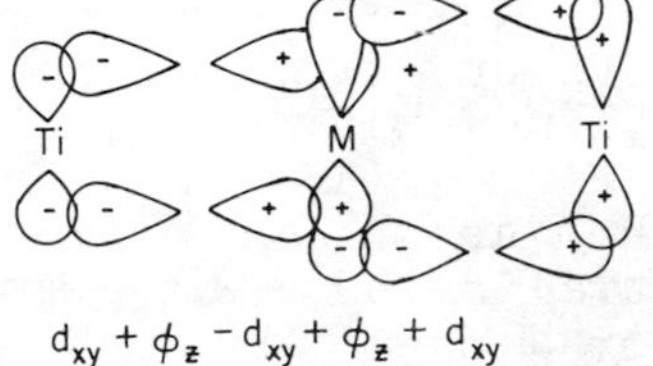

Figure 5. Exchange pathways for trimetallic Ti(III) complexes

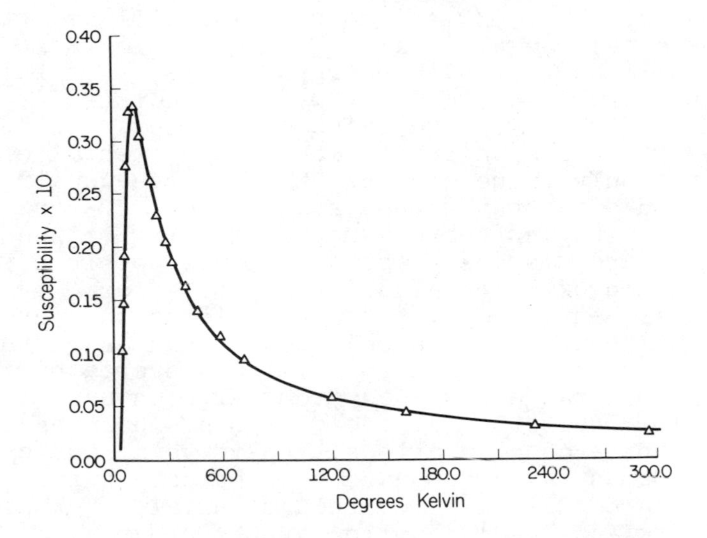

Figure 6. Experimental data (Δ) and theoretical curve for temperature dependence of molar susceptibility of $[Cp_2TiCl]_2$-$BeCl_2 \cdot 2C_6H_6$. Parameters used in curve generation: $J = -6.89$ cm^{-1}, $g = 1.91$, $\theta = 1.87°K$, $N\alpha = 260 \times 10^{-6}$ c.g.s.

be used rather than a p orbital giving a molecular orbital of a_1 symmetry (D_{2d}). We note in passing that the combination at the bottom of the figure which involves the more highly populated d_{xy} orbital (d_{z^2} in the C_{2v} notation of Dahl and Petersen) is only possible if an s or d orbital is available on the central atom. In order to determine if the presence of a d orbital is essential for the presence of magnetic exchange in this system, the compound $[Cp_2TiCl]_2BeCl_2 \cdot 2C_6H_6$ was prepared and recrystallized from a benzene solution. It is isostructural with the zinc compounds and has a temperature dependent magnetic susceptibility, which is shown in figure 6. The value of the coupling constant ($-6.89\ cm^{-1}$) is comparable to that of the zinc chloride compound. It appears that the availability of d orbitals on the central metal atom is not essential for 1,3 exchange.

A slightly different type of one dimensional chain system is obtained with larger metal ions in the central position. With Mn(II) in tetrahydrofuran (THF), the structure, as shown in figure 7, contains a six coordinate central metal atom. The spin state correlation diagram for a d^1-d^5-d^1 system (figure 8) suggests two possible ground states for $|J_{13}/J_{12}| \leq 1$, i.e., for $J_{13} < J_{12}$. Here, S' represents the total spin state of the system and S* is the spin state obtained by coupling the 1,3 atoms. J_{12} has been abbreviated as J. The (3/2,1) ground state will be obtained in the above range for all $J_{12} < 0$, while the (7/2,1) state will be lowest in energy for $J_{12} > 0$. The experimental μ_{eff} curve for $[Cp_2TiCl]_2MnCl_2(THF)_2$ is shown in figure 9. The value of μ_{eff} extrapolates to ~3.8 BM at 0°K which is within experimental error of the spin only value of 3.87 BM for three unpaired electrons and a quartet ground state, i.e., J_{12} must be negative with antiferromagnetic coupling between the titanium and manganese atoms. It was possible to obtain an approximate value for J_{12} of $-8\ cm^{-1}$ by fitting the experimental data, but J_{13} could not be determined with any accuracy since only small changes in the calculated magnetic susceptibilities over a narrow temperature range resulted from large changes in J_{13}. Zero field splitting at the d^5 ion has been neglected in the above considerations. Polycrystalline esr measurements have been made and indicate a zero field splitting of $0.1\ cm^{-1}$ for this material. Further studies of d^1-d^n-d^1 coupling are in progress.

In addition to the bridging groups and the central metal atom, another parameter which can influence the exchange coupling in biscyclopentadienyl d^1 complexes

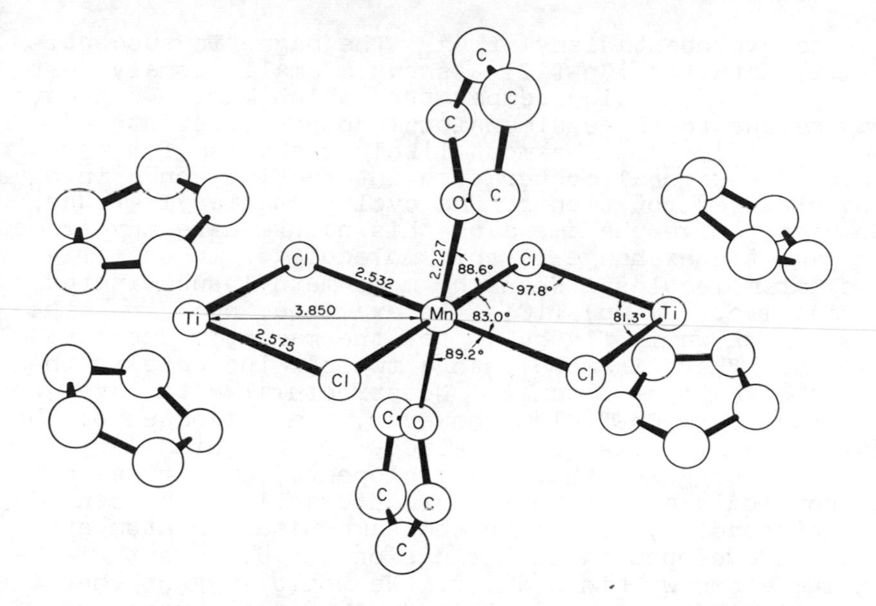

Figure 7. Molecular structure of $[Cp_2TiCl]_2MnCl_2(THF)_2$. *Thermal ellipsoids shown at 50% probability level.*

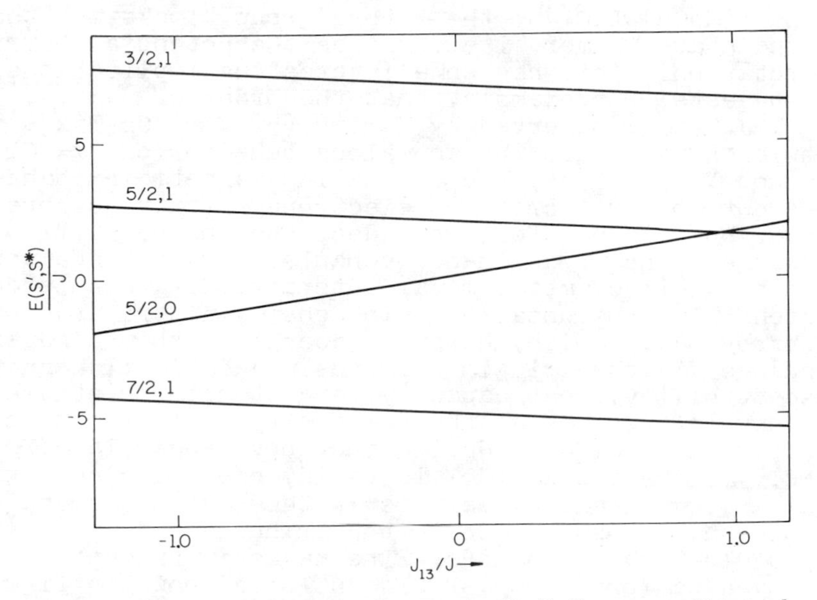

Figure 8. Spin state correlation diagram for d^1–d^5–d^1 *systems.* S' *is total spin state,* S* *is spin state obtained by coupling only the terminal paramagnetic centers,* J *is the exchange integral between terminal and central metals, and* J_{13} *is exchange integral between terminal metals.*

is the cyclopentadienyl ring. The magnetic susceptibility data for $[Cp_2TiCl]_2$ shows a small anomaly just above the transition temperature which Martin suggests may be due to three dimensional cooperative interactions (5),(9). The most likely mechanism for the three dimensional cooperative interaction would involve the hindered rotation of the cyclopentadienyl groups. Do cooperative phenomena of this nature have any effect on magnetic exchange in this molecule or in the trinuclear molecules? When the mono-methyl substituted derivative, $[(MeCp)_2TiCl]_2$ is examined, one finds the results shown in figure 10 for the magnetic susceptibility. There is more than a twofold increase in the value of J to $\sim$166 cm^{-1}. No structural data have been reported for $[Cp_2TiCl]_2$, however, a model based on the structural results for $[Cp_2TiCl_2]_2Zn$ (figure 11) strongly suggests that the cyclopentadienyl rings may be coupled via intramolecular interactions between cyclopentadienyl rings on the same titanium atom and between cyclopentadienyl rings on the different titanium atoms within a dimer. We would suggest that the mean Ti-Ti distance within a dimer for the various rotameric isomers which are obtained at high temperature will probably be greater than that for the ground state low temperature rotameric isomer. In the compound $[(MeCp)_2TiCl]_2$, the methyl group "locks in" the ground state isomer, i.e., raises the potential barrier to rotation. This has some interesting implications. It suggests, for example, that the order of $J_{Br} > J_I \sim J_{Cl}$ observed by Martin (5) for $[Cp_2TiX]_2$ complexes is due to the anomalous behavior of X = Cl and not X = I, i.e., the increasing magnetic exchange does not follow decreasing electronegativity as noted above, but instead follows a decrease in the Ti-Ti distance. The predominant mechanism for exchange would then be a direct metal-metal interaction. This is consistent with the unpaired spin density distribution found by Dahl, which, however, does not exclude superexchange via the bridging ligands. Additional magnetic susceptibility, heat capacity, and structural studies are now in progress at the University of Illinois in order to more clearly define the above and related systems. If the above model is valid, one would not expect a significant difference in the magnetic exchange upon methyl substitution in the trinuclear $[Cp_2TiCl]_2ZnCl_2$ compounds. The magnetic susceptibility data for $[(MeCp)_2TiCl]_2ZnCl_2$ is shown in figure 12 and gives a value of J of -7.3 cm^{-1} which is comparable to that of the unsubstituted material.

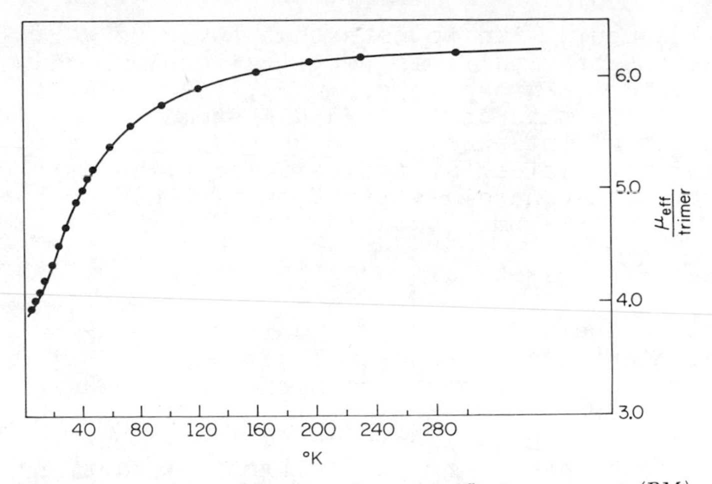

Figure 9. Temperature dependence of effective moment (BM) per trimer of $[Cp_2TiCl]_2MnCl_2(THF)_2$. Diamagnetic correction of -416×10^{-6} c.g.s. was applied.

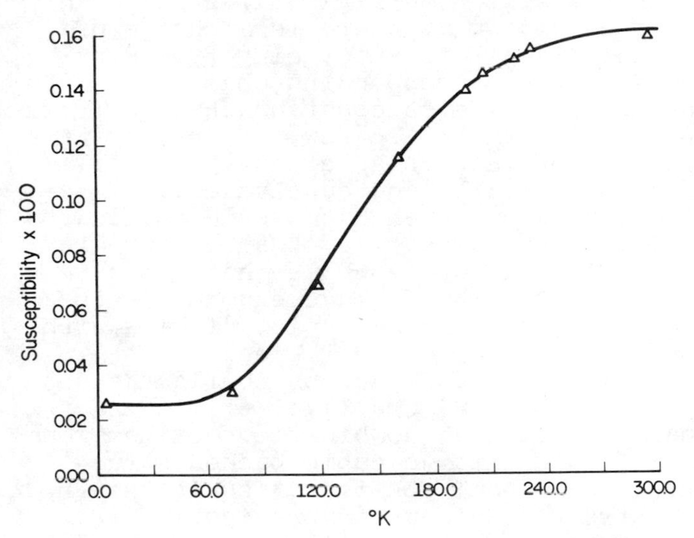

Figure 10. Temperature dependence of molar susceptibility of $[(MeCp)_2TiCl]_2$. Theoretical curve calculated using J = −166 cm^{-1}, $\theta = 0.87°K$, g = 2.06, $N\alpha = 260 \times 10^{-6}$ c.g.s.

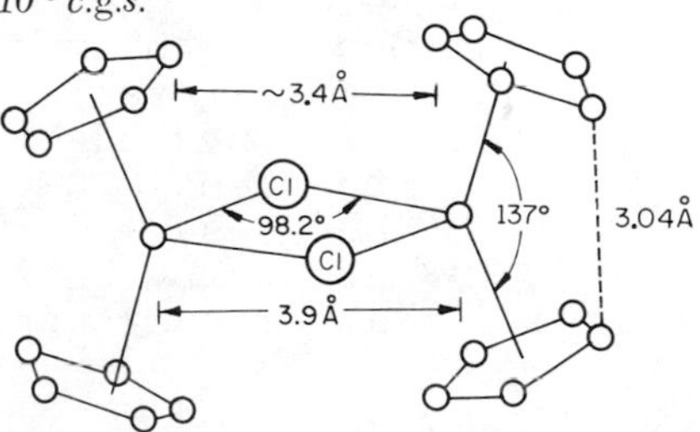

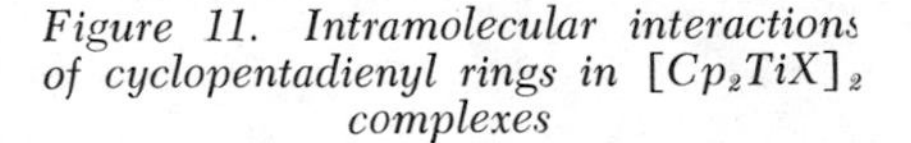

Figure 11. Intramolecular interactions of cyclopentadienyl rings in $[Cp_2TiX]_2$ complexes

Finally, we would like to note that numerous other combinations of metal ions (e.g., zirconium and vanadium) and bridging group (e.g., $-NR_2$, $-PR_2$, -OR, -SR) are possible in this series and a systematic experimental survey of the exchange in these compounds can be expected to provide, at the very least, an empirical understanding and classification of the electronic coupling of triatomic metal systems.

RMX_3 Complexes

Compounds of the type RMX_3 (R = unipositive cation, M = divalent transition metal cation, X = halogen) form a large class of one dimensional systems with the general structure shown in figure 13 (10). The one dimensional properties have been beautifully demonstrated by low temperature nmr and neutron diffraction studies (11). The fact that one is not dealing with a completely isolated one dimensional chain is evident in that the R group can be used to "tune" the electronic properties of the transition metal as illustrated in Table 3. Thus, one finds that there is a change of 0.14Å in the nickel-nickel distance going from R = $(CH_3)_4N^+$ to R = K^+. The corresponding change in 10 Dq is 750 cm^{-1} and the Curie-Weiss constant changes from 0 to -112°K ($RbNiCl_3$). There are few, if any, systems in which the first coordination sphere of the metal atom can be so systematically and subtly varied.

An important question which has become obvious to many people during the past year is, "Can one predict if a one dimensional structure such as that shown in figure 13 will be obtained?" We have examined this question from a strictly empirical point of view with the results shown in figure 14 and Table 4.

There are at least five different configurations adopted by RMX_3 compounds (figure 14): primitive cubic, hexagonal, and various combinations of hexagonal (one dimensional) and primitive cubic packed structures. The purely hexagonal configuration is the 2L, one dimensional structure discussed previously. 2L refers to the crystallographic repeat distance in the linear chain direction. The ratio of hexagonal (linear) to cubic packing increases from 6L to 4L to 9L. Remembering this, then, an examination of known RMX_3 structures reveals that the hexagonal 1-d configuration is favored by:

(1) Increased CFSE (crystal field stabilization energies)
(2) Larger X groups
(3) Larger R groups (to a point; $(C_5H_5)_4N^+$

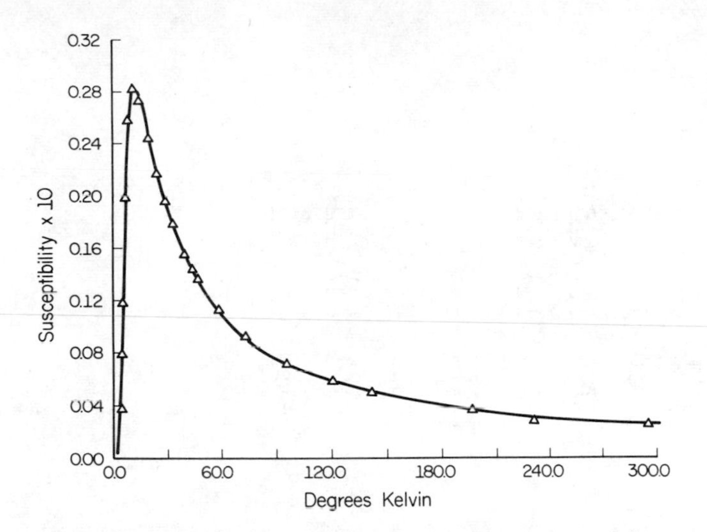

Figure 12. Temperature dependence of molar susceptibility of $[(MeCp)_2TiCl]_2ZnCl_2$. Theoretical curve calculated using $J = -7.26\ cm^{-1}$, $g = 1.92$, $\theta = 0.69°K$, $N\alpha = 260 \times 10^{-6}$ c.g.s.

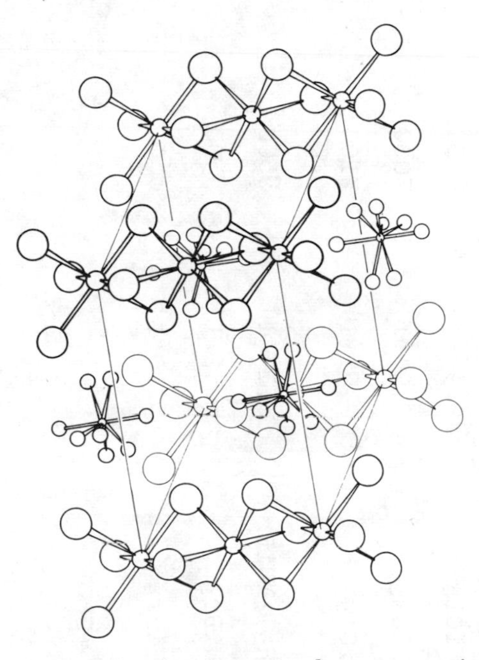

Figure 13. One-dimensional structure of RMX_3 compounds. $R = (CH_3)_4N^+$, $M = Ni$ and $X = Cl$ (10).

Table 3

	Ni-Ni distance Å	10Dq(cm^{-1})	15B'(cm^{-1})	Curie Weiss θ(°K)
$(CH_3)_4NNiCl_3$	3.054(5)	6350	13,450	0 (13)
$CsNiCl_3$	2.96(1)	6800	12,600	-76 (14),(15)
$RbNiCl_3$	2.94(1)	7000	12,400	-112 (14),(15)
$KNiCl_3$	2.91(1)	7100	12,300	---
$(CH_3)_4NNiBr_3$	3.17(1)	6300	11,600	---
$CsNiBr_3$	3.12(1)	6500	11,300	-101 (14),(15)
$RbNiBr_3$	3.10(1)	6600	11,300	-156 (14),(15)

Table 4

Summary of Structures of $CsMX_3$ Compounds

	F	Cl	Br	I
	(1.36)	(1.81)	(1.95)	(2.16)
Mg(0.65)[a]	--	2L (16)	2L (17)	2L (18)
V(0.87)	--	2L (19)	2L (20)	2L (18)
Cr(0.84)	--	2L (21)	2L (22)	2L (18)
Mn(0.80)	6L (23)	9L (24)	2L (20)	--
Fe(0.76)	6L (25)	2L (26)	2L (20)	--
Co(0.78)	9L (27)	2L (28)	--	--
Ni(0.78)	2L (29)	2L (30)	2L (31)	2L (18)
Cu(0.69)	--	2L (32)	4L (22)	
Cd(0.97)	--	6L (32)		

Summary of Structures of $RMCl_3$ Compounds

	$(Me)_4N$	Cs	Rb	K
	(2.60)	(1.69)	(1.48)	(1.33)
Vo(0.87)[a]	--	2L (19)	--	2L (19)
Cr(0.84)	--	2L (16)	--	--
Mn(0.80)	2L (34)	9L (18)	6L (35)	tetragonal
Fe(0.76)	--	2L (25)	2L (25)	--
Co(0.78)	--	2L (28)	2L (36)	--
Ni(0.78)	2L (11)	2L (30)	2L (32)	3L (20)
Cu(0.69)	--	2L (38)	--	4L (39)
Cd(0.97)	2L (33)	6L (33)	--	--

a. The numbers in parentheses are the ionic radius of each atom.

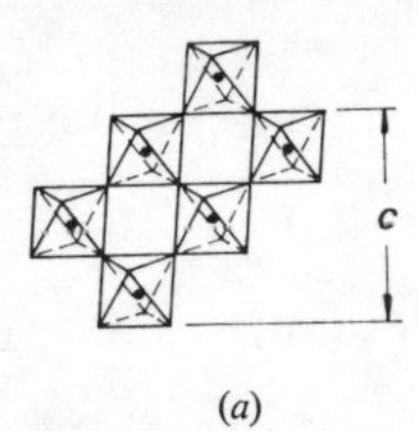

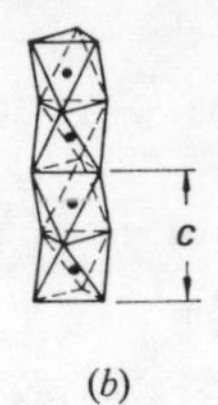

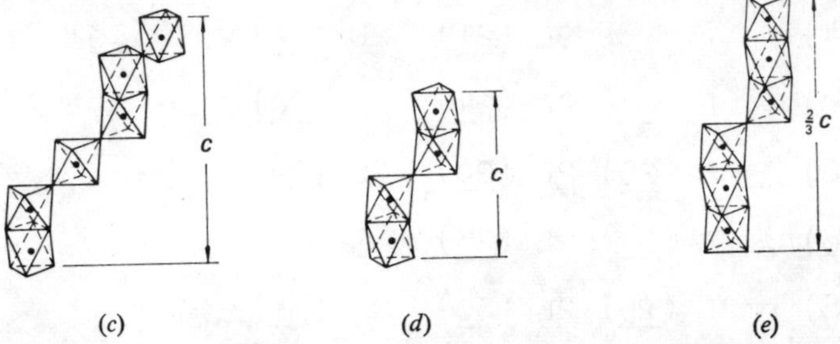

Acta Crystallographica

Figure 14. Perspective projections of octahedra network in a unit cell of RMX_3 complexes (15). *(a)* P. *(b)* 2L. *(c)* 6L. *(d)* 4L. *(e)* 9L.

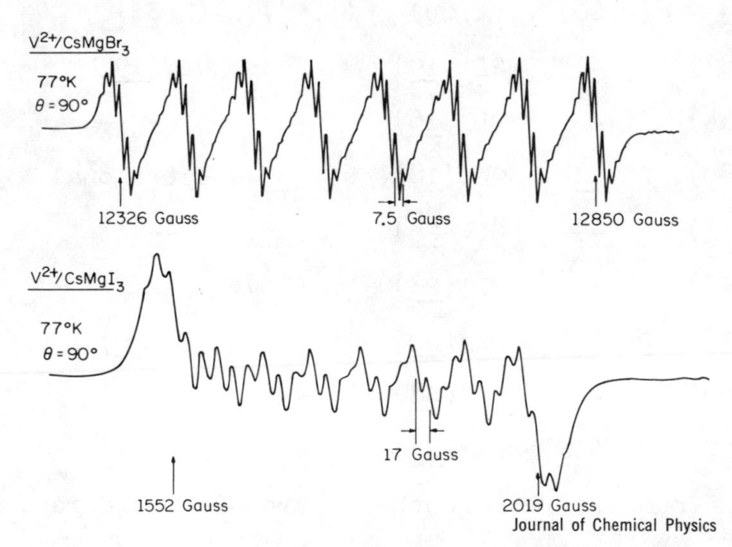

Journal of Chemical Physics

Figure 15. Ligand hyperfine in the esr spectra of V^{2+} doped into $CsMgBr_3$ and $CsMgI_3$ (41)

favors tetrahedral $NiCl_4^{2-}$.

We have examined the electronic properties of these complexes by a variety of methods, and some of the more recent work in this area will be described in the following two papers. In closing, two interesting observations which we have made will be briefly described.

$CsMgX_3$ salts provide a convenient isostructural diamagnetic matrix into which to dope the paramagnetic transition metal ions (16),(41). Covalency effects are strikingly apparent in the ligand hyperfine as it is shown superimposed on the $|-1/2> \rightarrow |+1/2>$ transition of the esr spectra of the V^{2+} bromine and iodine complexes (figure 15). The spin Hamiltonian parameters, excluding the ligand hyperfine tensors, are shown in Table 5. The increase in covalency $Cl^- \rightarrow Br^- \rightarrow I^-$ is evident in both the decrease in the metal hyperfine constants and in the changes in the g tensor. In fact, we observe for the iodine complex a very unusual example of an early transition element in a nearly octahedral field with both $g_{\parallel}$ and $g_{\perp}$ greater than the free electron g factor of 2.0023. This is qualitatively explicable in terms of McGarvey's theory (42),(43) of covalency for d^3 systems in which the ligand molecular orbital coefficients enter into the expression for the g tensor, weighted by the ligand spin orbital coupling constant.

The last experimental observation that will be described is illustrated by the single crystal electronic absorption spectrum of $CsCrCl_3$ (21) (figure 16). The concentration dependence of this spectrum in $CsMgCl_3$ is also shown. Two features of interest in the spectrum are (1) a band at 22,000 cm^{-1} which is not explained by ligand field calculations and (2) an enhanced intensity of spin forbidden transitions by approximately an order of magnitude. The polarization of the 22,000 cm^{-1} band is particularly strong and its temperature dependence is not that expected for a vibronic mechanism. The explanation for this effect was first proposed by Dexter (44) and later elaborated upon by Day (45) and others.

In its simplest form (figure 17), a single photon results in the formation of either two excitons or an exciton and a magnon. In both cases, the total spin symmetry is conserved. In $CsCrCl_3$, the former results in a band at approximately twice 10 Dq, the latter results in an enhanced allowedness for the spin forbidden quintet-triplet and quintet-singlet transition.

In summary, the infinite one dimensional complexes, RMX_3, provide an unusually broad and interesting class

Table 5

Spin Hamiltonian Parameters

		$CsMgCl_3$	$CsMgBr_3$	$CsMgI_3$
V^{2+}(297°K)	$g_{\parallel}$	= 1.9730 ± .0006	1.996 ± .002	2.04 ± .01
	$g_{\perp}$	= 1.9750 ± .0006	1.999 ± .002	2.04 ± .01
	D($x10^4 cm^{-1}$)	= ± 961. ± 4.	± 1320. ± 15.	± 2160 ± 50.
	A($x10^4 cm^{-1}$)	= ∓ 76. ± 1.	∓ 70. ± 2.	∓ 67. ± 2.
	B($x10^4 cm^{-1}$)	= ∓ 74. ± 1.	∓ 70 ± 2.	∓ 67. ± 2.
(77°K)	$g_{\parallel}$	= 1.9740 ± .0006	1.996 ± .002	2.038 ± .002
	$g_{\perp}$	= 1.9760 ± .001	1.999 ± .002	2.035 ± .002
	D($x10^4 cm^{-1}$)	= ± 858. ± 7.	± 1280 ± 15.	± 2080. ± 30
	A($x10^4 cm^{-1}$)	= ∓ 76. ± 1.	∓ 70. + 2.	∓ 65. ± 2
	B($x10^4 cm^{-1}$)	= ∓ 74. ± 1.	∓ 70 ± 2.	∓ 65. ± 2

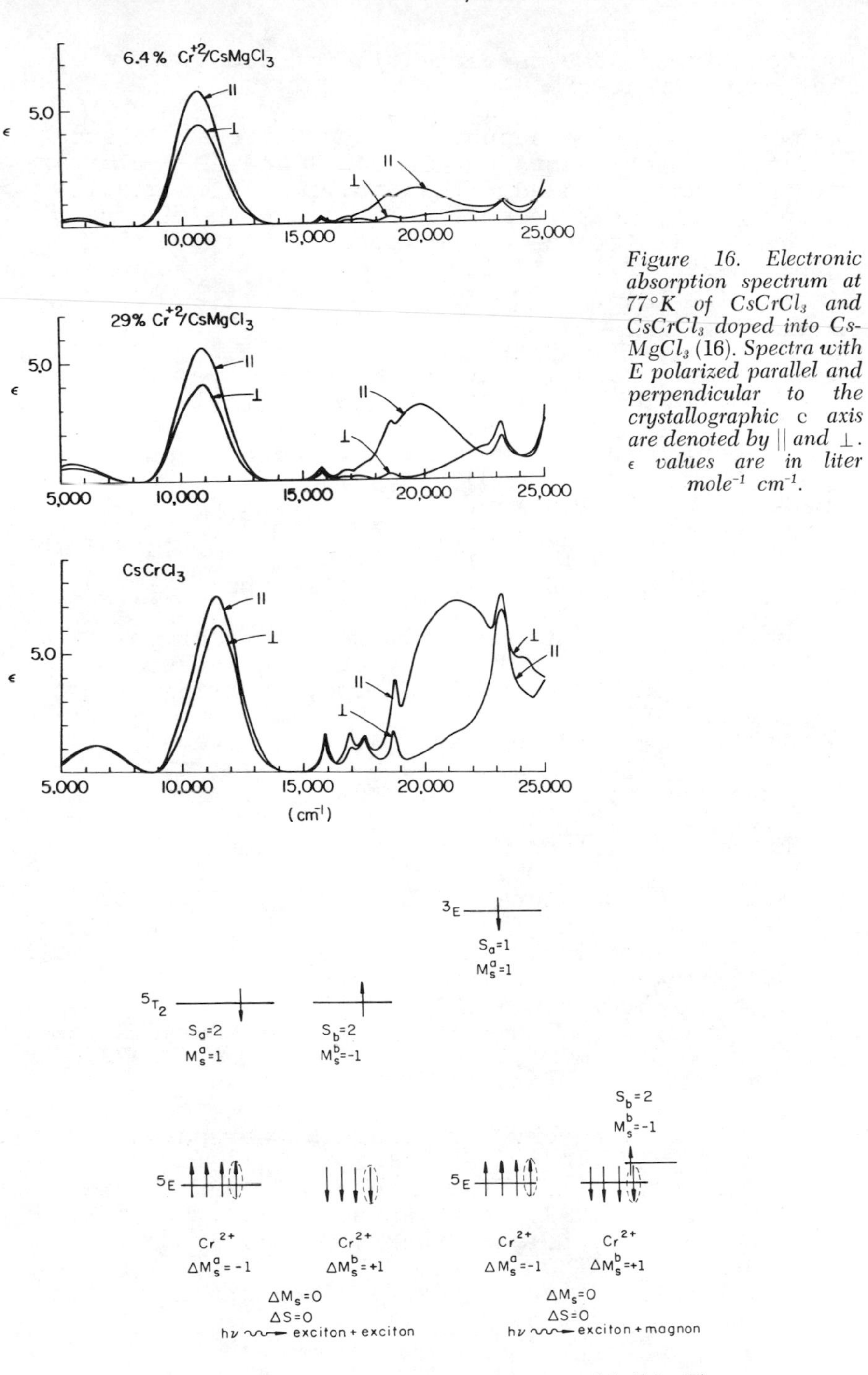

Figure 16. Electronic absorption spectrum at 77°K of $CsCrCl_3$ and $CsCrCl_3$ doped into $CsMgCl_3$ (16). Spectra with E polarized parallel and perpendicular to the crystallographic c axis are denoted by || and ⊥. ε values are in liter mole⁻¹ cm⁻¹.

Figure 17. Simultaneous pair excitation model (44, 45)

of materials for the investigation of cooperative electronic effects.

Acknowledgment. The support of the National Science Foundation under Grants NSF-GH-33634 and GP-31016X are gratefully acknowledged. The experimental magnetic susceptibility measurements were made with the greatly appreciated assistance of Professor David Hendrickson of the University of Illinois.

Literature Cited

1. Salzmann, J. J., Helv. Chim. Acta (1968), 51, 526.
2. Vonk, C. G., J. Cryst. Mol. Struct. (1973), 3, 201.
3. Gruber, S. J., Harris, C. M., and Sinn, E., J. Chem. Phys. (1968), 49, 2183.
4. Ginsberg, A. P., Martin, R. L., and Sherwood, R. C., Inorg. Chem. (1968), 7, 932.
5. Coutts, R. S. P., Wailes, P. C., and Martin, R. L., J. Organometallic Chem. (1973), 47, 375.
6. Martin, R. L., in "New Pathways in Inorganic Chemistry," Pages 175-231, E. A. U. Ebsworth, A. G. Maddock and A. G. Sharpe, Cambridge Press, London (1968).
7. Dahl, L. F. and Petersen, J. L., private communication.
8. Petersen, J. L. and Dahl, L. F., J. Amer. Chem. Soc. (1974), 96, 2248.
9. Martin, R. L. and Winter, G., J. Chem. Soc. (1965), 1965, 4709.
10. Stucky, G. D., Acta Cryst. (1968), B24, 330.
11. Yelon, W. B. and Cox, D. E., Phys. Rev. B. (1973), 7, 2024, and included references.
12. Goodgame, D. M. L. and Weeks, M. J., J. Chem. Soc. (1964), 1964, 5194.
13. Asmussen, R. W. and Soling, H., Z. Anorg. Allgem. Chem. (1965), 203, 3.
14. Asmussen, R. W. and Bostrup, E., Acta Chem. Scand. (1957), 11, 745.
15. Li, Ting-i, Stucky, G. D. and McPherson, G., Acta Cryst. (1973), B29, 1330.
16. McPherson, G. L., Kistenmacher, T. and Stucky, G. D., J. Chem. Phys. (1970), 52, 815.
17. McPherson, G. L. and Stucky, G. D., J. Chem. Phys. (1972), 57, 3780.
18. Li, Ting-i, Stucky, G. D. and McPherson, G. L., Acta Cryst. (1973), B29, 1330.

19. Seifert, H. J. and Ehrlich, P., Z. Anorg. Allgem. Chem. (1959), 302, 286.
20. Unpublished results.
21. McPherson, G. L., Kistenmacher, T. J., Folkers, J. B. and Stucky, G. D., J. Chem. Phys. (1972), 57, 3771.
22. Li, Ting-i and Stucky, G. D., Inorg. Chem. (1973), 12, 441.
23. Zalkin, A., Lee, K. and Templeton, D. H., J. Chem. Phys. (1962), 37, 697.
24. Goodyear, J. and Kennedy, D. J., Acta Cryst. (1972), B28, 1640.
25. Kestigian, M., Leipzig, F. D., Croft, W. J. and Guidoboni, R., Inorg. Chem (1966), 5, 1462.
26. Seifert, H. J. and Klatyk, K., Z. Anorg. Allgem. Chem. (1966), 342, 1.
27. Longo, J. M. and Kafales, J. A., J. Solid State Chem. (1969), 1, 103.
28. Seifert, H. J., Z. Anorg. Allgem. Chem. (1960), 307, 137.
29. Babel, D., Z. Naturforsch. (1965), 20A, 165.
30. Tishchenko, G. N., Tr. Inst. Krist. Akad. Nauk SSSR (1955), 11, 93.
31. Stucky, G. D., D'Agostino, S. and McPherson, G. L., J. Amer. Chem. Soc. (1966), 88, 4823.
32. Hexagonal, with a = 12.56, c = 11.56Å. See Ref. 27.
33. Siegel, J. and Gebert, E., Acta Cryst. (1964), 17, 790.
34. Morosin, B. and Graebner, E. J. Acta Cryst. (1967), 23, 766.
35. Seifert, H. J. and Koknat, F. W., Z. Anorg. Allgem. Chem. (1965), 341, 269.
36. Engberg, A. and Soling, H., Acta Chem. Scand. (1967), 21, 168.
37. Asmussen, R. W. and Soling, H., Z. Anorg. Allgem. Chem. (1956), 203, 3.
38. Schlueter, A. W., Jacobson, R. A. and Rundle, R. E., Inorg. Chem. (1966), 5, 277.
39. Willett, R. D., Dwiggins, C., Kruhard, R. and Rundle, R. E., J. Chem. Phys. (1963), 38, 2429.
40. Morosin, B., Acta Cryst. (1972), B28, 2303.
41. McPherson, G. L., Koch, R. C. and Stucky, G. D., J. Chem. Phys. (1974), 60, 1424.
42. McGarvey, B. R., J. Chem. Phys. (1964), 41, 3743.
43. McGarvey, B. R., J. Phys. Chem. (1967), 71, 51.
44. Dexter, D. L., Phys. Rev. (1962), 126, 1962.
45. Day, P., Inorganica Chimica Acta Reviews (1969), 3, 81.

12

Optical Properties of Linear Chains

S. L. HOLT

University of Wyoming, Laramie, Wyo. 82070

Compounds of the formulation RMX_3, where R is a heavy alkali metal ion (Cs^+ or Rb^+) or an organic cation such as $(CH_3)_4N^+$, M is a first row divalent transition metal ion, and X is Cl^-, Br^- or I^-, display unusual magnetic behavior.(1,2) This behavior is predominately one-dimensional and arises in large part because of the one-dimensional molecular structure of these materials. Figure 1 shows the chain-like constitution exhibited by all compounds of this type. These chains consist of face-sharing octahedra. The larger circles represent the bridging halide ions while the smaller circles represent the transition ions. As shown in Figure 2, these chains are physically separated by the cations, in this case, $(CD_3)_4N^+$. These cations provide the magnetic insulation between chains with the interchain distance and thus the strength of interchain interaction being dependent upon the size of cation, i.e., the larger the cation the smaller the interchain interaction.

Figure 3 shows the pathway of the exchange. As drawn, it suggests that the exchange involves primarily the overlap of the d_{z^2} orbital of Cation 1 with the $p\sigma$ orbital of the bridging anion followed by the interaction of the $p\sigma$ electrons with the d_{z^2} of Cation 2. Indeed, other orbitals may be involved as is schematically shown in Figure 4.

In this case, we have superexchange illustrated for both 90° bonded chromium(III) ions and 90° bonded iron(II) ions. In the upper part of the figure, we have the chromium(III) case and hence, are dealing with 3 d-electrons. The d-electrons on Cation 1 are found in t_{2g} orbitals, as are those on Cation 2. An electron from the $p\sigma$ orbital of the ligand is shown as being virtually exchanged into the e_g orbitals of Cation 1. The coupling process of lowest energy requires that the

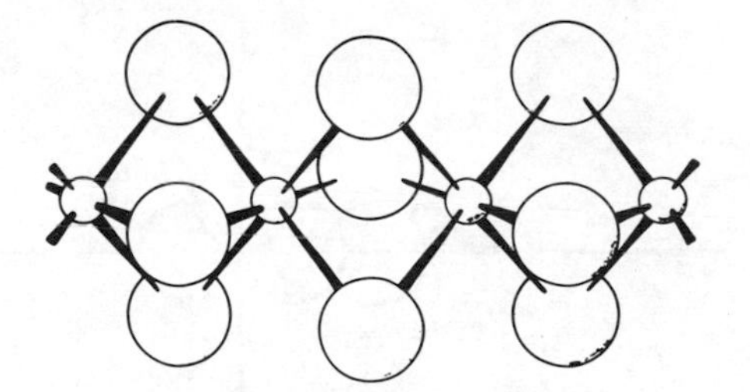

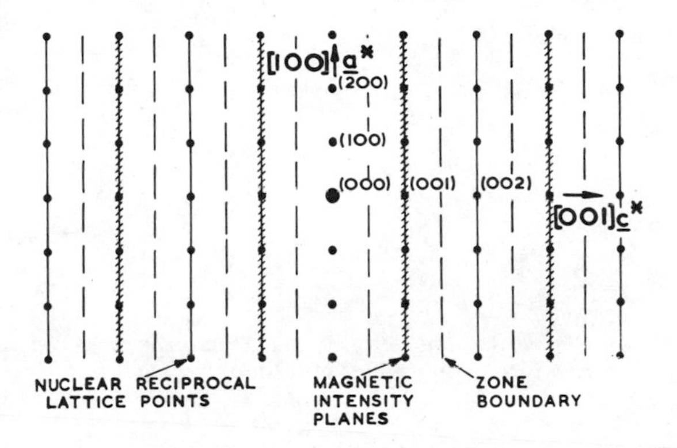

Figure 1. A portion of the linear chain of an RMX_3 compound (adopted from Ref. 1)

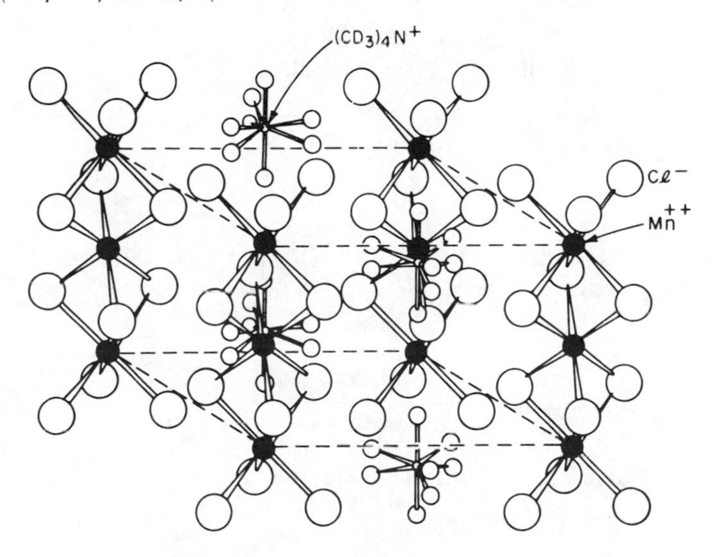

Figure 2. Crystal structure of $(CD_3)MnCl_3$ (adopted from Ref. 1)

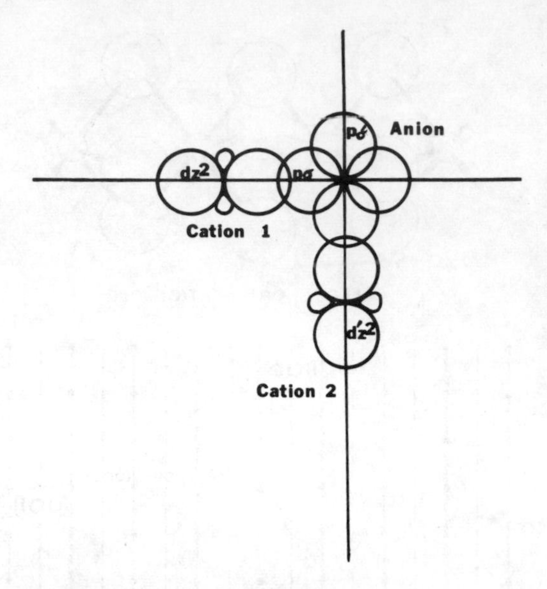

Figure 3. The 90° exchange pathway between two interacting metal ions (adopted from Ref. 2)

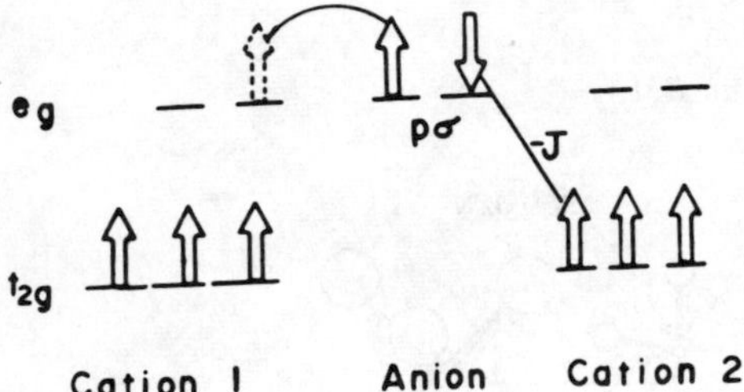

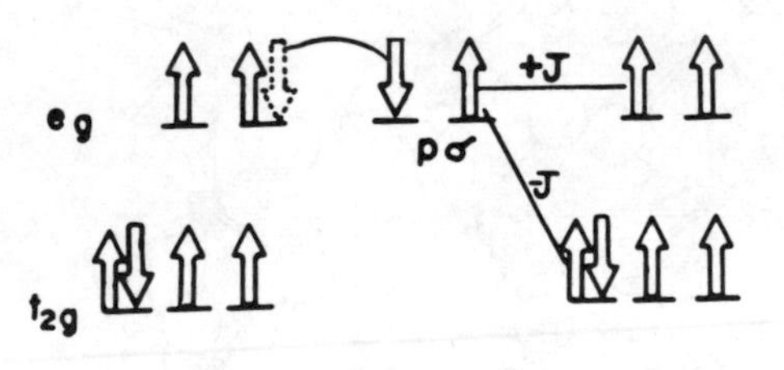

Figure 4. Superexchange pathway in (a) 90° bonded Cr^{3+}–Cr^{3+}; (b) 90° bonded Fe^{2+}–Fe^{2+} (adopted from Ref. 2).

spin of the virtually exchanged electron be the same as that in the t_{2g} set of Cation 1.

The Pauli Exclusion Principle then requires that the other electron in the pσ orbital of the ligand have a spin of opposite sign (in this case down). Because there are no electrons in the e_g orbitals of Cation 2, only one exchange process is possible, that process that gives rise to a -J, i.e., antiparallel exchange coupling between the pσ electron of the anion and t_{2g} electrons of Cation 2.

The net effect is to allow ferromagnetic coupling between Cation 1 and Cation 2 with the overall behavior shown by the compound being ferromagnetic.

In the lower part of the figure, we have the case of 90° exchange between two Fe^{2+} ions. In this case things are not so clear cut. Virtual exchange between the pσ electron of the anion and the electrons in the e_g orbitals of Cation 1 is antiferromagnetic in nature. This does not dictate the overall spin arrangement however. As we can see there exists the possibility for ferromagnetic as well as antiferromagnetic exchange between the electron in the pσ orbital of the anion and the electrons in the d-orbitals of Cation 2. The resultant magnetic behavior depends only upon which is stronger, the coupling between the electron in the pσ orbital and the e_g electrons or the coupling between the pσ electron and the t_{2g} electrons. If this coupling is sufficiently strong to produce preferential alignment and involves several cation centers then conditions have been fulfilled for the creation of spin-waves.

The drawing in Figure 5 is a schematic depiction of a spin-wave in a one dimensional ferromagnetically coupled system. As one can see, a spin deviation at one end of the chain causes spin deviations down the line producing the "spin-wave". As a quantum of energy called a phonon excites a vibration, a quantum of energy called a magnon causes a spin-wave. If one has isolated chains within a material, magnons will induce spin-waves in these chains independent of each other.

The presence of magnetic coupling and the inducement of spin-waves can be shown experimentally through inelastic-neutron-scattering techniques. Figure 6 shows the variation with temperature of the excitation which is associated with the presence of spin-waves in the compound $(CD_3)_4MnCl_3$. As can be seen the spin-wave is well defined at 1.9°K. This definition decreases as one increases the temperature to 40°K. This marked decrease in intensity is to be expected as the three-dimensional ordering temperature has been shown to be

Figure 5. Schematic of a spin wave. (a) View in the **ac** *plane. (b) View in the* **ab** *plane (adopted from Ref. 3).*

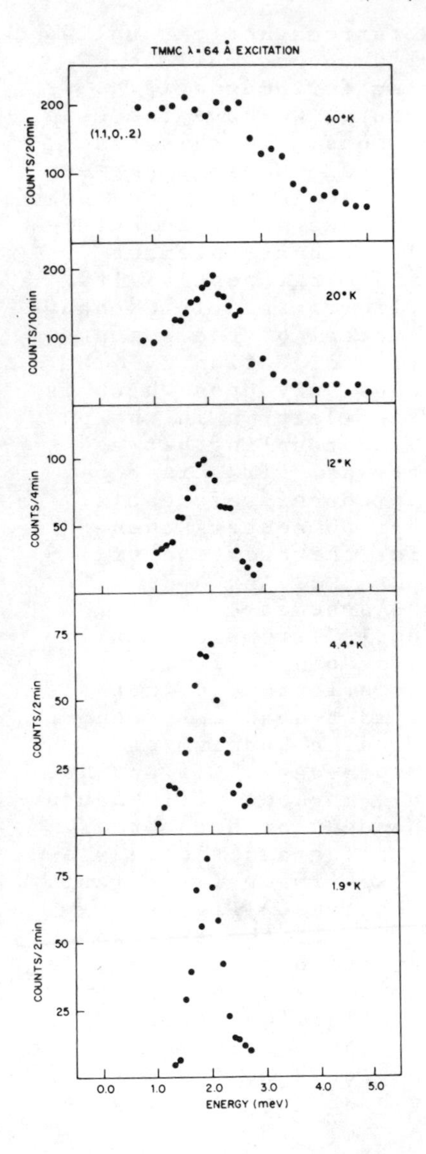

Figure 6. Variation with temperature of the excitation at (0.3, 0, 1.10), q_c^* = 0.10 reciprocal lattice units compound is $(CD_3)_4NMnCl_3$ (adopted from Ref. 1).

0.84°K. Thus, as one increases the temperature above this, continually greater disorder and decoupling along the chains will occur. One would expect, as a general phenomenon, the presence of spin-waves should be less evident the higher T is above T_N.

Our primary interest here will be with the optical manifestation of these spin-waves. Figure 7 shows the various types of electronic processes one is likely to encounter in a magnetically coupled transition metal containing material. We shall focus our attention primarily upon those transitions which are spin-forbidden in nature, i.e., those which undergo a change in spin-multiplicity, since it is here that the effects of the spin-wave phenomenon on the electronic transitions will be seen.

The first part of Figure 7 shows the process which occurs when you have a magnetically and optically dilute system in which the spin-multiplicity may change. This is a normal excitation caused by a photon where the sublattices act independently of each other. In the case shown here, we have only the excitation of an electron from the ground state on sublattice A into an excited state with the accompanying spin-flip. This will usually be seen in the spectrum of a compound as a weak absorption band. The weakness of such a transition arises because it is both parity and spin-forbidden, $\Delta S \neq 0$. Such is not the case in an optically concentrated system where cooperative excitations such as we see in the second part of Figure 7 may occur. In this case, by exciting an electron from the ground state in sublattice A and simultaneously exciting an electron with the opposite spin on sublattice B, we can conserve the spin-angular momentum, i.e., $\Delta S=0$. This is called an exciton·exciton transition. The consequences of the conservation of spin is to provide a band or transition which is relatively intense in comparison to a normal spin-forbidden transition. (In the case we have chosen both the excitation on sublattice A and the one on sublattice B are by themselves spin-forbidden. This is not an exclusive requirement for an exciton·exciton transition.) Note also that if this excitation of A and B are individually the same as the single exciton excitation shown in the first part of this figure, then the cooperative exciton·exciton transition will occur at approximately twice the energy of the exciton-only transition.

In the third part of the figure, we have an exciton+magnon transition. This is the first of these various phenomena which are strictly characteristic of a magnetically concentrated system. The first two

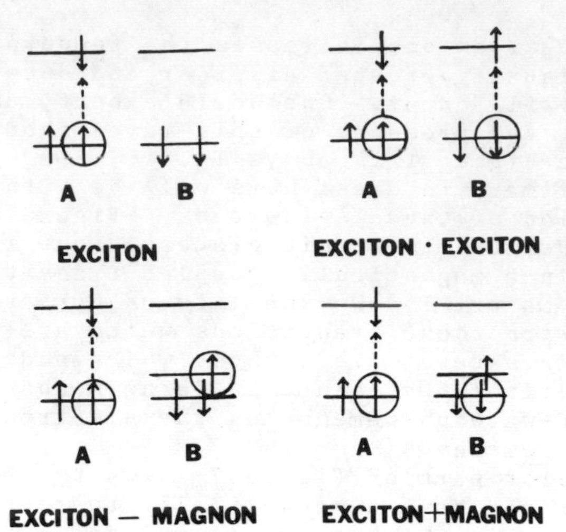

Figure 7. Exciton and exciton · magnon processes A and B refer to opposite sublattices (adopted from Ref. 2)

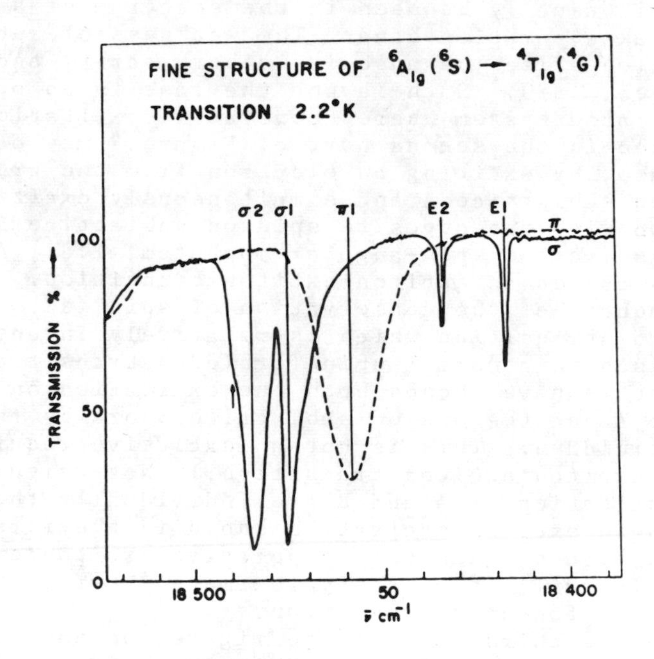

Figure 8. Exciton (E1, E2) and exciton + magnon (π1, σ1, σ2) transitions in MnF_2 (adopted from Ref. 4)

processes, the exciton and exciton·exciton transitions are characteristic of magnetically dilute systems as well as magnetically concentrated systems. The case of the exciton+magnon excitation requires a magnetically coupled system, however. The exciton+magnon transition is a combination of a normal spin-forbidden transition such as shown on sublattice A and a spin-deviation of low energy, such as shown on sublattice B. Here again, just as in the exciton·exciton transition we see that the selection rule $\Delta S=0$ is obeyed. In other words, we have an up-spin on A going to a down-spin simultaneously with a down-spin on B going to an up-spin. The difference between this and an exciton·exciton transition is that this process arises only in the case of magnetically coupled systems and the energy of the allowing excitation on B is of considerably less magnitude than that for the normal exciton·exciton transition.

The last diagram in Figure 7 shows a second process by which the presence of a cooperative magnetic interaction in the system helps to allow an electronic excitation. This is called the exciton-magnon. In this case, we have, as before, a spin-forbidden excitation on sublattice A but as opposed to the exciton+magnon, we have already created an excitation on sublattice B which then decays at the same time as the excitation occurs on sublattice A. As we can see, the spin is conserved in this system as well as in the previous two systems.

One observable difference between the exciton+magnon and exciton-magnon is their position relative to the pure exciton line. It should be clear from Figure 7 that the exciton+magnon will occur at higher energies than the pure exciton line, while the exciton-magnon will occur at lower energies than the pure exciton line. The case for the exciton+magnon is shown quite graphically in Figure 8.

These latter three excitations all have one thing in common and that is their unusual intensity. The variation of these intensities, as a function of T will differ, however. Figure 9 shows the calculated temperature dependence for both the exciton+magnon (cold band) and exciton-magnon (hot band) cases. This suggests that a major change in oscillator strength should occur below T_N. The magnitude of this change will depend upon the relative contribution of magnon hot and cold bands and phonon modes.

Figure 10 shows the calculated temperature dependence (solid line) for a two exciton transition in $RbMnF_3$. As can be seen from the experimental result

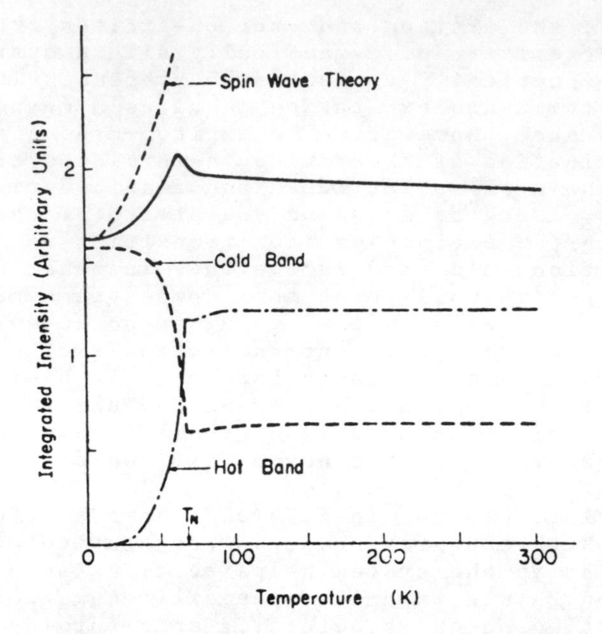

Figure 9. Calculated temperature dependence of magnon side bands in MnF_2 (adopted from Ref. 5)

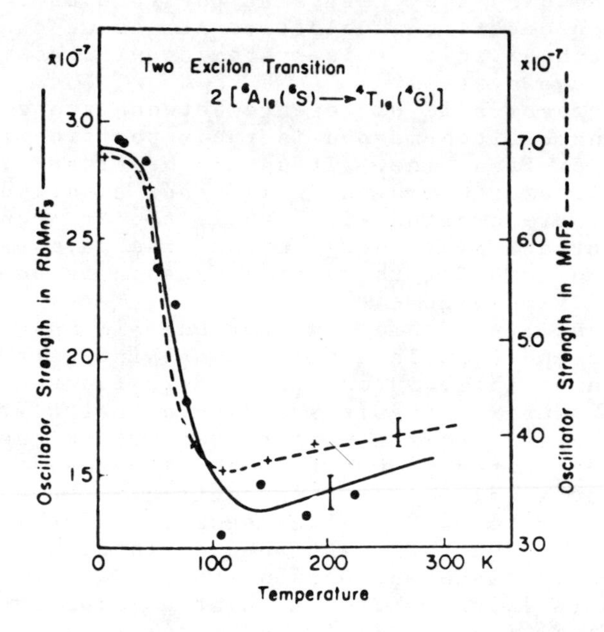

Figure 10. Temperature dependence of the intensity of two-exciton absorption in $RbMnF_3$ (adopted from Ref. 6)

(dashed line) the agreement is relatively good. Figure 11 provides us with the experimentally determined intensity variation of exciton-magnon transitions in $RbMnF_3$ and MnF_2.

The criteria then for identifying magnon assisted transitions are: 1) to look to energies slightly higher and slightly lower than the parent spin-forbidden transition for anomously intense manifolds and 2) ascertain if an anomolous intensity change occurs in the region of the Néel point. Criterion 1) should be qualified in that only the magnon assisted transitions may be visible, the parent line being too weak to be observed.

One of the RMX_3 systems, in which we have had considerable interest is that containing the transition metal ion nickel. Table I shows the various crystallographic parameters for a number of nickel-containing compounds. Also included are the room temperature magnetic moments, their Weiss constants and, where known, an indication of their Néel temperatures. As can be seen, the interchain distances vary from 9.35 Å in the case of $(CH_3)_4NiBr_3$, down to 6.85 Å in the case of $TlNiCl_3$. This then should give us a large range in which to look at the effect of inter- versus intrachain coupling.

To this moment, the measurements that we have made have been restricted to $(CH_3)_4NNiCl_3$, $CsNiCl_3$, $CsNiBr_3$, $RbNiCl_3$ and $RbNiBr_3$. Indeed, even in the case of the $(CH_3)_4NiCl_3$ we see from Table I that it has been found to be ferromagnetic, consequently should not and, in fact, does not exhibit any magnon assisted transitions in its optical spectrum. Consequently, our discussion is restricted to the alkali metal salts of the nickel halides.

In Figure 12 we see the spectrum of an ∿2% solid solution of $CsNiCl_3$ in the colorless, non-magnetic diluent $CsMgCl_3$. This is basically a normal spectrum for the Ni^{2+} ion. The absorption maximum occuring at approximately 7,000 cm^{-1} is the $^3A_{2g}\rightarrow{}^3T_{2g}(F)$ transition. This is followed by the spin-allowed $^3A_{2g}\rightarrow{}^3T_{1g}(F)$ transition at 12,000 wave numbers. Superimposed upon it is the spin-forbidden $^3A_{2g}\rightarrow{}^1E_g$ transition. The $^3A_{2g}\rightarrow{}^1T_{2g}$ transition then lies at ∿18000 cm^{-1} followed by the $^3A_{2g}\rightarrow{}^1A_{1g}$ transition at 19000 cm^{-1}. The spin-allowed $^3A_{2g}\rightarrow{}^3T_{1g}(P)$ transition is then observed at 22000 cm^{-1}. This in turn is followed by the spin-forbidden $^3A_{2g}\rightarrow{}^1T_{1g}(G)$ (∿24,000 cm^{-1}), $^3A_{2g}\rightarrow{}^1E_g(G)$ (25,700 cm^{-1}) and $^3A_{2g}\rightarrow{}^1T_{2g}(G)$ (26,000 cm^{-1}) transitions.

Figure 13, the spectrum of $CsNiCl_3$ at 5°K, shows

TABLE I

Structural and Magnetic Parameters of Some $RNiX_3$ Compounds

Compound	Space Group	Intrachain	Interchain	μ_{eff}, B.M.	θ, °K	TN, °K
$(CH_3)_4NNiCl_3$	$P6_3/m$	3.577(1)	7.268(8)	3.16 (80°)[a]	0	ferromagnetic
$(CH_3)_4NNiBr_3$	$P6_3$	3.17(1)	9.35(2)			
$CH_3NH_3NiCl_3$	Cmcm	3.12	6.94			
$TlNiCl_3$	$P6_3/mmc$		6.85			
$CsNiCl_3$	$P6_3/mmc$	2.97	7.18	3.41 (20°)[b]	-69	4.5
$CsNiBr_3$	$P6_3/mmc$	3.12	7.50	2.94 (20°)[c]	-101	
$CsNiI_3$	$P6_3/mmc$		7.66			
$RbNiCl_3$	$P6_3/mmc$	2.953(1)	6.955(1)	3.34 (20°)[b]	-100	11
$RbNiBr_3$	$P6_3/mmc$	3.104(4)	7.268(8)	2.84 (20°)[c]	-156	

a. B. C. Gerstein, F. D. Gehring and R. D. Willett, J. Appl. Phys., 43, 1932 (1972).

b. N. Achiwa, J. Phys. Soc. Japan, 27, 561 (1969).

c. R. W. Asmussen and H. Soling, Z. anorg. u. allgem. Chem., 283, 1 (1956).

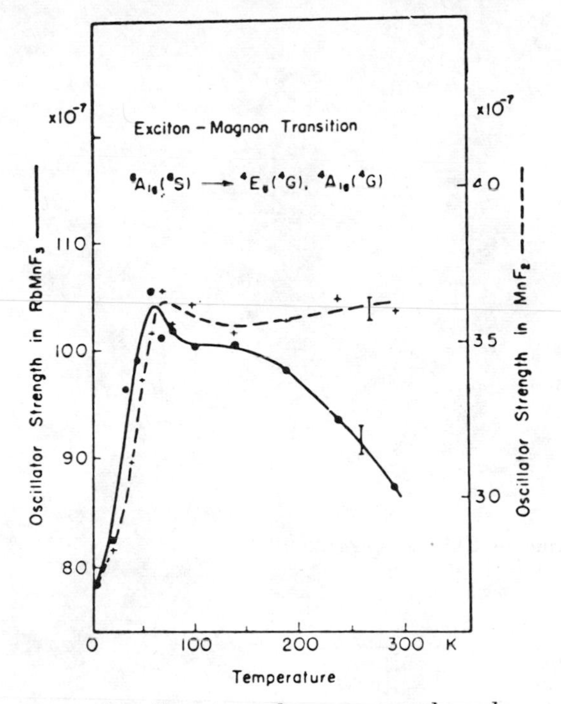

Figure 11. Experimental temperature dependence of oscillator strengths for exciton-magnon transition $^6A_{1g}(^6S) \rightarrow {}^4E_g(^4G), {}^4A_{1g}(^4G)$ *in* MnF_2 *and* $RbMnF_3$ *(adopted from Ref. 7)*

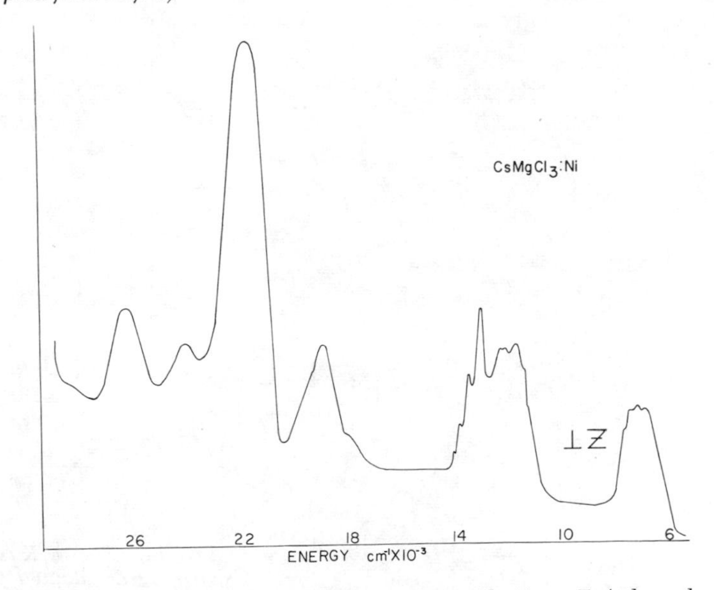

Figure 12. The 5°K spectrum of $CsMgCl_3$*:Ni,* $\perp Z$ *(adopted from Ref. 8)*

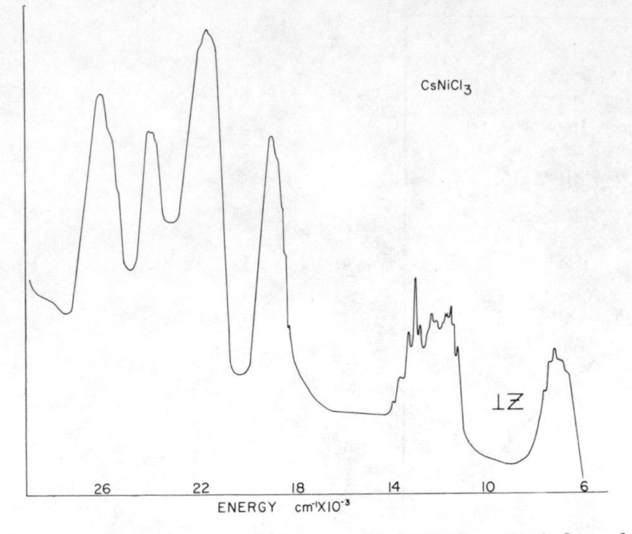

Figure 13. The 5°K spectrum of $CsNiCl_3$, ⊥Z (adopted from Ref. 8)

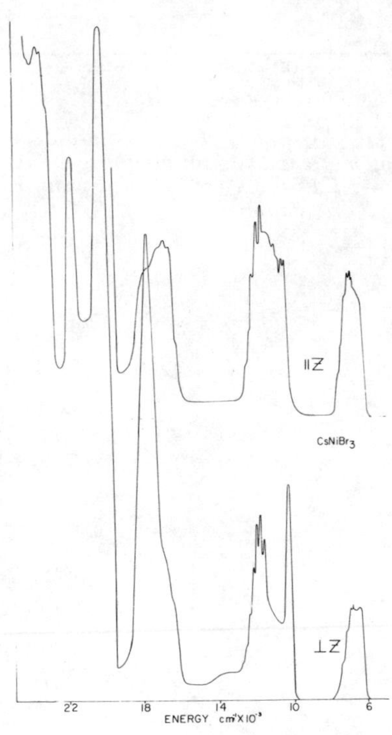

Figure 14. The 5°K spectrum of $CsNiBr_3$, ⊥Z (adopted from Ref. 8)

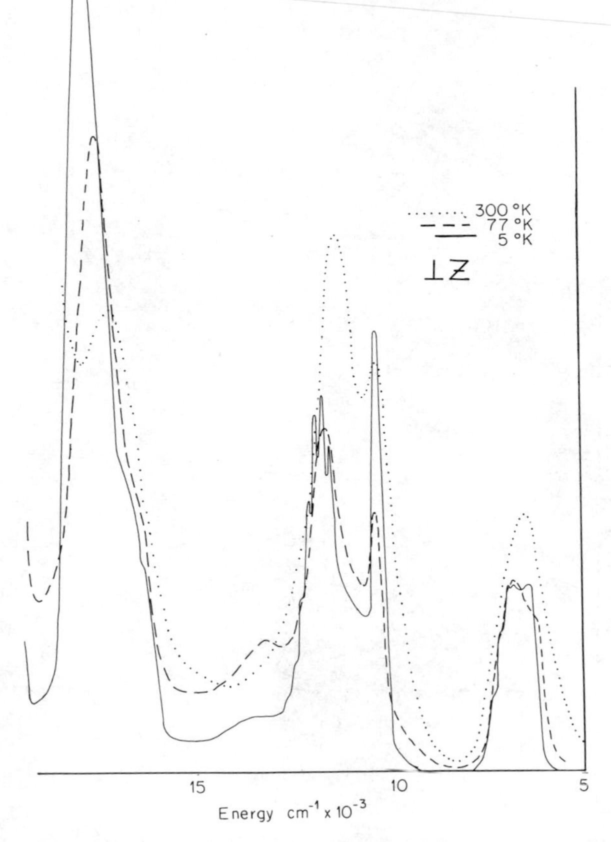

Figure 15. Temperature dependence of the spectrum of $CsNiBr_3$, $\perp Z$

TABLE II

Oscillator Strengths for Selected Transition in $RNiX_3$

Compound	Polarization	Transition	$f\times10^5$(300°K)	$f\times10^5$(77°K)	$f\times10^5$(4°K)
$CsNiCl_3$	⊥	${}^3A_{2g}\rightarrow{}^3T_{2g}$	1.8	1.1	1.4
		${}^3A_{2g}\rightarrow{}^3T_{1g}(F)$	4.7	1.8	1.8
		${}^3A_{2g}\rightarrow{}^1E_g(F)$		0.2	0.4
		${}^3A_{2g}\rightarrow{}^1A_{1g}(G)$	1.1	1.7	2.5
	∥	${}^3A_{2g}\rightarrow{}^3T_{2g}$	4.1	1.9	2.0
		${}^3A_{2g}\rightarrow{}^3T_{1g}(F)$	6.4	2.9	3.0
		${}^3A_{2g}\rightarrow{}^1E_g(D)$		0.4	0.6
		${}^3A_{2g}\rightarrow{}^1A_{1g}(G)$	0.4	0.7	0.8
$RbNiCl_3$	⊥	${}^3A_{2g}\rightarrow{}^3T_{2g}$	1.8	1.5	1.5
		${}^3A_{2g}\rightarrow{}^3T_{1g}(F)$	5.5	2.7	2.2
		${}^3A_{2g}\rightarrow{}^1E_g(D)$			0.3
		${}^3A_{2g}\rightarrow{}^1A_{1g}(G)$	1.2	2.2	2.5
	∥	${}^3A_{2g}\rightarrow{}^3T_{2g}$	4.2	2.5	2.1

Table II Continued

Compound	Polarization	Transition	$f \times 10^5$ (300°K)	$f \times 10^5$ (77°K)	$f \times 10^5$ (4°K)
		$^3A_{2g} \rightarrow {}^3T_{1g}(G)$	6.0	3.3	2.6
		$^3A_{2g} \rightarrow {}^1E_g(D)$			0.5
		$^3A_{2g} \rightarrow {}^1A_{1g}(G)$	0.9	0.9	1.0
$CsNiBr_3$	⊥	$^3A_{2g} \rightarrow {}^3T_{2g}$	3.0	1.9	1.8
		$^3A_{2g} \rightarrow {}^3T_{1g}(F)$	8.1	4.4	5.4
		$^3A_{2g} \rightarrow {}^1E_g(D)$			1.4
		$^3A_{2g} \rightarrow {}^1A_{1g}(G)$	3.7	5.4	8.5
	∥	$^3A_{2g} \rightarrow {}^3T_{2g}$	6.9	2.4	2.8
		$^3A_{2g} \rightarrow {}^3T_{1g}(F)$	11.7	5.2	4.8
		$^3A_{2g} \rightarrow {}^1E_g(D)$			0.7
		$^3A_{2g} \rightarrow {}^1A_{1g}$	1.9	3.6	4.8

a marked difference in the relative intensities of several of the bands when compared to Figure 12. Of principle interest are those bands which are above 18,000 wave numbers. That band at approximately 19,000 which has been identified as the spin-forbidden $^3A_{2g} \rightarrow {}^1A_{1g}$ transition is seen to have gained intensity relative to the spin-allowed transitions both above and below it. The same may be said for the higher energy spin-forbidden transitions which lie above the allowed $^3A_{2g} \rightarrow {}^3T_{1g}(F)$ transition.

Turning now to the spectrum of $CsNiBr_3$, Figure 14, one can again see that the same intensity pattern is repeated. Figure 15 shows the temperature dependence in the lower energy region for both the spin-allowed and spin-forbidden bands in $CsNiBr_3$. As can be seen, the primary feature of the temperature dependence is a rapid increase in the intensity of the formally spin-forbidden bands with decreasing temperature. This is in contrast to the decrease in intensity of the spin allowed transitions. This rapid increase in intensity at $T \gg T_N$ is in contrast to both work reported by Lohr and McClure(9) on some manganese salts and to the work of Fujiwara *et al* (7) cited earlier, Figure 11. That this behavior is common to all of the antiferromagnetic RMX_3 studied is graphically shown in Table II where the oscillator strengths, both in the $\perp$C and $\parallel$C directions, are shown as a function of three temperatures; room, 80° and 5°K. As may be seen, in all cases a rapid increase of the intensity of the magnon-assisted exciton lines in the alkali halides is noted. This suggests that the intensity producing mechanism is similar in all cases and perhaps of a different nature than that seen for $RbMnF_3$ and MnF_2. This latter point is one which deserves more study. Unfortunately, the limited amount of data available do not allow us to sort out inter- versus intrachain effects. Close inspection also shows that the parent exciton line does not appear to be discernable in the cases shown. The exciton+magnon are clearly seen, however.

I would like to express my thanks to the National Science Foundation for support of this research and to J. Ackerman, G. M. Cole, and E. M. Holt who collaborated in this work.

Work partially supported by the National Science Foundation grants GP-15432A1 and GP-41506.

Literature Cited

1. Hutchings, M. T., Shirane, G., Birgeneau, R. J. and S. L. Holt, Phys. Rev., (1972), B5, 1999.

2. Ackerman, J. F., Cole, G. M. and Holt, S. L., Inorg. Chim. Acta, (1974), 8, 323.

3. Kittel, C., Introduction to Solid State Physics, 4th edition, Wiley, New York, (1971) 539.

4. Sell, D. D., Greene, R. L. and White, R. M., Phys. Rev., (1967), 158, 489.

5. Shinagawa, K. and Tanabe, Y., J. Phys. Soc. Japan, (1971), 30, 1280.

6. Motizuki, K. and Miyata, S., Sol. State Comm., (1972), 11, 167.

7. Fujiwara, T., Gebhardt, W., Pentanides, K. and Tanabe, Y., J. Phys. Sco., Japan (1972), 33, 39.

8. Ackerman, J., Holt, E. M. and Holt, S. L., J. Sol. State Chem., (1974), 9, 279.

9. Lohr, L. L. and McClure, D. S., J. Chem. Phys., (1968), 49, 3516.

13

Spectroscopic and Magnetic Properties of $CsMI_3$ Type Transition Metal Iodides

G. L. McPHERSON and L. J. SINDEL

Tulane University, New Orleans, La. 70118

As the two previous papers demonstrate, there has been a great deal of interest in transition metal salts of the general formula $M(I)M'(II)X_3$ (where M(I) is a large univalent cation, M′(II) a divalent transition metal ion, and X a halide ion). These materials often crystallize in hexagonal lattices in which the most prominent structural feature is a parallel array of infinite, linear chains of octahedra sharing faces. The chains run parallel to the crystallographic c-axis with the transition metal ions at the centers and the halide ions at the corners of the octahedra (see Figure 1). The magnetic properties of these types of salts approach those of a one-dimensional system of interacting spins. The hexagonal linear chain structure is observed with the widest variety of transition metals and halogens when M(I) is a cesium ion. This paper discusses the magnetic and spectroscopic properties of several cesium metal triiodides which adopt this structure. Although the properties of these salts are inherently interesting, it is especially informative to compare the iodides to the analogous chlorides and bromides.

The cesium metal triiodides, $CsMgI_3$, $CsVI_3$, $CsCrI_3$, $CsMnI_3$ and $CsNiI_3$ have been shown by X-ray studies to adopt the linear chain structure.[1,2] Table I contains a summary of the crystallographic data for these salts. Although the space groups are not unambiguously determined it is very likely that all of the materials except the chromium salt are isostructural with $CsNiCl_3$ (space group $P6_3/mmc$). Crystallographic studies of $CsCrCl_3$[3] and $CsCrBr_3$[4] suggest that the structures of the chromium salts differ somewhat from that of $CsNiCl_3$; however, the basic linear chain feature is still retained. Undoubtedly there are minor structural variations among the $CsMX_3$ salts which result from the differences in the size of the halide ion. Certainly, a very important structural parameter is the intrachain metal-metal separation. This separation is equal to half of the lattice dimension in the "c" direction and is expected to be largest in the iodide salts.

Table I. Structural Properties of $CsMI_3$ Salts

Crystal System: Hexagonal

Extinctions: $hh\ell$, $\ell \neq 2n$

Space Group: $P6_3/mmc$, $P6_3/mc$, or $P\bar{6}2c$

Mol./unit cell: Z = 2

Lattice Constants:

	a	c
$CsVI_3$ [a]	8.21	6.81
$CsCrI_3$ [a]	8.12	6.85
$CsMnI_3$ [b]	8.18	6.95
$CsNiI_3$ [a]	8.00	6.76
$CsMgI_3$ [a]	8.20	7.01

[a] Reference (1)
[b] Reference (2)

In addition to the fairly subtle structural variations, significant changes in the nature of the metal-halogen bond would be expected in going from a chloride to a bromide and finally to an iodide lattice. These bonding differences are dramatically demonstrated by electron spin resonance measurements. The epr spectra of V^{2+}, Mn^{2+}, and Ni^{2+} doped into the isostructural magnesium salts, $CsMgCl_3$, $CsMgBr_3$, and $CsMgI_3$, have been studied.[5,6] The g- and metal hyperfine tensors indicate a considerable variation in the metal-halogen bonding in the three lattices. Table II gives a summary of hyperfine constants and g-values for the three lattices.

Table II. Hyperfine Constants* and g-Values+

		$CsMgCl_3$	$CsMgBr_3$	$CsMgI_3$
V^{2+}(77°K)				
	g_{obs}	1.975	1.994	2.036
	g_{calc}	1.957	1.950	1.942
	A	75.	70.	65.
Mn^{2+}(77°K)				
	g_{obs}	2.002	2.004	2.008
	A	80.	77.	75.
Ni^{2+}(77°K)				
	g_{obs}	2.25	2.23	2.16
	g_{calc}	2.38	2.40	2.40

*The hyperfine constants represent the average of the parallel and perpendicular components and are in units of 10^{-4} cm^{-1}. (Data taken from (6))

+The g_{obs} values represent the average of the parallel and perpendicular components of observed g-tensor. The g_{calc} values for the d^3 and d^8 systems are calculated from simple crystal field theory. (Data taken from (6))

One notices that there are considerable discrepancies between the observed g-values of V^{2+} and Ni^{2+} and those calculated from the following simple crystal field expression.

$$(g = 2.0023 - \frac{8\lambda}{\Delta})$$

Furthermore, the disagreement becomes more pronounced in going from chloride to bromide to iodide. Presumably the disagreement between the observed and calculated values arises from a ligand contribution to the g-value. The ligand contribution increases as the spin orbit constant of the ligand increases and also as the delocalization of the unpaired electrons from the metal to the ligands increases.[7-10] In view of the observed g-values, it appears that the metal-halogen bonding becomes more covalent proceeding through the series from chloride to iodide. The ^{55}Mn and ^{51}V hyperfine constants support this conclusion, since the constants show a steady decrease in going from chloride to iodide. A decrease in the metal hyperfine constant suggests an increase in the metal to ligand delocalization. (Reference (6) gives a more thorough discussion of the epr parameters.) Although the trend in the nature of the metal-halogen bonding in the $CsMX_3$ series is perhaps intuitively obvious, the epr studies provide a very satisfying experimental verification.

Plots of the reciprocal of the molar susceptibility versus the absolute temperature for $CsNiI_3$, $CsMnI_3$, and $CsCrI_3$ are shown in Figures 2, 3, and 4, respectively. Both $CsNiI_3$ and $CsMnI_3$ obey the Curie-Weiss law above 190°K, but show significant deviation at 77°K. The low temperature deviations and the large negative Weiss constants indicate that these salts are antiferromagnetic. The chromium salt, $CsCrI_3$, obeys the Curie-Weiss law throughout the 77° to 300°K region. This material, however, has a large negative Weiss constant which indicates that it is also antiferromagnetic. The magnetic properties of $CsVI_3$ differ from those of the three previously mentioned salts. The vanadium salt has a small paramagnetic susceptibility (2.3×10^{-3} esu/mole) which is essentially independent of temperature. This observation suggests that the antiferromagnetic interactions in this material are significantly stronger than those of the other salts. These interactions are effective even at room temperature. Although these susceptibility studies do not completely characterize the magnetic behavior of the iodides, there is little doubt about the antiferromagnetic nature of these materials.

The magnetic susceptibilities of a number of the analogous bromides and chlorides have also been studied. Data has been reported for $CsVCl_3$,[11] $CsCrCl_3$,[12] $CsMnBr_3$,[13] $CsNiBr_3$,[14] and $CsNiCl_3$.[14-16] A fundamental question to be considered when discussing the magnetic properties of the $CsMX_3$ salts is whether the magnetic exchange interactions are direct (through space) or indirect (through ligand). In principle both mechanisms are possible, since the metal-metal separations within a chain are fairly short (~3Å) and each metal ion shares three halide ligands with the neighboring metal ions in the chain. The comparison of the susceptibility data for the $CsMX_3$ salts shown in Table III gives some qualitative insight into this question.

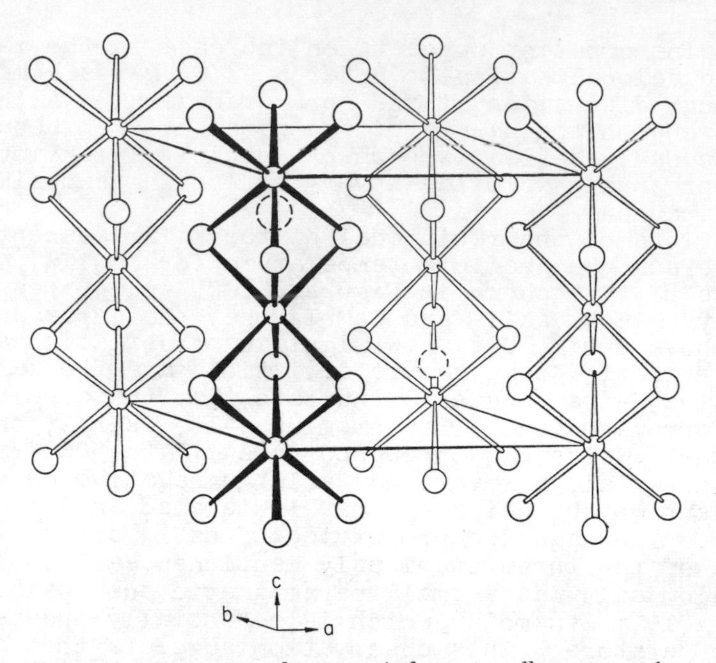

Figure 1. Perspective drawing of the unit cell contents of a $CsMX_3$ salt. Small open circles are metal ions, larger open circles are halide ions, dotted circles are Cs^+ ions.

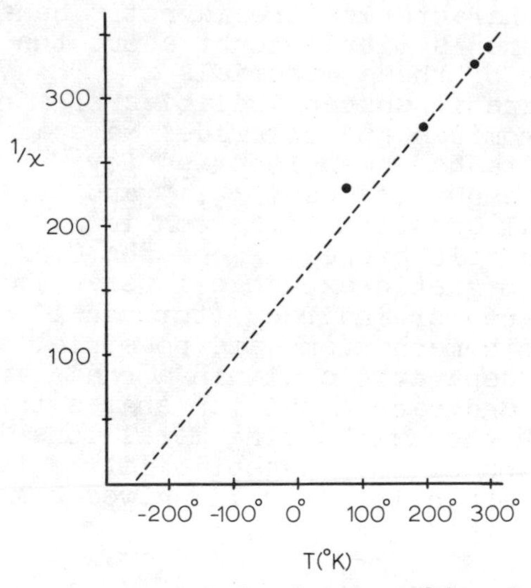

Figure 2. Reciprocal molar susceptibility of $CsNiI_3$ vs. absolute temperature. Curie-Weiss constants for the linear region: $C = 1.62$; $\theta = -250°$

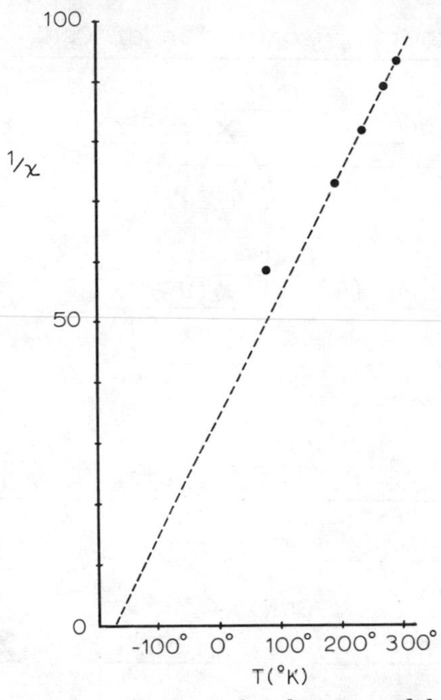

Figure 3. Reciprocal molar susceptibility of $CsMnI_3$ vs. absolute temperature. Curie-Weiss constants for the linear region: $C = 4.97$; $\theta = -165°$.

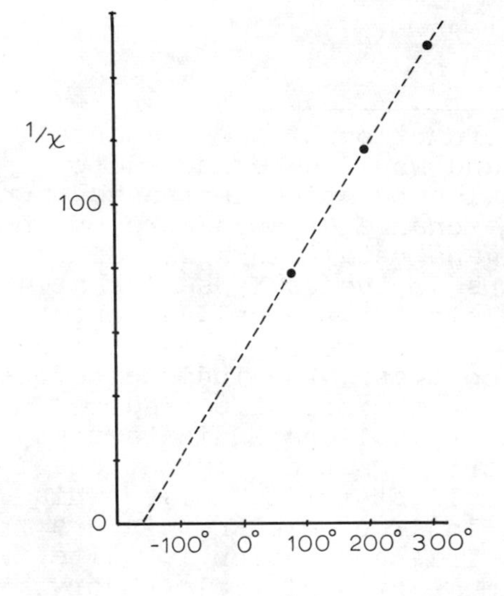

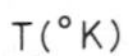

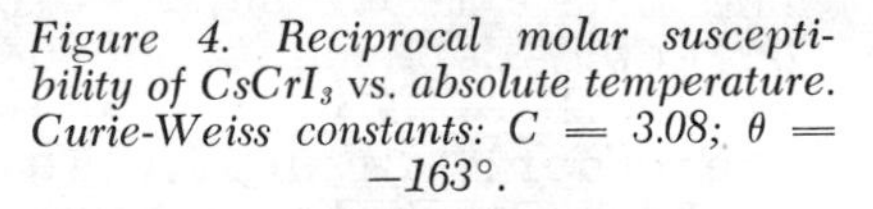

Figure 4. Reciprocal molar susceptibility of $CsCrI_3$ vs. absolute temperature. Curie-Weiss constants: $C = 3.08$; $\theta = -163°$.

Table III. Magnetic Properties of $CsMX_3$ Salts

	M-M Distance (Å)	χ(297°K)	χ(77°K)
	$CsVX_3$		
Cl	3.01	1370[a]	1440[a]
I	3.40	2220	2360
	$CsCrX_3$		
Cl	3.11	4500[b]	5300[b]
Br	3.25	--	--
I	3.42	6700	12800
	$CsMnI_3$		
Br	3.26	11000[c]	18500[c]
I	3.47	10750	17200
	$CsNiI_3$		
Cl	2.98	3800[d]	9200[d]
Br	3.12	3650[d]	8150[d]
I	3.38	2940	4350

χ values are in units of 10^{-6} esu/mole
[a]Reference (11)
[b]Reference (12)
[c]Reference (13)
[d]Reference (14)

Direct exchange is a function of the distance between interacting ions and would be expected to diminish as the metal-metal separation increases. On the other hand, indirect exchange depends more on the covalency of the metal-ligand-metal linkage. Since the metal-metal separations in the $CsMX_3$ series are directly dependent on the size of the halide ion, the strength of direct effects would be expected to follow the order: Cl>Br>I. In contrast, the indirect effects would be expected to exhibit the opposite order. The data for the cesium nickel trihalides indicate that the strength of the antiferromagnetic interactions is greatest for $CsNiI_3$ and smallest for $CsNiCl_3$. This observation suggests that indirect exchange is predominant in these salts. This conclusion is quite reasonable in light of simple crystal field theory.

The unpaired electrons of a d^8 system in an octahedral complex occupy the e_g set of orbitals which are directed toward the ligands. For a d^5 system such as Mn^{2+} direct as well as indirect interactions might be expected since the unpaired electrons occupy the t_{2g} and e_g orbitals. The data indicate that the coupling in $CsMnBr_3$ is perhaps a little stronger than in $CsMnI_3$, but the susceptibilities of the two salts are very similar. It appears that there are considerably stronger interactions in $CsCrCl_3$ than in $CsCrI_3$. It is possible that direct exchange is dominant in a d^4 system since the majority of the unpaired electrons occupy orbitals (t_{2g}) which are directed away from the ligands. We hesitate to speculate on the $CsVX_3$ salts since the susceptibilities are rather small and essentially independent of temperature. These small susceptibilities may result from a temperature independent paramagnetism which has nothing to do with the normal paramagnetism associated with the unpaired electrons of the V^{2+} ion. Clearly, rather strong antiferromagnetic interactions are present in these vanadium salts.

One very important point has been neglected in the qualitative discussion of the magnetic properties of the $CsMX_3$ salts. It has been firmly established that the interaction between two paramagnetic ions is critically dependent on the metal-ligand-metal angle. While the structures of the salts that have been discussed are all similar, this critical angle undoubtedly varies to some extent from lattice to lattice. Unfortunately, sufficient precise crystallographic data are not presently available to discuss this important point.

The electronic spectra of $CsVI_3$, $CsCrI_3$, $CsMnI_3$, and $CsNiI_3$ are shown in Figures 5 and 6. In general, the spectra show the ligand field transitions that would be expected from octahedral complexes of these transition metal ions. The spectrum of $CsCrI_3$ shows an intense absorption edge at approximately 10,000 cm^{-1}. The material absorbs strongly throughout the visible region. This intense absorption may be due to charge-transfer transitions. Charge-transfer absorption would be expected to appear at lower energies in these iodides than in similar chloride or bromide complexes. The shoulder on the absorption edge of the $CsCrI_3$ spectrum has been tentatively assigned to the spin allowed, $^5E \rightarrow {}^5T_2$, ligand field transition. Similarly, the $CsNiI_3$ spectrum has an intense absorption edge which appears at approximately 13000 cm^{-1} and presumably results from charge-transfer trans-

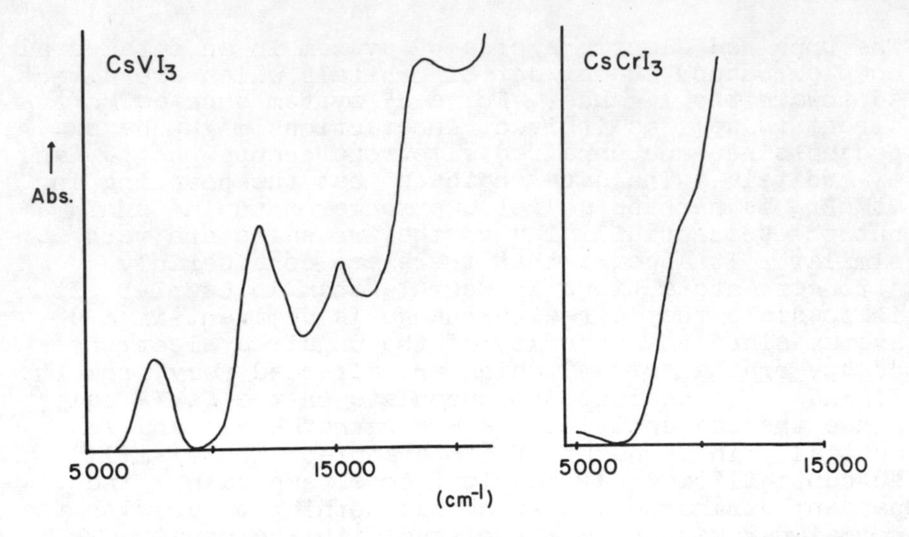

Figure 5. Absorption spectra of mulls of $CsVI_3$ and $CsCrI_3$ recorded at 77°K

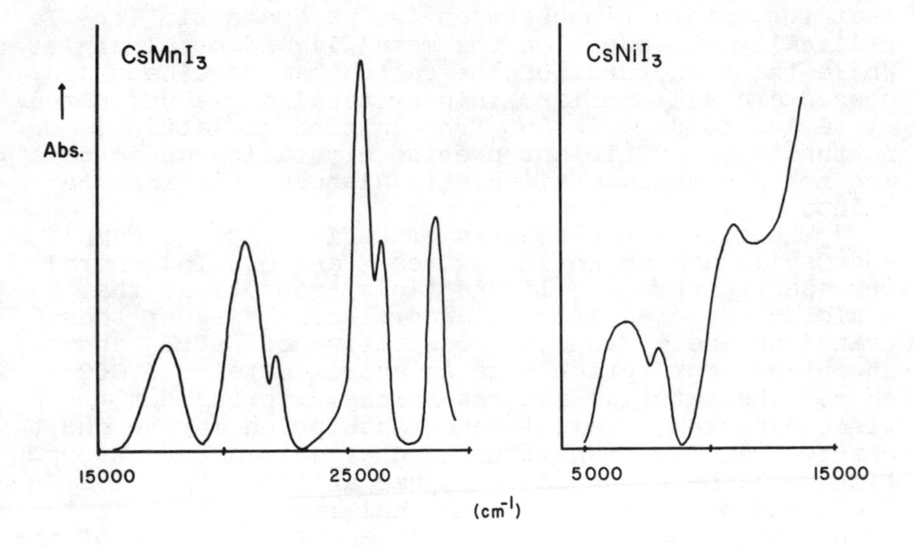

Figure 6. Absorption spectra of mulls of $CsMnI_3$ and $CsNiI_3$ recorded at 77°K

itions. The spectra of $CsVI_3$ and $CsMnI_3$ do not seem to be particularly unusual. Table IV gives the band assignments for the iodide salts based on an octahedral ligand field.

Table IV. Spectroscopic Assignments

$CsVI_3$		$CsCrI_3$	
Assignment	Energy (cm^{-1})	Assignment	Energy (cm^{-1})
$^4A_2 \rightarrow {}^4T_2$	7700	$^5E \rightarrow [^5T_2]$	9000 (sh)
$\rightarrow {}^4T_1(F)$	12000		
$\rightarrow [^2E_1{}^2T_1]$	13000(sh)		
$\rightarrow [^2T_2]$	15300		
$\rightarrow {}^4T_1(P)$	18700		

$CsMnI_3$		$CsNiI_3$	
Assignment	Energy (cm^{-1})	Assignment	Energy (cm^{-1})
$^6A_1 \rightarrow {}^4T_1(G)$	17800	$^3A_2 \rightarrow {}^3T_2$	6500
$\rightarrow {}^4T_2(G)$	20900	$\rightarrow [^1E]$	8000
$\rightarrow {}^4E,{}^4A_1(G)$	22100	$\rightarrow {}^3T_1(F)$	10900
$\rightarrow {}^4T_2(D)$	25500		
$\rightarrow {}^4E(D)$	26400		
$\rightarrow {}^4T_1(P)$	28600		

sh = shoulder
Brackets designate assignments which are uncertain.

Spectroscopic studies of $CsCrCl_3$[3], $CsCrBr_3$[4], $CsMnBr_3$[13], $CsNiCl_3$ and $CsNiBr_3$[17,18] have been reported. A comparison of the Dq values of the $CsMX_3$ salts is presented in Table V.

Table V. Dq Values for the $CsMX_3$ Salts

	$CsVX_3$	$CsCrX_3$	$CsMnX_3$	$CsNiX_3$
Cl	1000	1145[a]	--	695[d]
Br	--	1150[b]	680[c]	655[d]
I	770	900	605	650

[a]Reference (3)
[b]Reference (4)
[c]Reference (13)
[d]Reference (17)

The trends in the Dq values appear to follow that which would be predicted by the spectrochemical series. It should be mentioned that the Dq value for $CsMnI_3$ was derived following the procedure presented in reference (13).

The electrical resistivities of single crystals of $CsNiI_3$ have been studied as a function of temperature.[19] The material appears to be a semiconductor with a room temperature resistivity of 10^7 to 10^8 ohm cm and an energy of activation of approximately 0.7 ev.

For a material that is an intrinsic semiconductor the band gap (the energy separating the valence and conduction bands) should be equal to twice the energy of activation of conduction. The intense absorption edge appears in the spectrum of $CsNiI_3$ at approximately 1.5 ev which is about twice the observed energy of activation. This suggests that at room temperature the material is an intrinsic semiconductor. In spite of the linear chain structure, the resistivity of the material is essentially isotropic.

In conclusion, we hope that we have shown that the cesium metal triiodides have rather interesting solid state properties and that these compounds will be useful for further studies into the nature of the linear chain $M(I)M'(II)X_3$ compounds.

Literature Cited

1. Li, Ting-i, Stucky, G. D., and McPherson, G. L., Acta Crystallogr., (1973), B29, 1330.

2. McPherson, G. L. and Quarls, H. F., Unpublished Data.

3. McPherson, G. L., Kistenmacher, T. K. Folkers, J. B., and Stucky, G. D., J. Chem. Phys., (1972), 57, 3771.

4. Li, Ting-i and Stucky, G. D., Acta Crystallogr., (1973), B29, 1529.

5. McPherson, G. L., Kistenmacher, T. K., and Stucky, G. D., J. Chem. Phys., (1970), 52, 815.

6. McPherson, G. L., Koch, R. C., and Stucky, G. D., J. Chem. Phys., (1974), 60, 1424.

7. McGarvey, B. R., J. Chem. Phys., (1964), 41, 3743.

8. Garrett, B. B., DeArmond, K., and Gutowsky, H. S., J. Chem. Phys., (1966), 44, 3393.

9. Mesetich, A. A. and Buch, T., J. Chem. Phys., (1964), 41, 2524.

10. Mesetich, A. A. and Watson, R. E., Phys. Rev., (1966), 143, 335.

11. Seifert, H. J., Fink, H., and Just. E., Naturwiss., (1968), 55, 297.

12. Larkworthy, L. F. and Trigg. J. K., Chem. Commun., (1970), 1221.

13. McPherson, G. L., Aldrich, H. S., and Chang, J. R., J. Chem. Phys., (1974), 60, 534.

14. Asmussen, R. W. and Soling, H., Z. anorg. allgem. Chem., (1956), 283, 1.

15. Achiwa, N., J. Phys. Soc. (Japan), (1969), 27, 561.

16. Smith J., Gerstein, B. C., Liu, S. H., and Stucky, G., J. Chem. Phys., (1970), 53, 418.

17. McPherson, G. L. and Stucky, G. D., J. Chem. Phys., (1972), 57, 3780.

18. Ackerman, J., Holt, E. M., and Holt, S. L., J. Solid State Chem., in press.

19. McPherson, G. L., Wall, J. E., Jr., and Hermann, A. M., Inorg. Chem., in press.

14

Magnetic and Thermal Properties of the Linear Chain Series $[(CH_3)_3NH]\ MX_3 \cdot 2H_2O$

J. N. McELEARNEY, G. E. SHANKLE, D. B. LOSEE, S. MERCHANT, and R. L. CARLIN

University of Illinois, Chicago, Ill. 60680

Introduction

Several previous papers in this symposium have discussed properties of members of the linear chain series RMX_3 (where M=transition metal, X=halide and R=$(CH_3)_4N$, Cs or Rb). This paper will be concerned with the new linear chain series $[(CH_3)_3NH]MX_3 \cdot 2H_2O$, where M=(Mn, Co, Ni, Fe or Cu) and X=(Br or Cl). Large single crystals suitable for optical and oriented magnetic field studies may be easily obtained for most members of the series. Single crystal near-zero-field magnetic susceptibility data (measured using a mutual inductance technique) and heat capacity data (measured using standard heat pulse techniques) are presented here which show several of the interesting features of this series: anisotropic magnetic behavior greater than normal with the presence of low-dimensional characteristics as well as spin-canting in the ordered state.

Crystal Structures

To a large extent the magnetic behavior of these compounds is quite clearly related to their structure. Not all the members of the series are isomorphic, although they probably are all isostructural, as inferred from their magnetic properties. Structures have been obtained only for the non-isomorphic (M=Co, Cu; X=Cl) compounds (1,2), although X-ray studies indicate that the (M=Co, Mn; X=Cl) compounds both crystalize in the space group Pnma and are probably isomorphic.

The most important structural characteristics of these compounds may be seen by considering the projections of the (M=Co; X=Cl) prototype shown in Figs. 1 and 2. The structure consists of chains of edge-sharing trans-$[CoCl_4(OH_2)_2]$ octahedra. The cobalt atoms

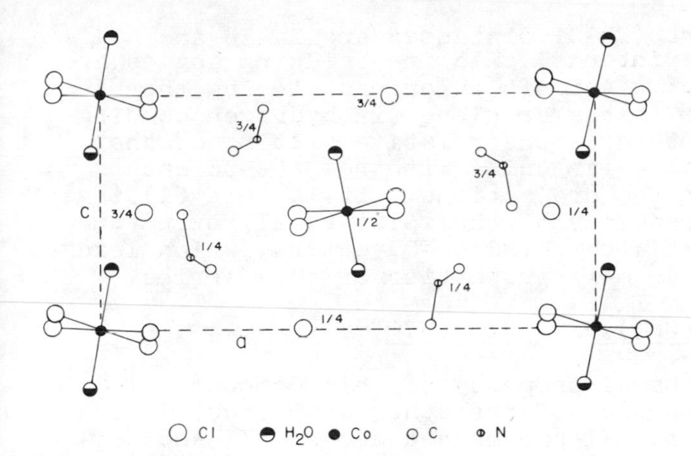

Figure 1. *Projection of the unit cell of* $[(CH_3)_3NH]CoCl_3 \cdot 2H_2O$ *onto the ac plane. Height above this plane of several of the atoms is indicated. (All cobalt atoms are at the same height.)*

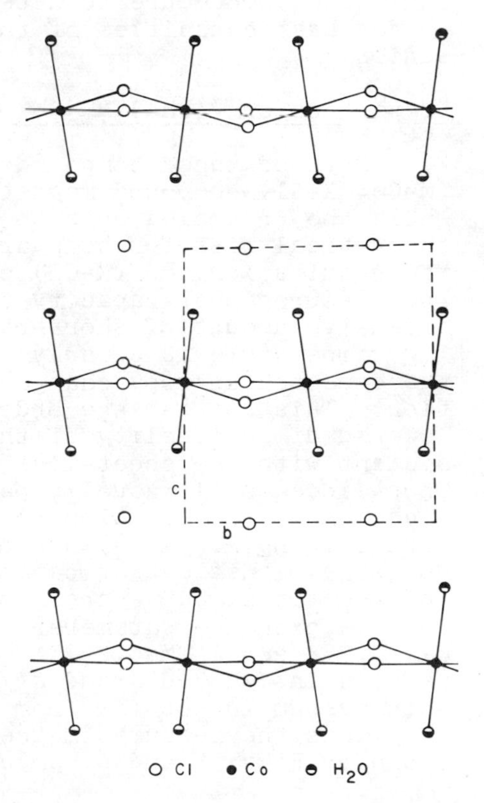

Figure 2. *Projection of a portion of the crystal structure of* $[(CH_3)_3NH]CoCl_3 \cdot 2H_2O$ *onto the bc plane. Portion used was a 3.33-A thick layer taken parallel to the bc plane and centered about the cobalt atoms. Dashed lines give unit cell boundaries.*

are 3.637 A apart (Co-Cl distances are 2.456 and 2.503 A) and the internal chlorine bridging angles are 93.14° and 95.52°. Anionic chlorines tie together chains which lie in the bc plane via hydrogen bonding. It is important to note the relative tilting of the $[CoCl_4(OH_2)_2]$ molecular units with respect to each other as seen in the projections. It is this tilting, taken with the tendency of the spins to align in some manner consistent with the O-Co-O vectors, which leads to the unusual magnetic properties of this series.

Properties of $[(CH_3)_3NH]CuCl_3 \cdot 2H_2O$

The most unusual property of this compound (in view of the properties of the other compounds) is that it behaves as a normal paramagnet above 2°K, showing only slight signs of magnetic exchange at 1°K prior to ordering at 0.15°K (3) and that thus intrachain magnetic exchange in this material must be quite small. Thus its measured heat capacity (which very nearly follows a T^2 law) has been used in conjunction with a corresponding states procedure to determine lattice contributions to the heat capacities of the other members of the series.

Properties of $[(CH_3)_3NH]CoX_3 \cdot 2H_2O$ (X=Cl, Br)

Both of these compounds behave very similarly. The (M=Co; X=Cl) compound magnetically orders at 4.14°K, while the Br analog does so at 3.86°K. The data and theoretical fits for both are nearly identical so only the results for the (X=Cl) compound will be given here. The measured heat capacity is shown in Fig. 3 and the extensive amount of short-range order above the ordering temperature is clearly evident. More than 90% of the expected entropy change occurs above the transition. This short-range order is indicative of the lowered dimensionality of the spin system and is consistent with the sheet-like nature of the structure. Thus since Co(II) usually has Ising-like characteristics, Onsager's solution to the anisotropic two-dimensional Ising model (4) has been used to fit the data. Both the magnetic heat capacity derived from the data and the fitted curve are shown in Fig. 4. The values for the exchange parameters determined from the fit are J/k=7.7°K and J'/k=0.09°K. (The fit of the Br analog results in J/k=7.0°K and J'/k=0.09°K.) Clearly the spin system is not far from being one-dimensional.

Thus the measured magnetic susceptibilities, shown in Figs. 5 and 6, should be nearly one-dimension-

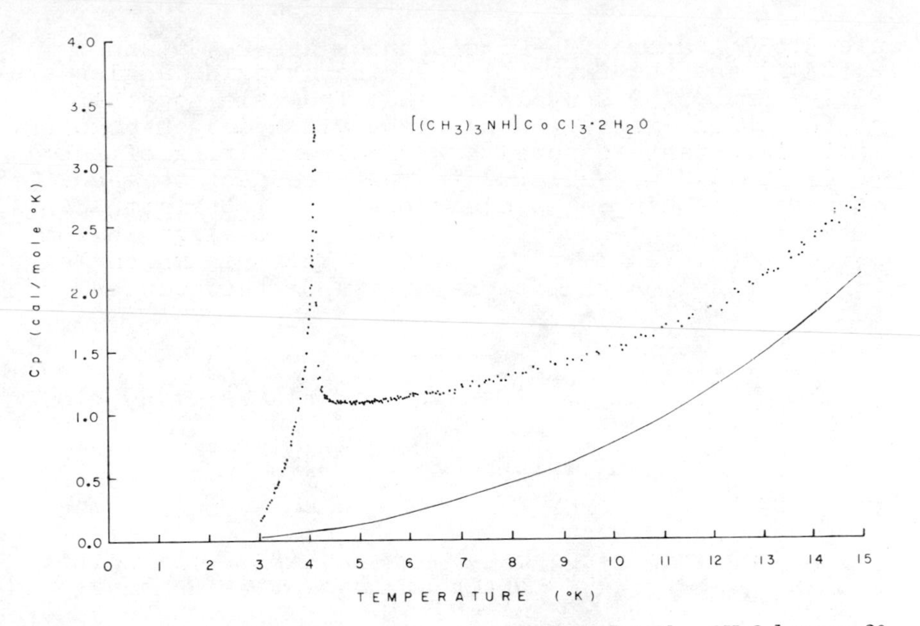

Figure 3. Zero-field heat capacity data for $[(CH_3)_3NH]CoCl_3 \cdot 2H_2O$ *between 3° and 15°K. Solid line represents the lattice contribution.*

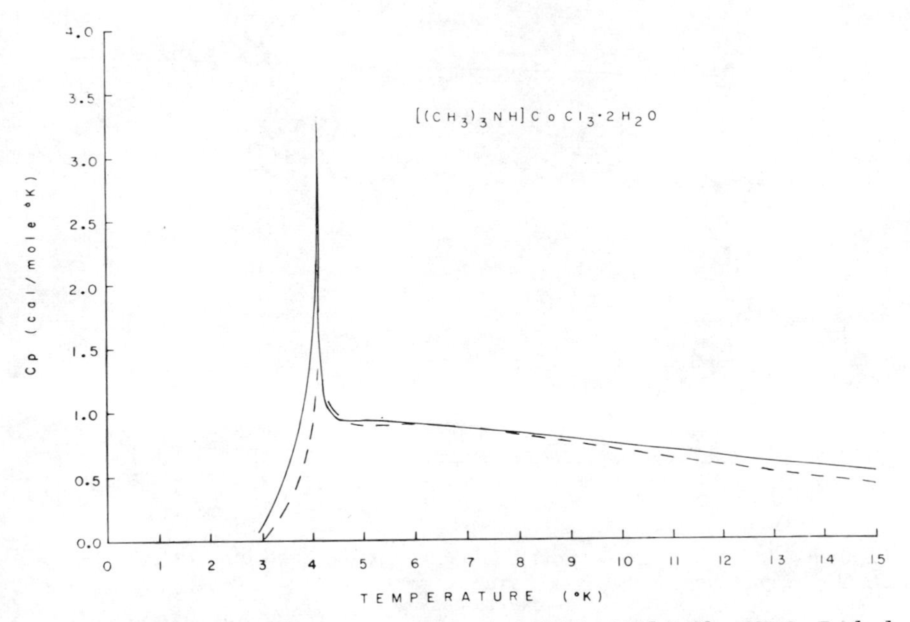

Figure 4. Magnetic heat capacity (solid line) of $[(CH_3)_3NH]CoCl_3 \cdot 2H_2O$*. Dashed line is the fit to Onsager's two-dimensional anisotropic Ising model solution.*

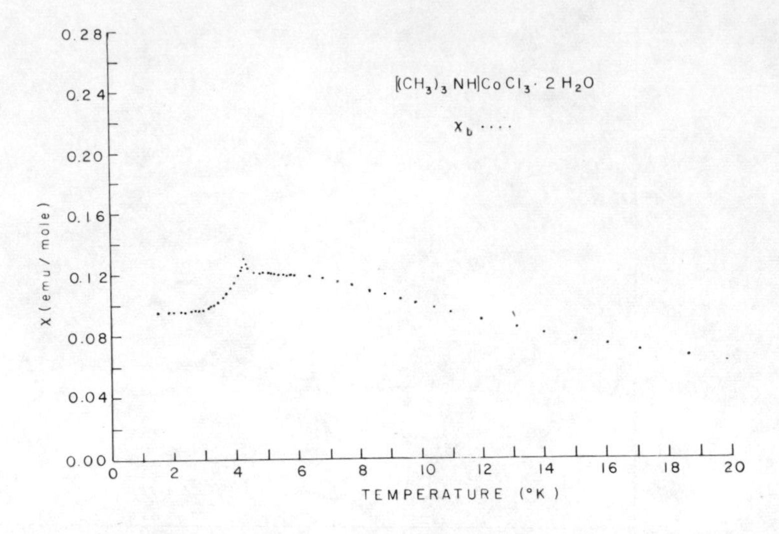

Figure 5. Zero-field magnetic susceptibility of $[(CH_3)_3NH]CoCl_3 \cdot 2H_2O$ measured parallel to the b axis

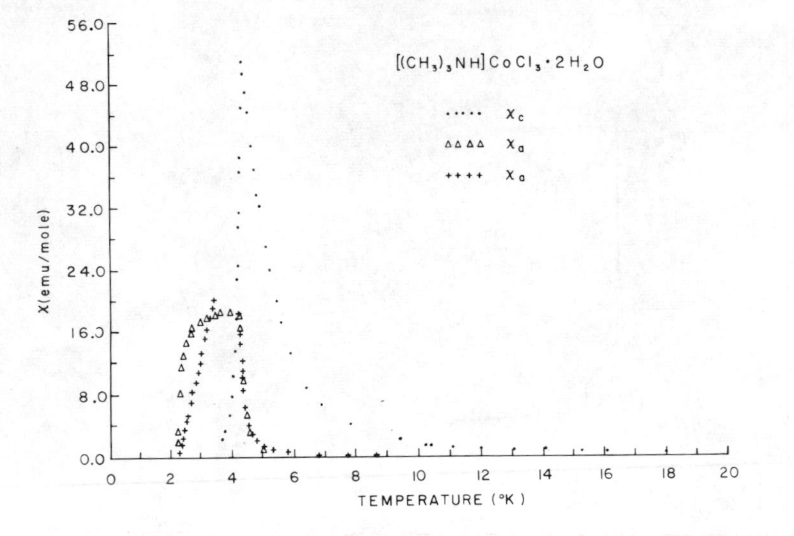

Figure 6. Zero-field magnetic susceptibilities of $[(CH_3)_3NH]CoCl_3 \cdot 2H_2O$ measured along a and c axes. Results of measurements along the a axis on two different crystals are shown.

al in character. The nearly discontinuous behavior near the transition temperature should be noted. It is evident that there is an extreme amount of anisotropy, χ_b being generally very much smaller than χ_a and χ_c, and that there are indications of a ferromagnetic moment parallel to the a axis immediately below the transition. Additionally, at the lowest temperatures χ_a and χ_b are constant, while χ_c decreases with decreasing temperature, as might be expected for an antiferromagnetic material for which c is the easy axis.

These facts lead to a spin arrangement model in which spins lie nearly along the c axis with some canting towards the a axis. All spins are perpendicular to the b axis. All the spins in each bc sheet of Co ions are aligned parallel with ferromagnetic coupling along and between chains in the sheet. Spins in any neighboring pair of bc sheets are aligned so as to give a net moment along the a axis with no net moment along the c axis. Thus the spin arrangement is that of a canted antiferromagnet or a weak ferromagnet. The intrachain exchange value determined from the heat capacity data has been used with the equations derived (5) for a one-dimensional Ising model to fit the χ_b and χ_c data. To take account of the a axis canting, the results derived by Moriya (6) for the susceptibility of a canted antiferromagnet have been applied to the χ_a data. The results of the fits are shown in Fig. 7 and are seen to be extremely good.

Properties of $[(CH_3)_3NH]MnX_3 \cdot 2H_2O$ (X=Cl, Br)

Both of these compounds behave similarly, with the (X=Cl) compound exhibiting a transition at 0.98°K, while the (X=Br) compound orders at 1.58°K. Because of the similarity of the data for these compounds only the latter will be discussed here. The measured heat capacity is shown in Fig. 8. The sharp spike corresponds to the magnetic ordering while the broad hump is indicative of the short-range order expected for a low-dimensional system. More than 80% of the magnetic entropy change expected for an S=5/2 system is gained above 1.58°K. Likewise, the single crystal susceptibilities measured for this compound, shown in Fig. 9, show the effects of extensive short-range order. The broad maximum in χ_c significantly above the ordering temperature is especially indicative of the behavior which should be expected for a linear chain. It is also important to note the amount of anisotropy below 10°K and that χ_c and χ_a behave as parallel and perpendicular antiferromagnetic susceptibilities, respec-

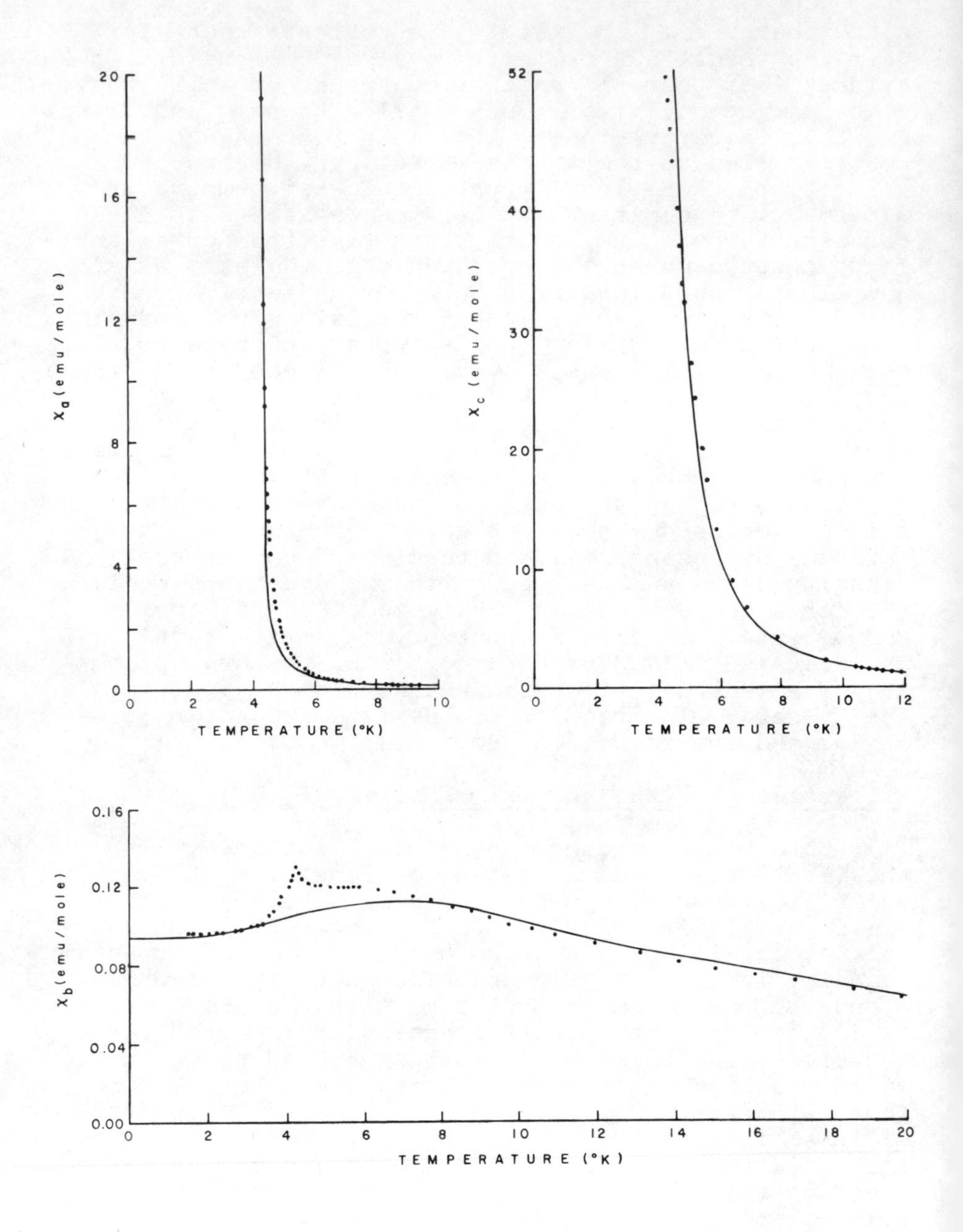

Figure 7. Results of the fits (solid lines) to the $[(CH_3)_3NH]CoCl_3 \cdot 2H_2O$ principal axis susceptibility data

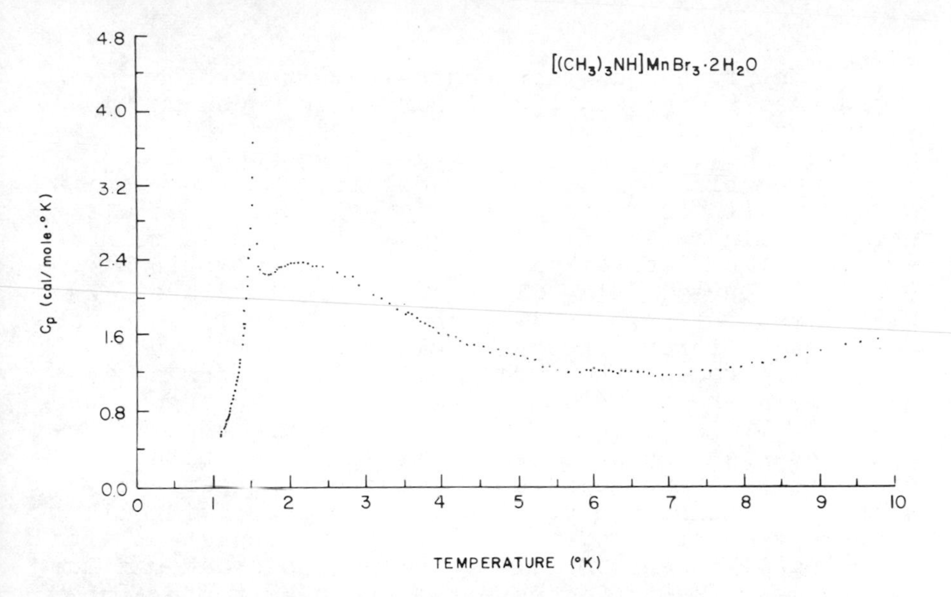

Figure 8. Zero-field heat capacity data for $[(CH_3)_3NH]MnBr_3 \cdot 2H_2O$

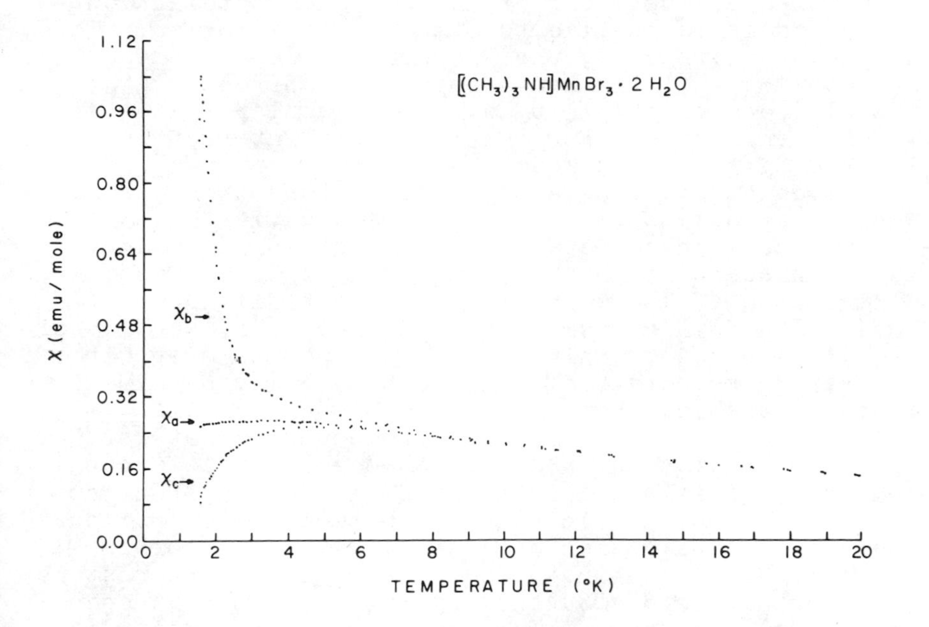

Figure 9. Zero-field magnetic susceptibilities of $[(CH_3)_3NH]MnBr_3 \cdot 2H_2O$ *measured along the a, b, and c axes*

tively.

As Mn(II) has an S ground state, it might be expected that the Heisenberg linear chain model previously applied to other Mn(II) chains (7,8) could describe the data. However, a rather significant molecular field correction (9) to that model is required when it is used to fit the data. The fit of χ_a achieved with an intrachain interaction of J/k=-0.41°K and a molecular field parameter of zJ'/k=-0.74°K is shown in Fig. 10. (The corresponding Cl analog parameters are J/k= -0.36°K and zJ'/k=-0.55°K.) The molecular field parameter must be interpreted as including all the interactions ignored by the model Hamiltonian, such as interchain interactions and non-isotropic exchange interactions. It is obvious from the structure that interchain interactions must be small, so the results indicate that the spin Hamiltonian must not be perfectly isotropic but instead must include a large Ising-like anisotropy. Actually, this is totally consistent with the Ising-like behavior of χ_c in which a broad maximum appears more than 100% above the ordering temperature. This is also consistent with recent results (10) which suggest anisotropic dipolar interactions must be present in a linear chain compound which possesses isotropic nearest-neighbor exchange. Such interactions would add an Ising-like term to the Hamiltonian. Thus, as might be expected, the χ_c data can be well described by assuming an Ising Hamiltonian and correcting the calculated susceptibility with a molecular field term which then accounts for the isotropic interactions ignored by the Ising model.

Even considering anisotropic interactions, special consideration must be given to the behavior of χ_b. The sharp rise with decreasing temperature of χ_b below 3°K is reminiscent of the behavior of χ_a in the (M=Co; X= Cl) compound. There are two important differences: these data do not rise to very high values, and immediately below the transition the data points drop with decreasing temperature. This behavior may be explained if it is assumed that there is a hidden canting of the spins along the b axis in this compound. That is, the spins in a given chain are aligned more or less parallel to the c axis and are canted in the ±b direction with antiferromagnetic coupling between them. Since there is no indication of canting-like behavior in χ_a, the spins must lie in the bc plane. It is interesting to note that this situation is opposite to that seen in the (M=Co) compounds. The supposition of canting is supported by the good fit to the χ_b data, shown in Fig. 11, obtained when the molecular field modified linear

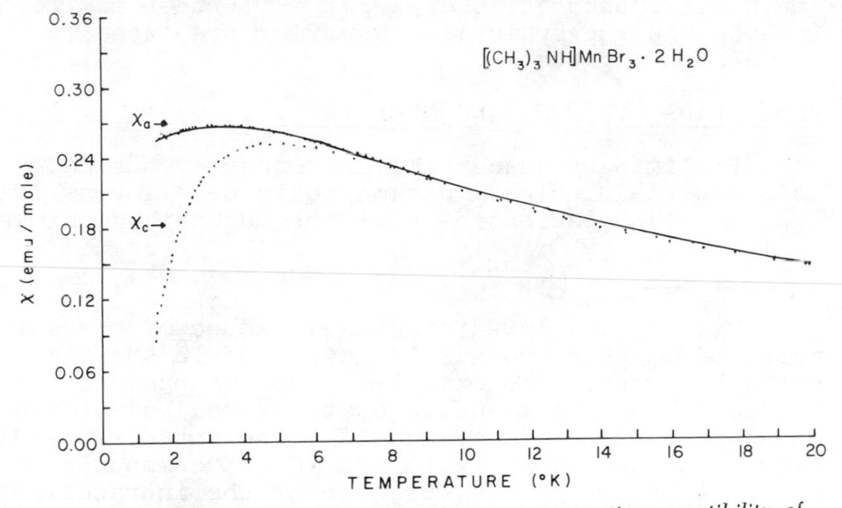

Figure 10. Fitted results (solid line) for the a axis magnetic susceptibility of $[(CH_3)_3NH]MnBr_3 \cdot 2H_2O$

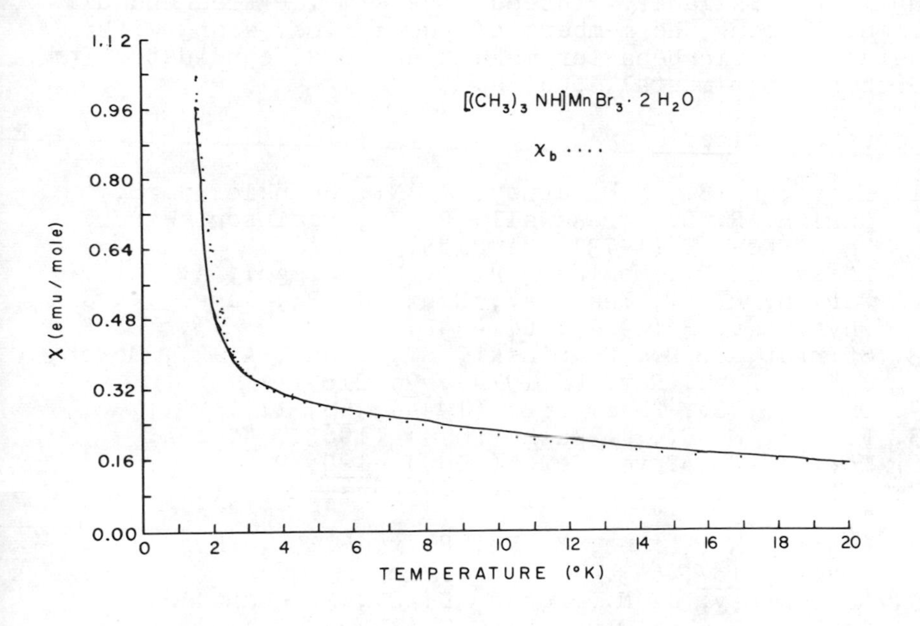

Figure 11. Fitted result (solid line) for the b axis magnetic susceptibility of $[(CH_3)_3NH]MnBr_3 \cdot 2H_2O$

chain model used for the χ_a fit is further modified to include the susceptibility behavior predicted by Moriya for canted magnets.

Properties of $[(CH_3)_3NH]NiBr_3 \cdot 2H_2O$

Preliminary susceptibility measurements made on this material indicate paramagnetic behavior as low as 1°K, with indications of antiferromagnetic exchange.

Conclusions

Clearly the low-dimensional nature of the structures of these compounds is reflected in their magnetic behavior. The extensive short-range order observed in the paramagnetic state is well described by one- or two-dimensional models. One interesting difference between the (M=Co) and (M=Mn) compounds is the relative change in the strength of the intrachain exchange when Cl is replaced by Br - increasing for (M=Mn) and decreasing for (M=Co). Unfortunately, no interpretation is possible until detailed crystal structures are available. Indeed, the similarities and differences among the members of the series, along with the anisotropic behavior make them ideal candidates for further experimental studies.

Literature Cited

1. Losee, D. B., McElearney, J. N., Shankle, G. E., Carlin, R. L., Cresswell, P. J., Robinson, Ward T., Phys. Rev. B (1973), 8, 2185.
2. Losee, D. B., McElearney, J. N., Siegel, A. E., Carlin, R. L., Khan, A., Roux, J. P., James, W. J., Phys. Rev. B (1972), 6, 4342.
3. Stirrat, C. R., Dudzinski, S., Owens, A. H., Cowen, J. A., Phys. Rev. B (1974), 9, 2183.
4. Onsager, L., Phys. Rev. (1944), 65, 117.
5. Fisher, M. E., J. Math. Phys. (1963), 4, 124.
6. Moriya, T., Phys. Rev. (1960), 120, 91.
7. Smith, T., Friedberg, S. A., Phys. Rev. (1968), 176, 660.
8. Dingle, R., Lines, M. E., Holt, S. L., Phys. Rev. (1969), 187, 643.
9. McElearney, J. N., Losee, D. B., Merchant, S., Carlin, R. L., Phys. Rev. B (1973), 7, 3314.
10. Walker, L. R., Dietz, R. E., Andres, K., Darack, S. Solid State Commun. (1972), 11, 593.

15

The Ferromagnetic Ordering of Ferrous Chloride Polymers of Diimine Ligands

W. M. REIFF*, B. DOCKUM, and C. TORARDI
Northeastern University, Boston, Mass. 02115

S. FONER, R. B. FRANKEL, and M. A. WEBER
Francis Bitter National Magnet Laboratory, Massachusetts Institute of Technology, Cambridge, Mass. 02139

ABSTRACT

Magnetization and Mössbauer studies show that Fe(bipyridine)Cl_2 and Fe(phenanthroline)Cl_2 order ferromagnetically at $T_c \sim 5$ K. Methyl substituted bipyridine derivatives appear to be slowly relaxing paramagnets in the range of 12 to 2 K. These results are correlated with near- and far-infrared and x-ray spectroscopic data which suggest six-coordinate chloro-bridged polymeric structures for the unsubstituted diimine systems, and five-coordinate dimeric structures for the methyl substituted compounds.

Introduction

In recent years there have been extensive studies of magnetic behavior of simple dimeric or small multimetal cluster compounds containing a variety of organic ligands. The driving force for such studies is the hope of gaining a better understanding of exchange interactions in magnetically condensed inorganic salts such as anhydrous and hydrated metal halides which exhibit extended cooperative magnetic behavior. For instance, anhydrous ferrous chloride has the cadmium chloride structure with intra- and inter-chain chloro-bridging. The intra-chain interaction for this compound is ferromagnetic ($J > O$) whereas the inter-chain interaction is weaker and antiferromagnetic. As a consequence, anhydrous ferrous chloride exhibits[1,2] a "meta-magnetic" phase transition from an antiferromagnetic to paramagnetic state in an external field of about 11 kG.

Extended (lattice) ferromagnetic interaction in transition metal-organic ligand systems is much less common than for simple inorganic

* Please address correspondence to this author.

salts. In this article we present a Mössbauer and magnetic susceptibility study of such behavior in polymeric octahedral complexes of the type Fe(di-imine)Cl_2 where the di-imine is either 2,2'-bipyridine, hereafter bipyridine or 1,10-phenanthroline and substituted derivatives. We also present x-ray and spectroscopic data bearing on the molecular structure of these compounds. It is of interest to see if there are any significant magnetic dilution effects on placing organic ligands "in" ferrous chloride and at the same time maintaining a polymeric structure. The weakening of inter-chain interaction by the ligand dilution allows the possibility of lower dimensionality magnetic interaction and weak meta-magnetic behavior. In this connection we make comparisons of the magnetic behavior of Fe(bipyridine)Cl_2 and Fe(phenanthroline)Cl_2 to our preliminary results for Fe(pyridine)$_2Cl_2$. The latter system is also a chloro-bridged polymer(3) but contains the trans-FeN_2Cl_4 chromophore rather than the analogous cis chromophore as in Fe(bipyridine)Cl_2. The effects of preparative technique are also considered in the discussion of Fe(5,5'-di-CH_3-bipyridine)Cl_2 prepared by high vacuum thermolysis.

Chemical Preparation

Analytical data for the compounds studied are given in Table I. All of the complexes were studied as powders as the preparations do not yield appropriate single crystals. The synthetic methods used consisted of rapid precipitation from aqueous-hydrochloric acid solution or vacuum thermolysis of [Fe(di-imine)$_3$] Cl_2 yielding fine dust-like powders. Attempts at conductivity or molecular weight measurements by the usual solution methods results in disproportionation to the thermodynamically more stable Fe(di-imine)$_3^{2+}$.

Table I: Analytical Data

Compound	Calculated				Observed			
	C	H	N	Fe	C	H	N	Fe
Fe(phenanthroline)Cl_2	46.95	2.63	9.12	18.20	46.77	2.66	9.34	18.60
Zn(phenanthroline)Cl_2	45.50	2.55	8.85		45.55	2.41	8.79	
Fe(bipyridine)Cl_2	42.42	2.86	9.91		42.26	2.86	10.10	
Fe(4,4'-di-CH_3-bipyridine)Cl_2	46.31	3.90	9.01		46.36	3.86	8.69	
Fe(5,5'-di-CH_3-bipyridine)Cl_2	46.31	3.90	9.01		45.81	3.87	8.35	

Susceptibility and Magnetization Studies

A preliminary study of the magnetic susceptibility[4] showed that a solution preparation of Fe(phenanthroline)Cl_2 orders ferromagnetically with $T_c = 8 \pm 2$ K. Fits with a Curie-Weiss law correspond to a paramagnetic Curie temperature $\theta = 12 \pm 4$ K and the Curie-Weiss constant C = 3.81 emu/mole. In this section we present magnetic data for Fe(bipyridine)Cl_2 also prepared in solution and the substituted derivative Fe(5,5'-di-CH_3-bipyridine)Cl_2 obtained by vacuum thermolysis. Fe(bipyridine)Cl_2 clearly is similar to Fe(phenanthroline)Cl_2 but has a stronger ferromagnetic interaction. Evidence for a ferromagnetic interaction is seen (Fig. 1) in the χ_g^{-1} vs T plot for which the intercept is large ($\geqslant$ 20 K) and positive. The plot of χ_g^{-1} vs T is linear from ~60 K to 200 K and a fit with the Curie-Weiss law yields $\theta = +25$ K and C = 5.26 emu/mole. The temperature dependence of the dc magnetization σ at low field (Fig. 2) shows the expected rapid rise in the vicinity of T_c which is estimated to be $\leqslant$ 8 K. A more precise estimation of T_c will be discussed in connection with the Mössbauer data. Between 0 and $\simeq$10 kG at 4.2 K there is a rapid rise in the magnetic moment per gram, σ. However, above ~10 kG there is a gradual increase in σ and the compound is clearly not saturated for applied field B_o as large as 200 kG, Fig. 3. It is interesting to point out that the sharp rise in σ vs T seen in Fig. 2 is quite similar to that found[5] for the linear chain polymer Co(pyridine)$_2Cl_2$. In the latter compound the sharp rise in σ is attributed to strong one-dimensional ferromagnetic intra-chain interaction (correlation) above the Néel temperature at which the complex undergoes three-dimensional antiferromagnetic ordering. A similar positive intrachain correlation may well be occurring in Fe(bipyridine)Cl_2 and Mössbauer data bearing on this possibility will be discussed subsequently.

The field and temperature dependence of the magnetic properties for Fe(5,5'-di-CH_3-bipyridine) Cl_2 are somewhat different from those of Fe(bipyridine)Cl_2. A plot of χ_g^{-1} vs T (Fig. 4) appears to have a near zero or small negative temperature intercept. A fit of the data yields $\theta \simeq 0$ K or slightly negative with C = 4.3 emu/mole. Thus this material appears to exhibit very weak, possibly antiferromagnetic interactions. A weak exchange interaction is also indicated by the gradual rise in σ vs B_o for this compound. The difference in magnetic behavior for the two bipyridine systems is related to a difference in molecular structure.

Magnetic Moments. Effective magnetic moments in Bohr magnetons (μ_B) are calculated using the relation

$$\mu_{eff} = \sqrt{3k/N_o}\ \sqrt{\chi_M(T-\theta)} = 2.828\ \sqrt{C}$$

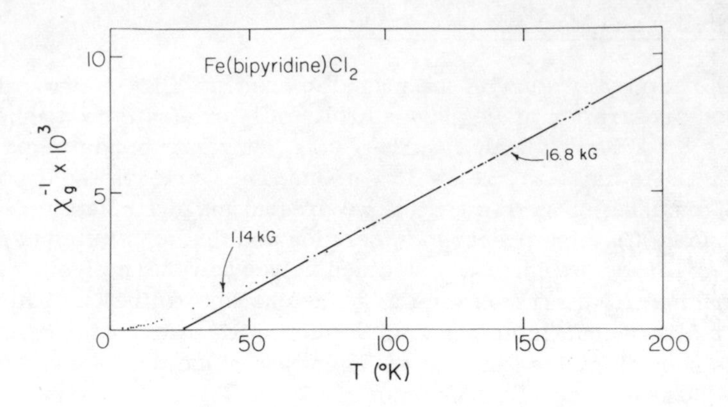

Figure 1. χ_g^{-1} vs. T *for Fe(bipyridine)Cl₂, applied field* B_o = *1.14 kG for* T ≤ *80 K,* B_o = *16.8 kG for* T > *80 K*

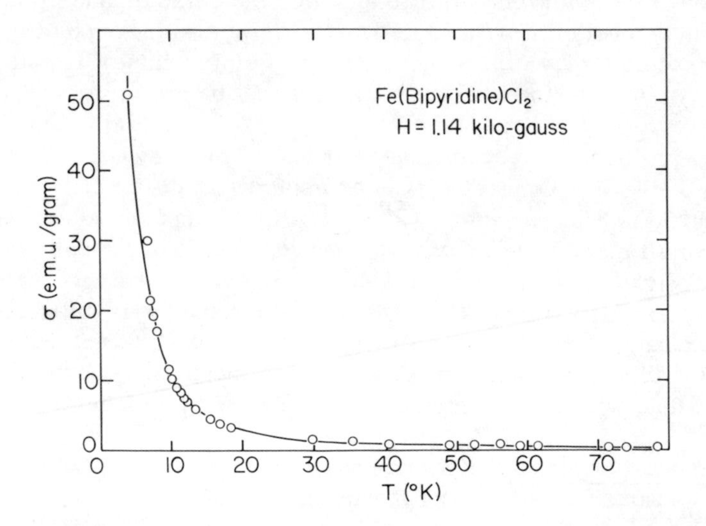

Figure 2. Magnetic moment per gram, σ, vs. T *at 1.14 kG for Fe(bipyridine)Cl₂*

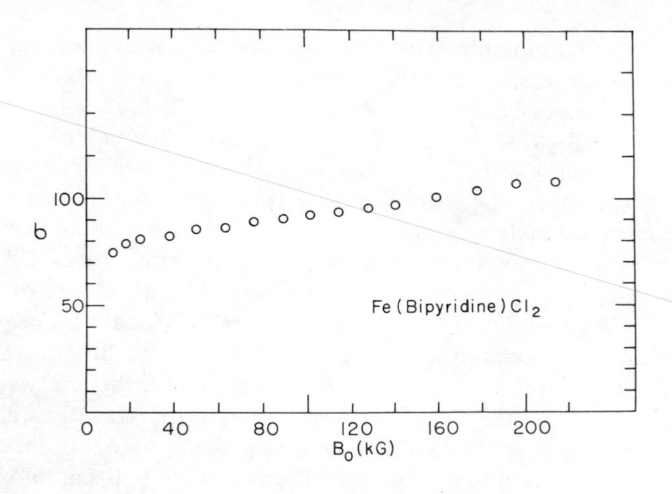

Figure 3. σ vs. *applied field,* B_o, *at 4.2 K for Fe(bipyridine)*Cl_2

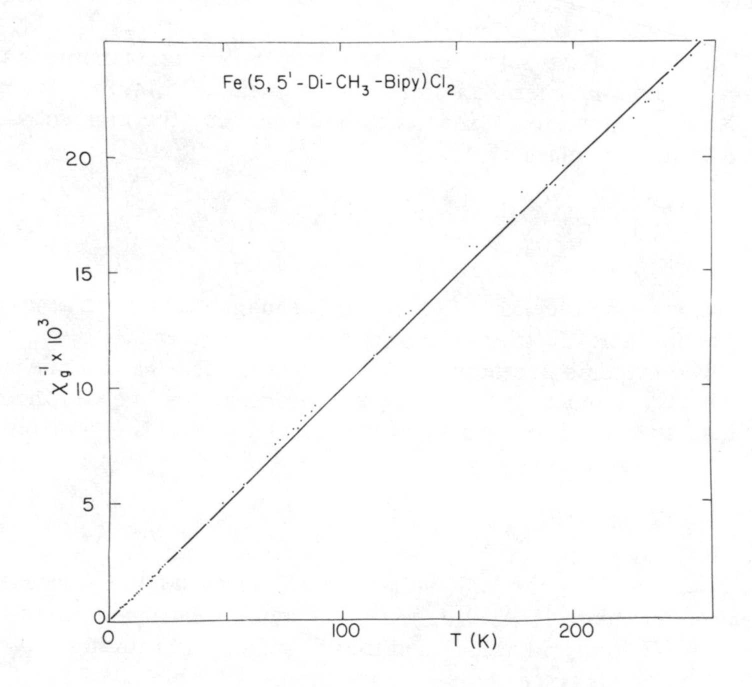

Figure 4. χ_g^{-1} vs. T *for Fe(5,5′-di-*CH_3*–bipyridine)*Cl_2, B_o = *16.8 kG*

where k is the Boltzmann constant, N_o Avogadro's number and χ_M is the corrected molar susceptibility. The values of μ_{eff} obtained for the preceding values of C are: Fe(phenanthroline)Cl_2 (5.5μ_B), Fe(bipyridine)Cl_2 (6.4 μ_B) and Fe(5,5'-di-CH_3-bipyridine)Cl_2 (5.8 μ_B). Since the spin-only value of the magnetic moment for high-spin ferrous is 4.9 μ_B, the foregoing values of μ_{eff} represent in part unusually large orbital contributions to the moment, in particular for the last two compounds. There is no evidence of any high-spin ferric or other impurity that could account for the magnetic moment values we observe. For instance, an high-spin iron III impurity of ~30% would be required to increase the room temperature moment from 5.1 to 5.4 μ_B. Such an impurity content would be readily observed in both the analytical data and Mössbauer spectra (Fig. 5). No evidence for iron III is detected. The Mössbauer spectrum corresponds to a pure material containing a single iron II environment. Furthermore, for iron III complexes of diimine ligands, there is generally instability and a strong tendency for reduction of ferric to diamagnetic ferrous even on exposure to ordinary light. It is interesting to note that in a study of Fe(bipyridine)Cl_2 and Fe(phenanthroline)Cl_2 at 300 K, Dwyer et al[6] also observed large moments, 5.79 μ_B and 5.72 μ_B respectively. Thus we believe that the observed large effective moments are "real" and probably reflect, in part, an orbital contribution.

Difficulty in saturating the compounds at low temperatures (Fig. 3) suggests a large magnetic anisotropy consistent with large, anisotropic orbital contributions to the moment. For instance, the moment μ calculated using the relation:

$$\mu = \frac{M\sigma}{N_o\mu_B}$$

where M is the molecular weight and σ the magnetization in emu/gram are given in Table II. It is evident that with increasing B_o the moments of all three systems are increasing. However, the high field behavior shown in Fig. 3 and a similar data for Fe(phenanthroline)Cl_2 indicate that these systems are not saturated even at 200 kG. The spin-only value of μ is given by

$$\mu_S = g\mu_B S.$$

When g = 2 and S = 2 for high-spin ferrous, $\mu = 4\mu_B$, the expected saturation moment. There is, thus, substantial enhancement of the moment for Fe(bipyridine)Cl_2 and this is reflected in its high μ_{eff} and μ. The somewhat larger value for the moment (Table II) of Fe(bipyridine)Cl_2 than for Fe (phenanthroline)Cl_2 is consistent with the

Table II: Magnetic Moments

Compound	T(K)	B_o(kG)	μ_B
Fe(bipyridine)Cl_2	4.2	16.8	4.7
Fe(phenanthroline)Cl_2	3.0	48.4	3.1
	4.2	215	3.9
Fe(5,5'-di-CH_3-bipyridine)Cl_2	4.2	14.7	1.9
	4.2	16.8	2.1
	3.0	48.4	3.4

larger ferromagnetic interaction of the former; a ratio of θ's for these complexes is ~2:1.

To conclude this section it is important to mention that the second order Zeeman effect and spin-orbit coupling can also contribute to the enhancement of magnetic moments to values greater than the spin-only. However, this problem has been studied[7] in some detail for simple tetrahedral iron complexes such as $[(C_2H_5)_4N]_2 FeCl_4$ whose effective moments (~5.4 to 5.6 μ_B) are high. It is found that the combination of the foregoing effects increases the moment to only ~5.2 μ_B. For the rather distorted systems of this investigation these effects are expected to be less important. Thus the effective moments for the present compounds are unusually high and we have no simple explanation for their origin. It is difficult to envision a large direct orbital contribution. A small amount of configuration interaction of the ground $3d^6$ with a nearby $4s^1 3d^5$ having a much higher spin-only moment could greatly enhance the moments but this would be difficult to demonstrate.

Mössbauer Studies

The Mössbauer spectra of Fe(bipyridine)Cl_2 and Fe(phenanthroline)Cl_2 made in solution are very similar. The onset of magnetic order is easily seen in the Mössbauer spectra, as illustrated in Fig. 6 for Fe(phenanthroline)Cl_2. The Curie points of Fe(bipyridine)Cl_2 and Fe(phenanthroline)Cl_2 are T_c = 3.8 K and T_c = 5.0 K respectively. Magnetization data previously reported for the Fe(phenanthroline)Cl_2 suggested a transition temperature of about ~8 K. The higher apparent T_c observed by the magnetization measurements

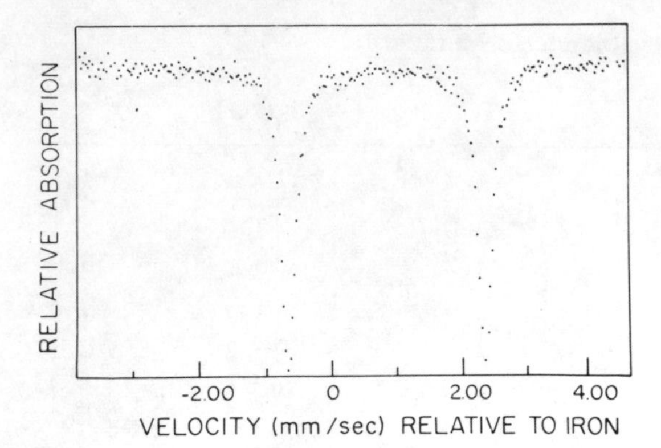

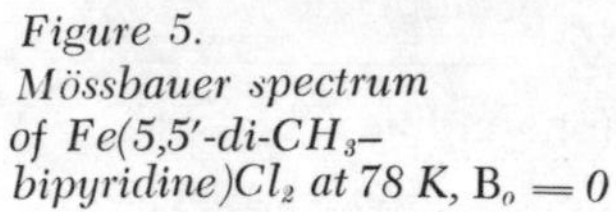
Figure 5. Mössbauer spectrum of Fe(5,5′-di-CH_3–bipyridine)Cl_2 at 78 K, $B_o = 0$

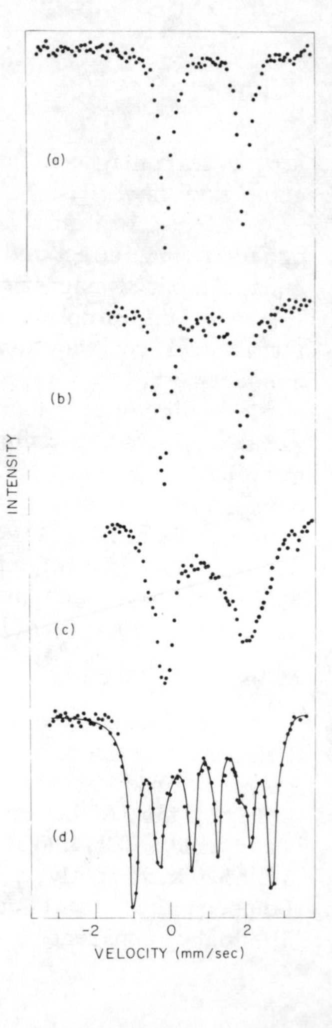

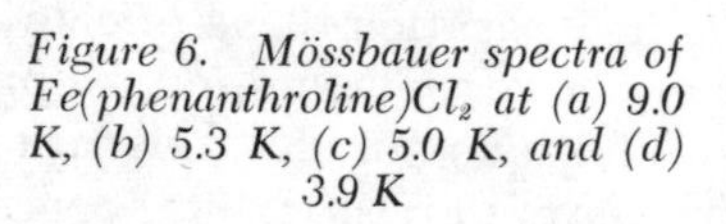
Figure 6. Mössbauer spectra of Fe(phenanthroline)Cl_2 at (a) 9.0 K, (b) 5.3 K, (c) 5.0 K, and (d) 3.9 K

may be ascribed to the difficulty in defining the magnetic transition for the irregularly shaped powder sample. The large susceptibility above T_c is illustrated in Fig. 7 where spectra of Fe(bipyridine)Cl_2 (T_c = 3.8 K) at 4.2 K are shown at zero magnetic field and in longitudinal magnetic fields of 4 and 35 kG. It is seen that the small external field of 4 kG induces a hyperfine field without polarizing the moment along the external field direction. This is indicated by the presence of $\Delta m = 0$ lines in the spectrum and by the fact that the angle β between the principal component of the electric field gradient and the magnetic hyperfine field as deduced from the spectrum is unique, and close to that observed in the ordered state below T_c in zero external field. At B_o = 35 kG, the appearance of the spectrum is considerably altered due to the polarization of the ferromagnetic moment by the external magnetic field, with a consequent randomization of β

Below their respective Curie temperatures Fe(bipyridine)Cl_2 and Fe(phenanthroline)Cl_2 have essentially similar spectra. For the former at 1.5 K, H_n = -60 kG, ΔE = +1.70 mm/sec and $\beta \approx 60^o$ while for the latter H_n = - 75 kG, ΔE = +2.03 mm/sec and $\beta \approx 60^o$.

As discussed previously the temperature and field dependence of magnetization of Fe(5, 5'-di-CH_3-bipyridine)Cl_2 (prepared by thermolysis) suggests weak, possibly negative, magnetic exchange interactions. The temperature dependence (in zero field) of the Mössbauer spectra for this material also suggests weak magnetic interactions. Instead of a sharp "ferromagnetic" transition over a small ($\leq$ 0.5 K) interval as in Fig. 6, Fe(5, 5'-di-CH_3-bipyridine)Cl_2 exhibits gradual changes of the magnetic hyperfine splitting over a much larger temperature range. The transitions of the quadrupole doublet start broadening at ~12 K and a fully resolved Zeeman spectrum is not observed until ~2 K indicating slow paramagnetic relaxation rather than a cooperative ordering process. It will be shown that this compound contains high-spin iron II in a highly distorted 5-coordination environment. For such a low symmetry, longer spin-lattice relaxation times and slow paramagnetic relaxation are not unexpected. The observation of this phenomenon is far less common for high-spin ferrous than for ferric complexes.

Molecular Structure Studies

Far Infrared Spectra and X-ray Data. The difference in the magnetic behavior of Fe(bipyridine)Cl_2 and Fe(5, 5'-di-CH_3-bipyridine)Cl_2 is probably due to the preparative method and a difference of basic molecular structure resulting therefrom, rather than from just simply a substituent effect. Figure 8 shows the far-infrared spectra of Fe(phenanthroline)Cl_2 (prepared in solution) and Zn(phenanthroline)Cl_2. The latter zinc complex is known by a single crystal x-ray study[8] to be a pseudo-tetrahedral monomer. We have compared the x-ray powder

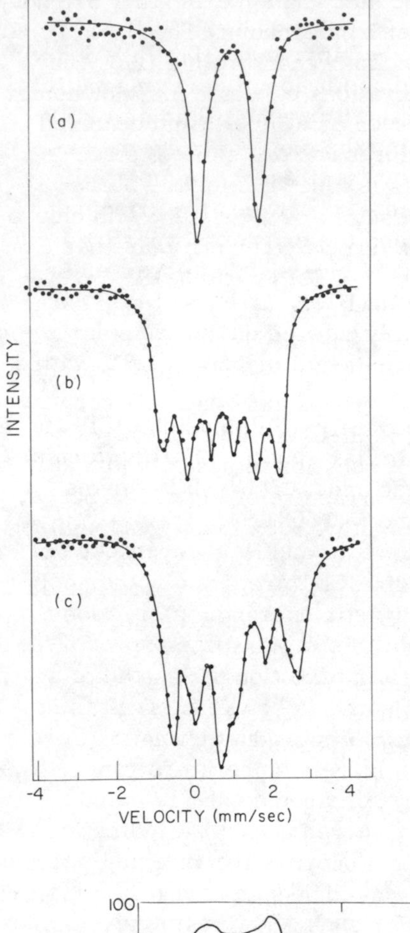

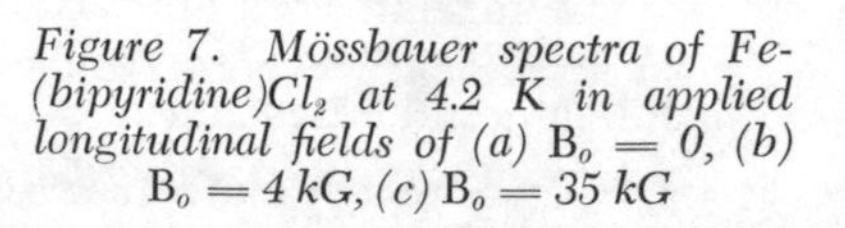

Figure 7. Mössbauer spectra of Fe-(bipyridine)Cl_2 at 4.2 K in applied longitudinal fields of (a) $B_o = 0$, (b) $B_o = 4$ kG, (c) $B_o = 35$ kG

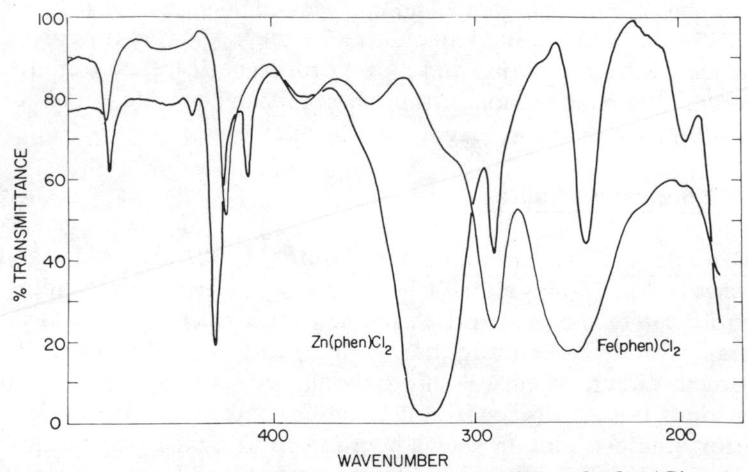

Figure 8. Far-infrared spectra of Fe and Zn (phenanthroline)Cl_2–mineral oil mull on polyethylene at 300 K

patterns of Fe(phenanthroline)Cl_2 and the zinc analogue and they are not similar, indicating these systems are not isomorphous. This is reflected in the far-infrared spectrum of Zn(phenanthroline)Cl_2. The two strong terminal Zn-Cl stretching vibrations expected for a monomer having approximate C_{2V} symmetry are seen as a broad band centered ~325 cm^{-1}. In the iron analogue these vibrations are shifted to considerably lower energy (~255 cm^{-1}); this is also true for the zinc and iron bipyridine complexes. These results are consistent with chloro-bridging[9, 10] as shown in Fig. 9. A similar polymer structure has been proposed as part of a recent study[11] of Sn(bipyridine)Cl_2, i.e., an infinite linear polymer chain with chloro-bridging.

The far infrared spectrum of Fe(5, 5'-di-CH_3-bipyridine)Cl_2 is shown in Fig. 10. Strong bands at 322 and 240 cm^{-1} indicate the presence of both terminal and bridging chloro groups and hence a structure involving five-coordinate iron II. Fe(4, 4'-di-CH_3-bipyridine)Cl_2 exhibits a similar spectrum and the di-CH_3 substituted systems probably have a dimeric structure as shown in the structure of Fig. 11. That is, a combination of steric effects from the methyl substituents and high vacuum thermolysis preparation results in dimeric rather than the extended polymeric structure proposed for the solution preparation of Fe(bipyridine)Cl_2. A dimeric structure similar to that presented in Fig. 11 has been found in a single crystal x-ray study[12] of [Ni(2, 9-di-CH_3-phenanthroline)Cl_2]$_2$. Recent[13] magnetic studies of other similar chloro-bridged nickel II dimers suggest weak (intradimer) antiferromagnetic exchange and is consistent with our results for the methyl substituted derivative.

Near Infrared Spectra. The near-infrared spectrum of Fe(phenanthroline)Cl_2 given in Fig. 12 is typical[14, 15] of a pseudo-octahedral FeN_2Cl_4 chromophore and strongly supports the proposed structure. Fe(pyridine)$_2Cl_2$ is an octahedral polymer[3] containing the same chromophore but with trans-nitrogens. As expected, it exhibits a near-infrared spectrum quite similar to that of Fe(phenanthroline)Cl_2 and Fe(bipyridine)Cl_2. The sign of the quadrupole interaction for these systems is positive and consistent with a d_{xy} ground orbital. We thus tentatively assign the transitions observed at 6000 and 10, 300 cm^{-1} as $^5B_2(d_{xy}) \rightarrow {}^5B_1(d_{x^2-y^2})$ and $^5B_2(d_{xy}) \rightarrow {}^5A_1(d_{z^2})$ respectively.

For a five-coordinate FeN_2Cl_3 chromophore as suggested for the methyl substituted complexes, one expects the d-d transition at somewhat lower energies. We observe a ligand-field band for Fe(4, 4'-di-CH_3-bipyridine)Cl_2 at 9100 and 5000 cm^{-1}.

Mössbauer Isomer Shifts. In Table III we present some isomer

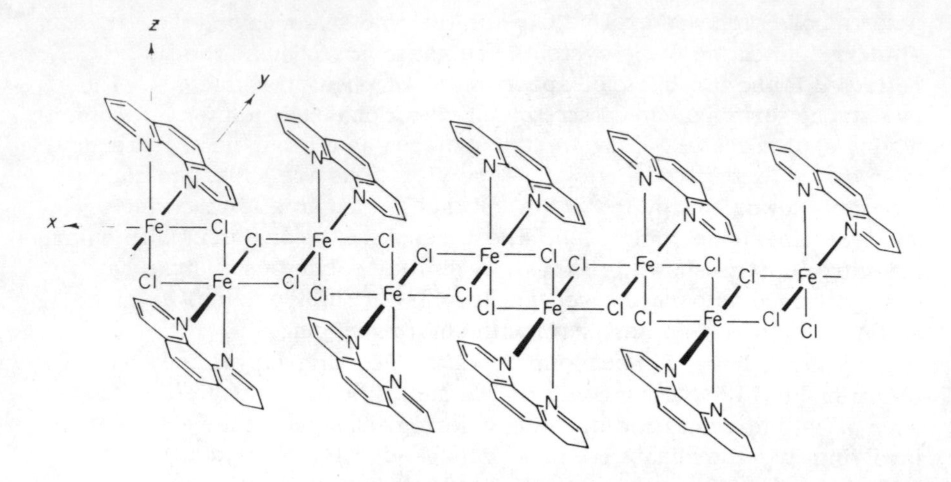

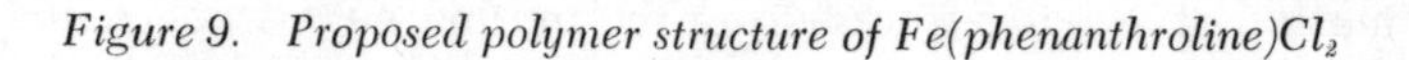

Figure 9. Proposed polymer structure of Fe(phenanthroline)Cl_2

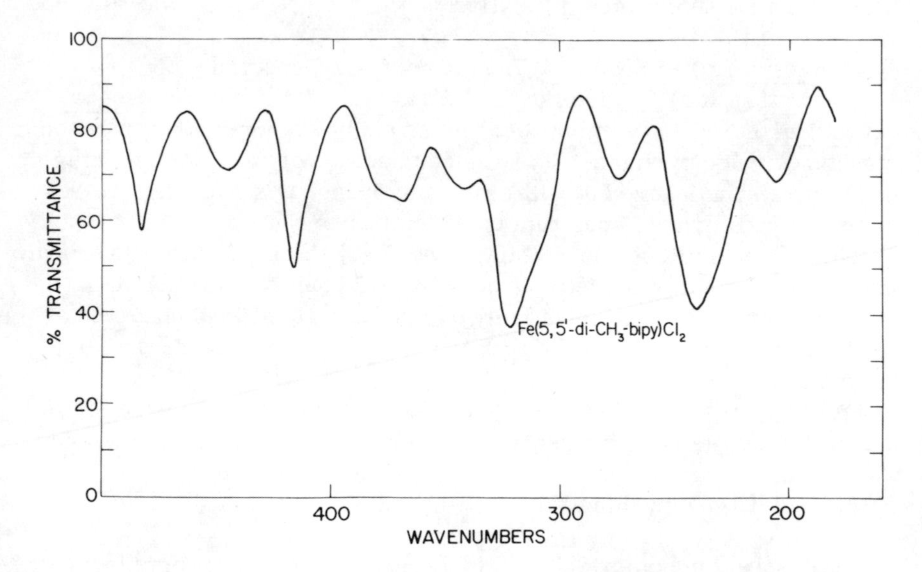

Figure 10. Far-infrared spectrum of Fe(5,5'-di-CH_3–bipyridine)Cl_2–mineral oil mull on polyethylene at 300 K

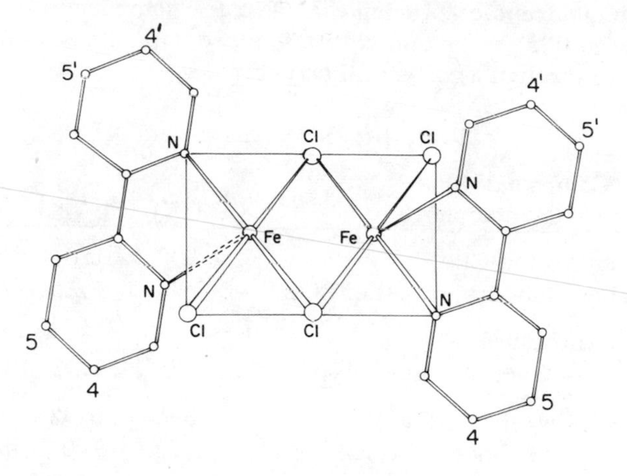

Figure 11. Proposed dimer structure for Fe(5,5′-di-CH_3–(bipyridine)Cl_2

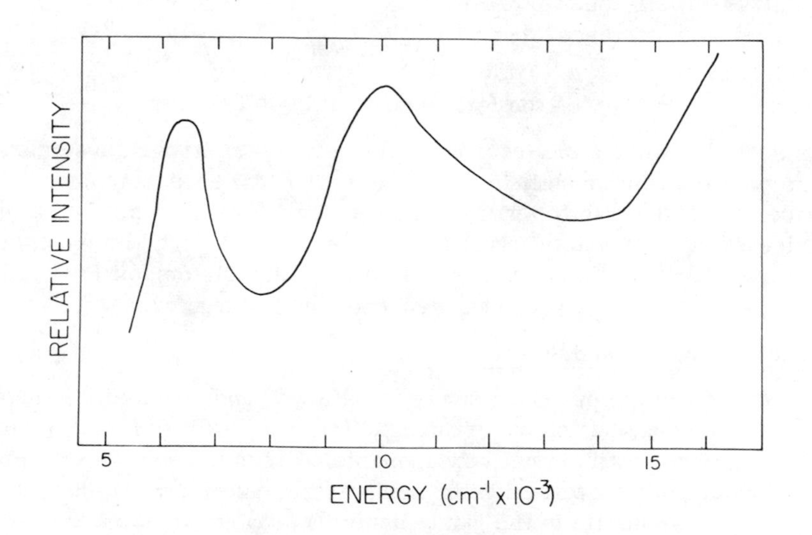

Figure 12. Near-infrared spectrum of Fe(phenanthroline)Cl_2, fluorocarbon grease mull at 300 K

shift (δ) and quadrupole splitting (ΔE) data for compounds (1,2,3,4) investigated in this work and compare these data to those for other compounds (5 through 8) of known structure.

Table III: Mössbauer Data

Compound Structure-Chromophore	T(K)	δ†	ΔE	Reference
1. $Fe(phenanthroline)Cl_2$ polymeric-octahedral-cis-FeN_2Cl_4	300	1.04	1.45	*
	78	1.16	2.17	*
2. $Fe(phenanthroline)Br_2$ polymeric-octahedral-cis-FeN_2Cl_4	78	1.13	1.98	*
3. $Fe(4,4'\text{-di-}CH_3\text{-bipyridine})Cl_2$ dimer-five-coordinate FeN_2Cl_3	300	0.92	3.55	*
	78	1.04	3.73	*
4. $Fe(5,5'\text{-di-}CH_3\text{-bipyridine})Cl_2$ dimer-five-coordinate FeN_2Cl_3	300	0.93	3.51	*
	78	1.06	3.72	*
5. $Fe(pyridine)_2Cl_2$ polymeric-octahedral-trans-FeN_2Cl_4	300	1.07	0.56	15
	78	1.20	1.28	
6. $Fe(pyridine)_4Cl_2$ monomeric-octahedral-trans-Fe Cl_2N_4	300	1.05	3.15	15
	78	1.14	3.35	
7. $Fe(quinoline)_2Cl_2$ monomeric-tetrahedral-FeN_2Cl_2	300	0.86	2.76	16
	78	0.98	3.07	
8. $Fe(2,9\text{-di-}CH_3\text{-phenanthroline})Cl_2$ monomeric-tetrahedral-FeN_2Cl_2	300	0.80	2.66	*
	78	0.91	2.77	*

* This work

† mm/sec, relative to iron foil, source at 300 K.

It is seen that for the six-coordinate $FeCl_2$-imine nitrogen systems a more positive isomer shift is observed.(17) With decreasing coordination number the isomer shift generally decreases(18) as evidenced by table entries 7 and 8 for known tetrahedral complexes.(19) The intermediate isomer shift(20) and large quadrupole interaction of compounds 3 and 4 are reasonable for the proposed five-coordinate structure.

Magnetic Behavior and Structure

In view of the proposed structure (Fig. 9) and considerable supporting data, the observation of ferromagnetism for $Fe(bipyridine)Cl_2$ and $Fe(phenanthroline)Cl_2$ is not unreasonable. Various models(21,22) predict ferromagnetic interaction for such a structure when (a) the metal and bridging atoms lie in the same plane; (b) bridge angles are ~90°; (c) the exchange involves bridge atom p- and metal d-orbitals;

and (d) the exchange is between metal atoms in oxidation states corresponding to a more than half-filled t_{2g} manifold. A study of the ferromagnetic behavior for other bridging anions (F^-, Br^-, I^-) may enable assessment of the contribution of direct metal-metal exchange as opposed to "super-exchange" via the bridging ligands since the metal-metal distance can be varied considerably. This work is now in progress. Also interesting is the absence of meta-magnetic behavior for the systems of this investigation in studies to as low as 1.4 K and for fields up to 200 kG. We have observed metamagnetic behavior in Fe(pyridine)$_2$Cl$_2$ and Fe(pyridine)$_2$(NCS)$_2$. This work will be discussed in a forthcoming publication.[23]

Literature Cited

1. Wilkinson, M.K., Cable, J.W., Wollan, E.O., and Koehler, W.C., Phys. Rev., (1959) 113, 497.
2. Simkin, D.J., Phys. Rev., (1968) 177, 1008.
3. Dunitz, J.D., Acta Cryst. (1957) 10, 307.
4. Reiff, W.M., and Foner, S., J. Amer. Chem. Soc., (1973) 95, 260.
5. Takeda, K., Matsukawa, S., and Haseda, T., J. Phys. Soc. Japan, (1971) 30, 1330.
6. Broomhead, J.A., Dwyer, F.P., Austr. J. Chem., (1961) 14, 250.
7. Clark, R.J.H., Nyholm, R.S. and Taylor, F.B., J. Chem. Soc., (A) (1967) 1802.
8. Reimann, C.W., Block, S., Perloff, A., Inorg. Chem., (1966) 5, 1185.
9. Postmus, C., Ferraro, J.R., and Wozniak, W. Inorg. Chem., (1967) 6, 2030.
10. Wilde, R.F., Srinivasan, T.K.K., Ghosh, S.N., J. Inorg. Nucl. Chem., (1973) 35, 1017.
11. Fowles, W.A., and Khan, I.A., J. Less Common Metals, (1968) 15, 209.
12. Preston, H.S., and Kennard, C.H.L., J. Chem. Soc. (A) (1969), 2682.
13. Hendrickson, D. private communication.
14. Goodgame, D.M.L., Goodgame, M., Hitchman, M.A., and Weeks, M.J., Inorg. Chem., (1966) 5, 635.

15. Long, G.J., Whitney, D.L., and Kennedy, J.E., Inorg. Chem., (1971) 10, 1406.

16. Long, G.J., and Whitney, D.L., J. Inorg. Nucl. Chem., (1971) 33, 1196.

17. Burbridge, C.D., Goodgame, D.M.L., Goodgame, M., J. Chem. Soc. (A) (1967), 349.

18. Erickson, N.E. in, "The Mössbauer Effect and Its Application in Chemistry," Advances in Chemistry Series, No. 88, American Chemical Society, Washington, D.C. (1967).

19. Edwards, P.R., Johnson, C.E., and Williams, R.J., J. Chem. Phys., (1967) 47, 2074.

20. Reiff, W.M., Erickson, N.E., and Baker, W.A., Inorg. Chem., (1969) 8, 2019.

21. Anderson, P.W., "Magnetism", (1963) Volume 1, Ch. 2, Rado, G.T., and Suhl, H., Ed., Academic Press, New York, N.Y.

22. Goodenough, J.B., "Magnetism and the Chemical Bond", Interscience, New York (1963), pp. 165-185.

23. Reiff, W.M., Long, G.J., Little, B.F., Foner, S., Frankel, R.B., in preparation.

Acknowledgements

W.M. Reiff is pleased to acknowledge the partial support of the Research Corporation and the Petroleum Research Fund administered by the American Chemical Society. He thanks the National Science Foundation (NSF Grant No. GH39010) for the major support of the investigation. Finally, he thanks Dr. Graham Hunt of AFCRL for use of the Perkin-Elmer 180 IR-spectrometer. M.A. Weber was supported by an Organization of American States Fellowship. The Francis Bitter National Magnet Laboratory is supported by the National Science Foundation.

16

The Unusual Magnetic Properties of an Imidazolate Bridged Polymer of an Iron(III) Hemin

IRWIN A. COHEN
Brooklyn College, The City University of New York, Brooklyn, N.Y. 11210

DAVID OSTFELD
Seton Hall University, South Orange, N.J. 07079

For the last several years we have been interested in electron transfer processes involving metalloporphyrins because of the important biological role of the cytochromes (1,2). This has led us to consider the stability of ligand bridging between metalloporphyrins and the utilization of such bridges in electron transfer and spin coupling processes.

The antiferromagnetic coupling of the two irons in $[TPPFe(III)]_2O$ due to the oxo bridge, shown schematically in Figure 1, is now well known (3). But other bridges, especially those involving ligands generally associated with heme proteins have not been as well investigated. Histidine has been found bound to the heme iron in hemoglobin (4) and cytochrome c (5) and this has generated a great deal of interest in imidazole adducts of iron porphyrins. Thus the $[\text{porphyrin Fe(III)(ImH)}_2]^+$ ion has been well studied by several research groups (6-13). The structure of that cation is shown schematically in Figure 2. Imidazolate complexes of porphyrins have not been well examined in spite of the interesting possibility of bridge formation and mediation of electron transfer.

Reported imidazolate complexes are generally assumed to be polymeric (14) as confirmed by single crystal structure determinations on the Zn(II) (15) and Cu(II) (16) complexes. The ability of this ligand to conduct electrons is not shown by the electrical conductivity of the imidazolates since crystalline samples of Co(II), Cu(II) and Zn(II) imidazolates have been found to possess greater resistivities than imidazole itself (17). However, the antiferromagnetic behavior found for copper imidazolate (18) has indicated the possible role of the anion in spin pairing between metal ions. Thus, when we observed that the addition of base to solutions of hemin fluoroborate and imidazole caused the formation of compounds with porphine to iron to imidazolate ratios of 1:1:1, we embarked on a study of their magnetic properties.

Addition of a solution of base to one containing $TPPFeBF_4$ and imidazole consistently produced an insoluble product.

Examination of the X-ray powder diffraction patterns of the products indicated that exact duplication of the reaction conditions reproduced identical products but that variations in the solvents lead to products with different solid phases. In addition, the powder patterns indicated that irreversible changes occurred upon grinding the solid phases, after which all the products failed to produce any observable reflections and thus appeared to become amorphorous. No common solvent (e.g. H_2O, CH_3OH, $(CH_3)_2CO$, DMSO, CH_3CN, CH_2Cl_2, C_6H_6) was found capable of dissolving any of the products, although long exposure to methanol generally caused decomposition of the solids and produced the oxo dimer, $(TPPFe)_2O$. In no case did the infrared (KBr) spectrum of the product reveal any absorption at 870^{-1}. This shows the absence of $(TPPFe)_2O$ in each of the products. The ir absorption bands characteristic of the TPPFe group were observed at the normal frequencies and the Raman spectrum of the one solid product examined was extremely similar to that observed for solid $[TPPFe(ImH)_2]Cl$.

Preparative reactions were carried out under two sets of conditions which differed mainly in the time allowed for addition of the basic solution to that of the hemin and imidazole. The reactions considered to be rapid involved addition of the basic solutions dropwise with stirring over a period of less than 30 min. The reactions considered to be slow involved two phase systems which were carefully added to the same flask so as to preserve an interface. The product formed at the interface over a period of more than two days. Variation of the composition of the solvent or changes in the time allowed for the reaction did not appear to change the overall stoichiometry of the product. Analytical data indicated that the reaction product was TPPFeIm · S where Im is the imidazolate anion and S is a small molecule, usually solvent, such as H_2O or CH_3OH. For one set of reaction conditions, S was observed to be a mole of ImH. In no case was any anion except imidazolate found to be present in any solid porphine reaction product. It was also found possible to convert a sample of TPPFeIm · ImH to TPPFeIm · H_2O simply by washing the insoluble solid with wet methanol and dichloromethane to remove trapped imidazole.

The structure of the product is proposed as indicated in Figure 3 and is justified by the following: A) the low solubility of the product indicates a high molecular weight; B) the absence of additional anions requires the presence of imidazolate anion (rather than imidazole) for charge compensation of the ferric hemin; C) the similarity of the Raman spectrum of the product to that of the $TPPFe(ImH)_2$ cation reveals the hexacoordination of the hemin iron; D) the overall stoichiometry of the solids is always TPP/Fe/Im = 1/1/1; E) the additional small molecule (eq. H_2O, CH_3OH) contained in the solids are easily removed without changing the ratio of TPP/Fe/Im; F) other metal imidazolates have been found to be polymeric and none have been proven

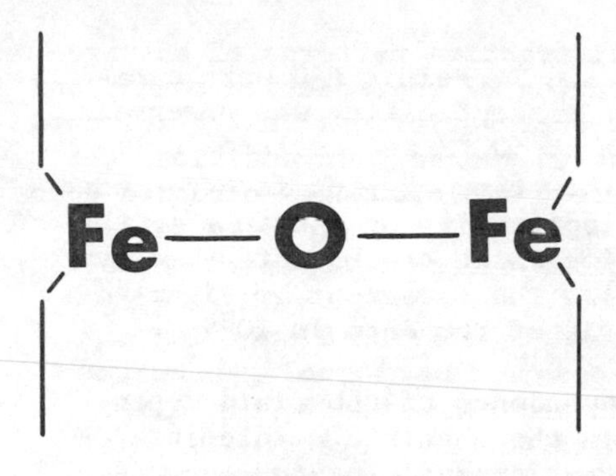

Figure 1. Schematic of the structure (3b) of the μ oxo bridge in $[TPPFe]_2O$. Vertical lines represent the plane of the porphine ring.

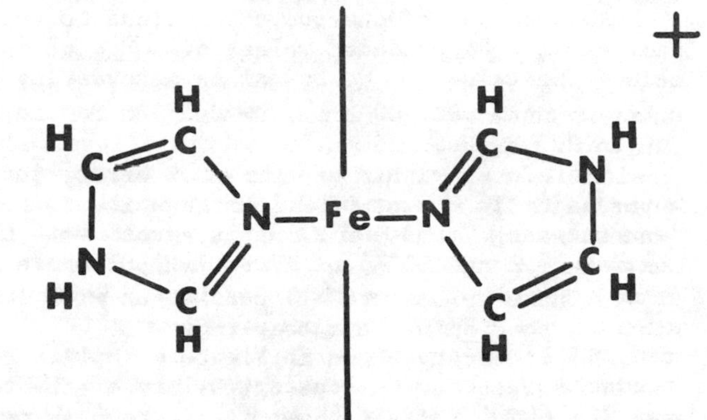

Figure 2. Schematic of the structure (12) of the cation, $TPPFe(ImH)_2^+$. Vertical lines represent the plane of the porphine ring.

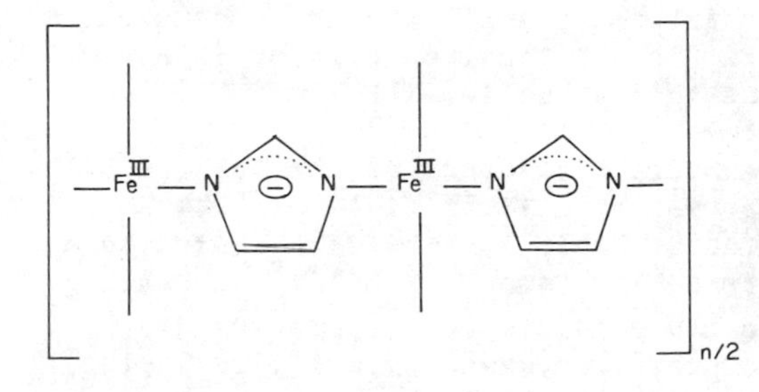

Figure 3. Proposed structure of $[TPPFeIm]_n$. Vertical lines represent the plane of the porphine ring.

to be momeric.

The magnetic properties of solid TPPFeIm · H_2O were examined and a field dependence of the susceptibility was observed. Figure 4 shows a plot of χ_M^c vs 1/H at 300°K and includes only those data points with H > 6kG. This type of plot is often used to determine the paramagnetic susceptibility of a sample in the presence of ferromagnetic impurities which can be saturated by the available magnetic fields (19). The intercept in Figure 4 as determined by a least squares fit of the data is $10^6\chi_i$ = 1980 ± 60 cgs/mole at 300°K.

A study of the temperature dependence of the field dependent susceptibility would ascertain the possible maintenance of the Curie Law for χ_i. This requires large fields to magnetically saturate the sample, but the use of a Cryostat restricts our available fields to a maximum of 5980 gauss. However, the use of a few data points allows a reasonable estimate of χ_i at 77°K.

Application of the above technique to two different solid samples at 77°K produced values of $10^6\chi_i$ of 6910 and 5990 cgs/mole. The values of χ_i reveal an interesting value of the magnetic moment. At 300°K μ_{eff} = 2.15 BM/Fe and at 77°K μ_{eff} = 1.9 to 2.1 BM/Fe.

Therefore, within experimental error, the polymers exhibit a paramagnetic moment of 2.1 BM/Fe which is temperature independent between 77 and 300°K. This agrees with the measured moment (between 4.2 and 50°K) of TPPFe(ImH)$_2$Cl where μ_{eff} = 2.36 BM/Fe (13). These values are all consistent with low spin Fe(III) exhibiting some spin orbit coupling.

The linearity shown in Figure 4 implies that the field dependent portion of the susceptibility of the sample is saturated when H > 6kG. This is shown in Figure 5 as molar magnetization vs field. In this case the magnetization (M = χ_dH) is determined from the observed χ_M^c less the field independent χ_i; taken as 1920 x 10^{-6} cgs/mole in accord with the result of Figure 4. The room temperature saturation magnetization, for TPPFeIm · H_2O is found as 11.5 cgs units/mole.

This corresponds to an extremely small moment at saturation (300°K) of 2.06 x 10^{-3} BM/Fe, if all the iron present is assumed to be involved. As small as this value is, TPPFeIm · H_2O shows magnetic ordering at room temperature.

The study of the temperature dependence of the susceptibility of TPPFeIm · H_2O was carried at a fixed field of 6.44 KGauss and the results are presented in Table I. The data indicates a non-linear relationship between $1/\chi_M^c$ and T with $1/\chi_M^c$ decreasing faster than the Curie-Weiss Law predicts as temperature decreases. This behavior (downward curvature of a $1/\chi$ vs T plot) is reminiscent of a ferrimagnetic material above its Neel temperature (20). However, that interpretation can be rejected since the Neel temperature according to this data would have to

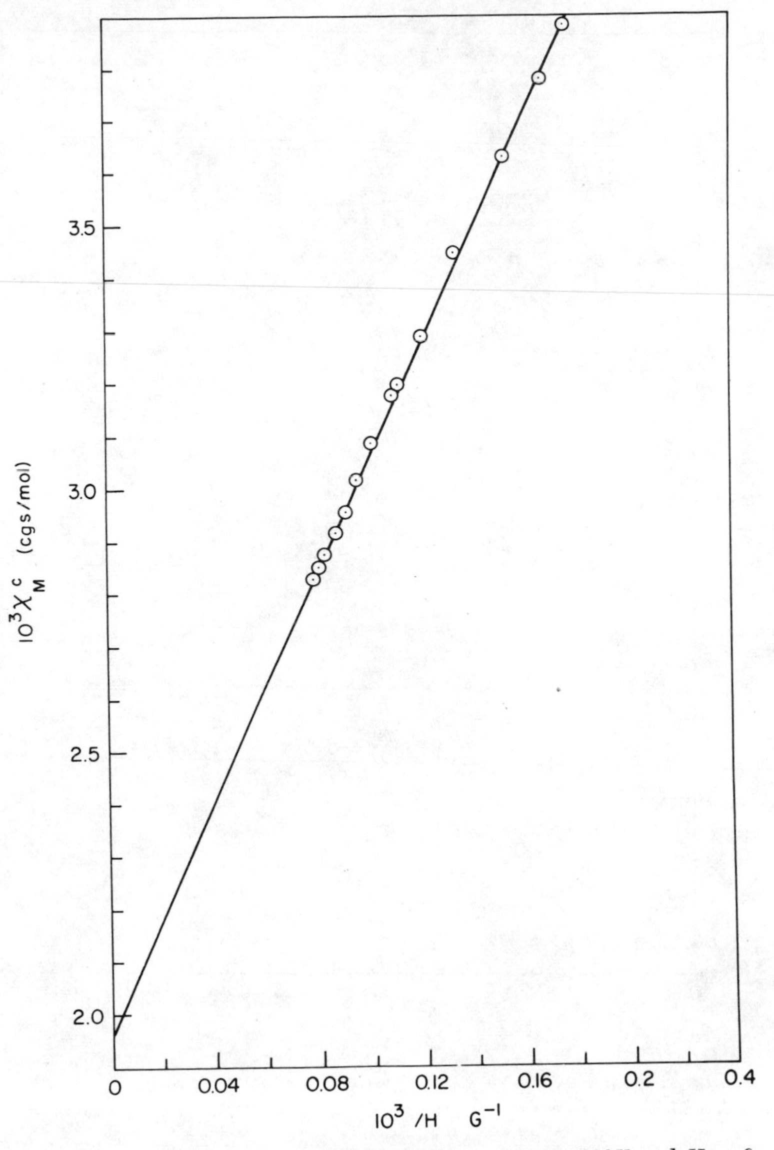

Figure 4. Observed χ_M^c vs. $1/H$ *for TPPFeIm* · H_2O *at 300°K with* H > 6 *KGauss*

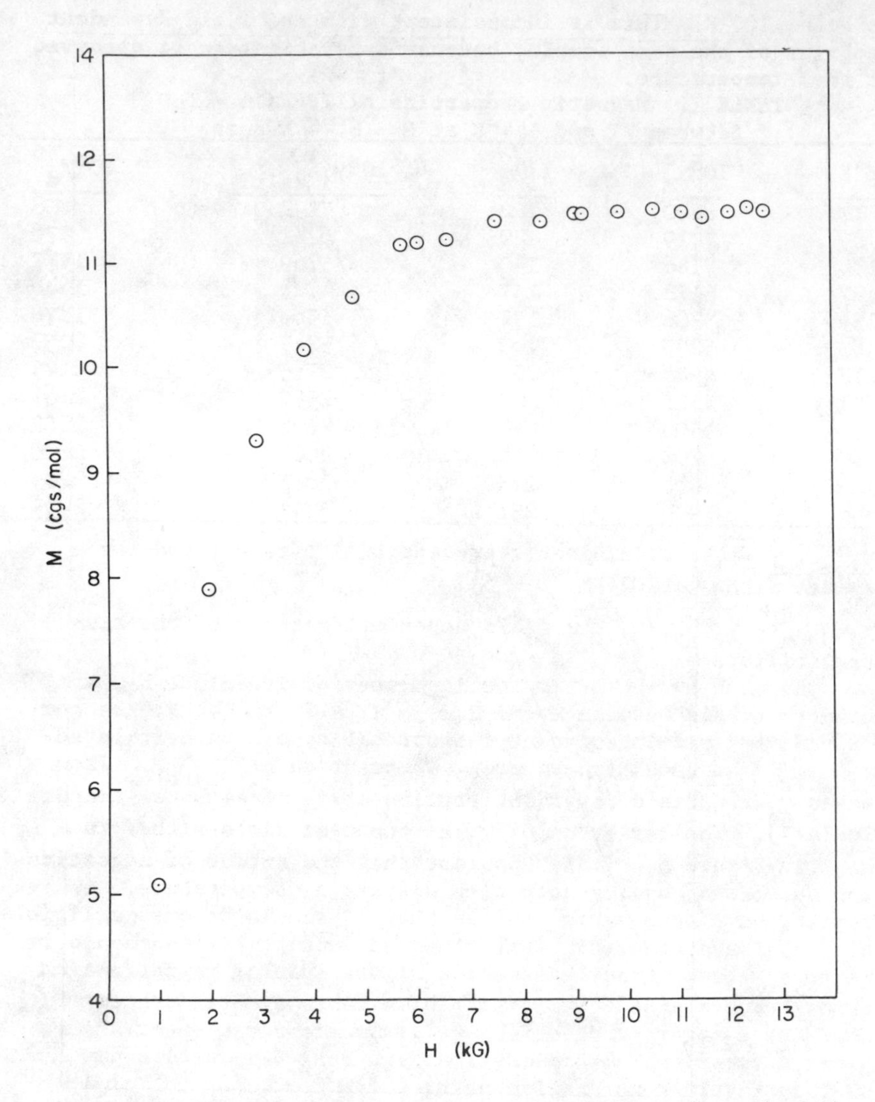

Figure 5. Molar magnetization at 300°K vs. H for TPPFeIm · H_2O. The field independent portion of the susceptibility, $\chi_i = 1920 \times 10^{-6}$ cgs/mole, was subtracted from the observed χ_M^c, $M = H(\chi_M^c - \chi_i)$.

be below 100°K. This is inconsistent with the field dependent behavior of the same sample, because saturation can be observed at room temperature.

TABLE I. Magnetic Properties of TPPFeIm · H_2O Between 77 and 344°K at H = 6.44 KGauss

T°K	$10^6\chi_M^c$	$1/\chi_M^c$	$10^6\chi_{i\ calc.}$ [a]	$10^6\chi_d$ [b]
77	8200	122	7792	408
89	7719	129	6741	978
104	7146	140	5769	1377
127	6372	157	4724	1648
154	5766	173	3986	1870
185	5295	189	3243	2052
217	4908	204	2765	2143
251	4551	220	2390	2161
286	4300	232	2098	2202
300	4200	238	2000	2200
319	4084	245	1881	2203
344	3978	251	1744	2234

[a] $\chi_{i\ calc.}$ = the paramagnetic susceptibility calculated for a system with μ = 2.2 BM.

[b] $\chi_d = \chi_M^c - \chi_{i\ calc.}$; the field dependent portion of the susceptibility.

Inasmuch as the paramagnetic moment of TPPFeIm · H_2O was shown to remain between 2 and 2.2 BM from 77 to 344°K; the corresponding field independent susceptibility can be calculated ($\chi_{i\ calc.}$) at each temperature. Subtraction of $\chi_{i\ calc.}$ from χ_M^c leaves χ_d the field dependent portion of χ_M^c at each temperature (Table I). The variation of χ_d at constant field with T is shown in Figure 6. It is apparent that the extent of magnetization decreases considerably with decreasing temperature. An extrapolation of the curve implies that χ_d should become negligible below 70°K and then the total observed susceptibility should be due to χ_i alone. The observation of the loss of magnetization at low temperature is indicative of a ferromagnetic system which undergoes a magnetic phase change at low temperature or a ferrimagnetic system which possesses a high Neel temperature and a low temperature compensation point (20).

The dependence of χ_M^c on T at low temperature and high field is shown in Figure 7. It is evidently quite complex and does not fit the Curie-Weiss Law in any region of T. The lack of knowledge of the value of χ_i below 77°K prohibits separation of χ_M^c into χ_i and χ_d at each temperature but the anomalous behavior of the sample at about 60°K is evident. It is also rather important to note that the total observed susceptibility at low temperature is only about 10% of the χ_i calculated on the basis of μ_{eff} = 2 BM. Clearly, the constant moment observed above 77°K is not

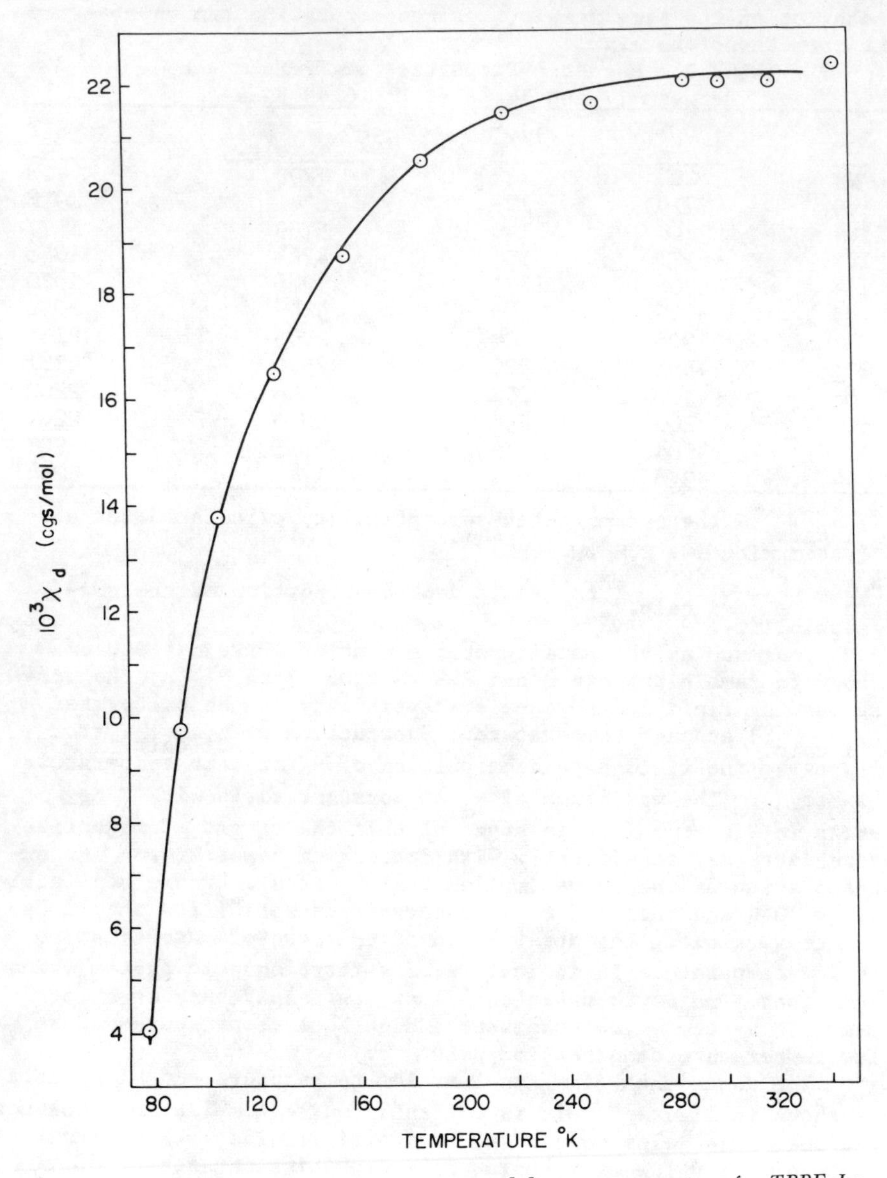

Figure 6. Field dependent portion of susceptibility vs. *temperature for TPPFeIm H_2O at* H = *6.44 KGauss. Value of $\chi_{i\,calc}$ was calculated at each temperature for a system with μ_{eff} = 2.2 BM and $\chi_d = \chi_M{}^c - \chi_{i\,calc}$.*

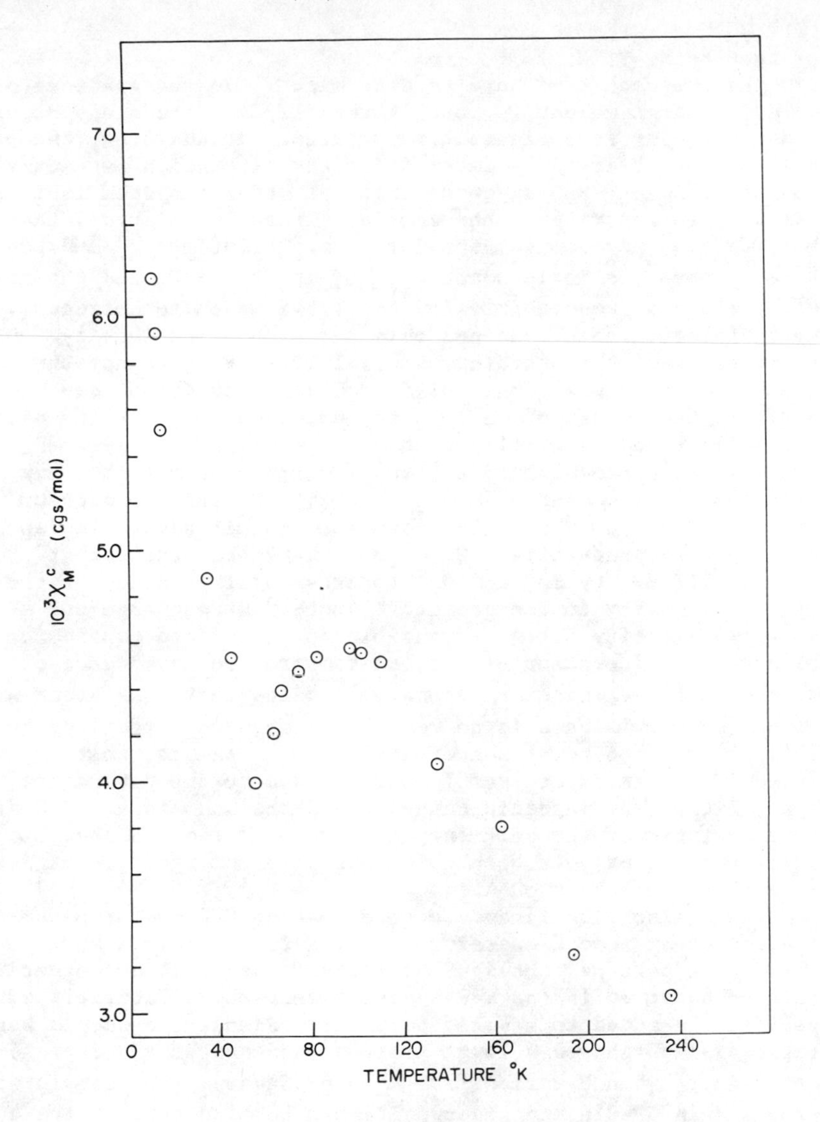

Figure 7. Observed χ_M^c for TPPFeIm · H_2O vs. *temperature at 15 KGauss*

constant below 77°K.

The approach used here to determine χ_i in the presence of a field dependent moment is usually required for the study of samples containing ferromagnetic impurities. In addition, the very small moment observed by determining the saturation magnetization of TPPFeIm · H_2O suggests that the ordered system is in fact only an impurity in the sample. It could be argued that the $[TPPFeIm]_n$ system is behaving like $[TPPFe(ImH)_2]^+$ and posses a temperature independent μ_{eff} of about 2.3 BM and the impurity alone is responsible for the field dependent effects. The analyses certainly do not rule out such a possibility. On the other hand, the starting material $TPPFeBF_4$ does not show any χ_d and simply grinding the solid product which is not expected to effect the amount of an impurity does change the value of χ_d. It is also rather significant that the treatment of $TPPFeBF_4$ with imidazole produces $TPPFe(ImH)_2{}^+$ which does not show any field dependent moment and the treatment of $TPPFeBF_4$ with OH^- produces $(TPPFe)_2O$ which also does not exhibit any field dependent magnetic properties. We do not therefore expect that treatment of $TPPFeBF_4$ by ImH and OH^- together will produce a ferromagnetic impurity in the product. Another strong argument against an impurity being responsible for the field dependence follows a consideration of the low temperature dependence of χ_M^c. Below 60°K the value of χ_M^c drops well below that consistent with a μ = 2 system. There is no way that a magnetic impurity can subtract from the total susceptibility of a sample; that is, the minimum χ for an impure sample must be due to the paramagnet itself. Either the magnetic behavior and the anomaly at 60°K is characteristic of the bulk sample or we must require that the impurity and the polymer both <u>coincidentally</u> undergo some magnetic phase change at about 60°K.

It is also significant to note that no EPR spectrum could be observed at room temperature or 77°K for TPPFeIm · H_2O. However even 1 part per thousand of either iron metal or magnetite could be observed in the EPR spectrum when those materials were specifically added to a solid porphine. Finally, a sample was prepared which showed a large field dependence of χ_M^c (i.e. 133% change in χ_M^c when H varied from 2 to 6 KGauss). The sample was suspended in CH_2Cl_2 and gaseous HCl was bubbled through the suspension for 1 min. The solid immediately dissolved. The solution was not disturbed for 5 min. and then the solvent was removed in a stream of N_2. The solid was dried <u>in vacuo</u> at room temperature and was <u>not purified</u> or additionally treated in any manner. The final solid as was obtained was examined magnetically and the χ_M^c was observed to change by only 6.3% between high and low field. Furthermore, the χ_M^c observed corresponded to a μ_{eff} between 5.6 and 5.8 BM/Fe. The visible spectrum indicated that the treatment with HCl had converted the polymer to TPPFeCl [known μ_{eff} = 5.9 BM (<u>21</u>)].

We conclude that the field dependent magnetic behavior of the sample is characteristic of the polymer because mild treatment which breaks up the polymer also eliminates the field dependence of χ_M^c. The very high susceptibilities and field dependence observed indicates that $[TPPFeIm]_n$ is capable of magnetic ordering. The fact that the field dependent magnetization can be saturated (at room temperature) eliminates antiferromagnetic coupling and implies that ferro- or ferri- magnetic coupling is occurring. The temperature dependence of χ_d does not allow the straightforward assignment of either ferro- or antiferromagnetism to this system. A ferrimagnetic model requires the presence of two different types of magnetic sites for the iron with an antiferromagnetic interaction between them (20). It is extremely difficult to envision how this could occur for this system. In particular, interchain interactions can be ignored because of the large separation between chains required by the size of the porphine ring. However, the unusual properties observed may be due to the linear structure of the polymer.

The problem of unidimensional magnetically ordered systems has been of considerable theoretical and experimental interest for several years. This is in part due to the proof by Mermin and Wagner (22) that an infinite one dimensional system at finite temperature cannot exhibit long range order and thus cannot act as a perfect ferro- or antiferromagnet. It is however still possible for short range interactions to occur which can lead to both types of magnetic ordering. Several groups have subsequently discussed the theory of these systems (23,24). Most of the experimental work has concerned antiferromagnetic systems (25) but $CsNiF_3$ has been found to act as one dimensional ferromagnet by susceptibility and neutron diffraction studies (26). The results observed for $[TPPFeIm]_n$ appear too complicated to attempt any quantitative fit to the theories at this time. However the very low saturation moment derived from the field dependent magnetization of $[TPPFeIm]_n$ is consistent with the occurrence of only weak, short range ordering of the ferromagnetic type. If only a small portion of the bulk sample is ordered then the remainder of the iron sites could continue to exhibit normal paramagnetism and contribute to a field independent moment of reasonable magnitude as observed. This model could also explain the extreme variability of χ_d between samples because the conditions of precipitation and presence of solvent could greatly change the extent of the small ordering that occurs.

In order to better ascertain the applicability of a linear ferromagnetic model for $[TPPFeIm]_n$ we will require considerably more information derived from low temperature susceptibility and Mossbauer studies. Knowledge regarding the μ_{eff} below 60°K and the nature of the anomaly about 60°K is currently being sought.

The facts that the polymer can be prepared and the imidazolate bridges allow metal-metal interactions to occur, support our interest in this system. Thus our current study of the role of imidazole in oxidation reduction reactions involving iron porphyrins may provide a model for part of the biological electron transport chain.

We thank Arthur Tauber at the U.S. Army Electronics Command for the measurement of susceptibility below 77°K and the Public Health Service for support of this work through Grant AM-11355.

Literature Cited

1. Cohen, I. A., Jung, C. and Governo, T., J. Amer. Chem. Soc. (1972) 94, 3003.
2. Abbreviations: TPP = the tetraphenylporphine dianion, $C_{44}H_{28}N_4^{=}$; Im = the imidazolate anion, $C_3H_3N_2^-$; χ_i = the field independent magnetic susceptibility; χ_d = the field dependent magnetic susceptibility; $\chi_{i\ calc.}$ = the field independent magnetic susceptibility calculated from an assumed moment; χ_M^c = the observed corrected molar magnetic susceptibility.
3. (a) Nicholas, M., Mustacich, R. and Jayne, D., J. Amer. Chem. Soc. (1972) 94, 4518. (b) Hoffman, A. B., Collins, D. M., Day, V. W., Fleischer, E. B., Srivastava, T. S. and Hoard, J. L., ibid (1972) 94, 3620. (c) Cohen, I. A., ibid (1969) 91, 1980. Also see references found within (a) and (b).
4. Bolton, W. and Perutz, M. F., Nature (1970) 228, 551.
5. Dickerson, R. E., Tarkano, T., Eisenberg, D., Kallai, O. B., Samson, L., Cooper, A. and Margoliash, E., J. Biol. Chem. (1971) 246, 1511.
6. Ogoshi, H., Watanabe, E., Yoshida, Z., Kincaid, J. and Nakamoto, K., J. Amer. Chem. Soc. (1973) 95, 2845.
7. LaMar, G. N. and Walker, F. A., ibid (1973) 95, 1782.
8. Duclos, J. M., Bioinorganic Chem. (1973) 2, 263.
9. Coyle, C. L., Rofson, P. A. and Abbott, E., Inorg. Chem. (1973) 12, 2007.
10. LaMar, G. N. and Walker, F. A., J. Amer. Chem. Soc. (1972) 94, 8607.
11. Kolshi, G. B. and Plane, R. A. ibid (1972) 94, 3740.
12. Collins, D. M., Countryman, R. and Hoard, J. L., ibid (1972) 94, 2066
13. Epstein, L. M., Straub, D. K. and Maricondi, C., Inorg. Chem. (1967) 6, 1720.
14. (a) Baraniak, E., Freeman, H. C. and Nockolds, C. E., J. Chem. Soc. (1970) A, 2558. (b) Eilbeck, W. J., Holmes, F. and Underhill, A. E., ibid (1967) 757. (c) Yamane, T. and Davidson, N., J. Amer. Chem. Soc. (1960) 82, 2118.
15. Freeman, H. C., Advan. Protein Chem. (1967) 22, 257.
16. Jarvis, J. A. J. and Wells, A. F., Acta Crystallogr. (1960) 13, 1028.
17. Brown, G. P. and Aftergut, S., J. Polym. Sci. (1964) 2, Part A-2, 1839.
18. Inuve, M., Kishita, M. and Kubo, M., Inorg. Chem. (1965) 4, 626.

19. Williams, D. E. G., "The Magnetic Properties of Matter", p. 99, Elsevier Publishing Co., Inc., New York, N. Y. 1966.
20. Morrish, A. H., "The Physical Principles of Magnetism", pp. 486-498, John Wiley and Sons, Inc., New York, N. Y. 1965.
21. Maricondi, C., Swift, W. and Straub, D. K., J. Amer. Chem. Soc. (1969) 91, 5205.
22. Mermin, N. D. and Wagner, H., Phys. Rev. Lett. (1966) 17, 1133.
23. Kondo, J. and Yamaji, K., Prog. Theor. Phys. (1972) 47, 807.
24. Bonner, J. C. and Fisher, M. F., Phys. Rev. (1964) 135, A640.
25. For a few examples see: (a) McElearney, J. N., Merchant, S. and Carlin, R. L., Inorg. Chem. (1973) 12, 906. (b) Jeter, D. Y. and Halfield, W. E., J. Inorg. Nucl. Chem. (1972) 34, 3055. (c) Hutchings, M. T., Shirane, G., Birgeneau, R. J. and Holt, S. L., Phys. Rev. B. (1972) 5, 1999. (d) Sweeney, W. V. and Coffman, R. E., Biochim. Biophys. Acta (1972) 286, 26. (e) Casey, A. T., Morris, B. S., Sinn, E. and Thackeray, J. R., Aust. J. Chem. (1972) 25, 1195. (f) Griffiths, R. B., Phys. Rev. (1964) 135, A659.
26. (a) Steiner, M. and Dorner, B., Solid State Comm. (1973) 12, 537. (b) Steiner, M., ibid (1972) 11, 73. (c) Steiner, M. and Dacks, H., ibid (1971) 9, 1603. (d) Steiner, M., Kruger, W. and Bakel, D., ibid (1971) 9, 227. (e) Steiner, M., Z. Angew. Physik (1971) 32, 116.

17

Local and Collective States in Single and Mixed Valency Chain Compounds

P. DAY

University of Oxford, South Parks Rd., Oxford OX1 3QR, England

Abstract

Compounds containing chains of metal atoms are found with four distinct classes of structure, each of which, as a result of the widely varying strength of the metal-metal interaction, has associated with it its own characteristic pattern of physical properties. Thus one may have either single or mixed valency phases with near neighbor contact between metal ions formed either directly or through anion bridges. The optical, X-ray photo electron, magnetic and transport properties of examples of each class are surveyed, with particular emphasis on work carried out at Oxford, to highlight the relative importance of single center and collective excited states in each category, in relation to their structures and the magnitude of the metal-metal interaction. It is pointed out that a collective description may be appropriate for magnetic and electronic excitations even without electron delocalization.

1. Introduction

Only in the last few years have inorganic chemists interested in the electronic structures of metal complexes paid much attention to the consequences of the interactions which can take place between molecules and ions placed next to each other in the solid state. At their weakest such interactions may manifest themselves only in relatively subtle shifts and splittings of electronic absorption bands, and in magnetic ordering at low temperatures. On the other hand, strong intermolecular interactions lead to spectacular color changes or the appearance of new absorption bands not found in the spectra of the isolated molecules or ions. Finally, in extreme cases the entire range of physical properties of the crystal may be profoundly different from those expected of ordinary ionic or molecular solids. Of the latter situation, the most famous instance is certainly the one-dimensional metallic conducting behavior of the partially oxidized platinum compounds such as $K_2Pt(CN)_4Br_{0.3}\cdot 3H_2O$. The exceptional properties of this material have served to focus attention on metal atom chain compounds in general.

For a number of years we ourselves have been interested in both magnetic and charge transfer interaction effects in inorganic crystals. The purpose of the present paper is to bring together some of these observations, primarily concentrating on one-dimensional examples, to exemplify how the optical, magnetic and electron transport properties are influenced by the strength of the metal-metal interaction. Metal atom chain compounds are found with a great variety of structures and are formed by a large number of different elements. It may be useful therefore to attempt some general classification of the pattern of physical behavior characteristic of each structure type, and of the strength of the metal-metal interaction. One reason why this class of materials is so interesting is that they span the entire range from localized to collective behavior and thus should be a good testing ground for theoretical models of the solid state.

2. Localized and Collective States

Before discussing the various classes of metal chain compound, a general point which needs to be touched on concerns the meaning of the words 'localized' and 'collective' as used in this context. Unlike solid state physicists, inorganic chemists are not usually accustomed to constructing wavefunctions which are invariant to the operations of translation within a periodic lattice. Nevertheless, if a crystal such as $K_2Pt(CN)_4$ absorbs a photon of such an energy that one of the constituent complex anions undergoes a ligand field transition, it is an inescapable fact that we do not know which of the 10^{23} anions in the crystal has been excited, and that as a result, the excited state wavefunction must allow an equal probability of each ion being excited: in other words the wavefunction is a Bloch function. In one sense the state which it describes is 'collective,' since the excitation belongs to the whole lattice. How far and how fast it can move in practice, however, depends on the magnitude of inter-ionic coupling or transfer integrals compared with either intra-ionic electron correlation or repulsion effects on the one hand, or electronic-vibrational interactions on the other. Crudely, if the excitation is capable of being transferred from molecule to molecule faster than each molecule can accommodate its geometry to the excitation, then the excitation, or 'exciton,' 'belongs' to the whole lattice.

Otherwise, its range of travel is limited to a few adjacent molecules (or finally, only to one), and it is localized.

Delocalization of excitation, however, is far from implying delocalization of electrons, or the occurrence of electronic conductivity. Exciton migration over hundreds of angstroms is familiar in aromatic molecular crystals, for example, yet in their ground states they are excellent insulators. For conductivity it is obvious that we need to form crystal states in which the electron occupancy at different sites has been altered, e.g., creating $Pt(CN)_4^{3-}$ and $Pt(CN)_4^-$ in $K_2Pt(CN)_4$. In ordinary ionic and molecular crystals such processes require a large energy input, and it is therefore inappropriate to talk of collective electron states in these materials. Interionic electron transfer states are only formed readily in solids containing an integral number of electrons per atom when the electron concentration is high enough to screen an excited electron from the positive hole left on the ion from which it originated. In contrast, if different sites in the crystal are already occupied by different numbers of electrons when the lattice is formed, much less energy is needed to form states in which these sites are merely interchanged. It may then be possible to form conducting states at lower electron concentrations, or with smaller overlap between ions, than one would need to generate energy bands of finite width in stoichiometric single valence solids. This is the significance of mixed valency, as we shall see in the survey of metal ion chain compounds which follows.

3. Types of Transition Metal Chain Compound

Transition metal compounds in which each cation has only two nearest neighbors are formed both in single and mixed valency situations. In turn, each of these may be formed either with or without anion bridges between the neighboring cations. Some examples of each type of compound are given in Table 1. The principal class of anion bridges single valence chains is that of the hexagonal perovskites ABX_3. When X is a halide ion and A a univalent ion the lattice can be thought of as made up of a close packed array of A and X, if they are of comparable size, e.g., Cs^+ and Cl^-. Now in a close packed assembly of ions the number of octahedral holes equals the number of close packed atoms, but in the system AX_3 only one quarter of the octahedral holes are surrounded exclusively by X^- ions and are hence available for occupation by the B^{2+} cations. The ABX_3 stoichiometry would then be generated by filling all the holes of this type. If the lattice of (A+3X) is cubic close packed we have the cubic perovskite structure formed by many oxides and fluorides. However another possibility is hexagonal close packing of (A+3X). In that case the octahedral BX_6 groups are stacked in columns sharing pairs of opposite faces so that the

Table 1. Examples of Transition Metal Chain Compounds

	Single Valence	Mixed Valence
Anion Bridged	Hexagonal perovskites, $ABX_3:3d^n$	$d^8:d^6$ square planar/octahedral halides
	$AMCl_3$(A = alkali metal or NR_4^+,	$PtenCl_3$
	M = V, Cr, Mn, Fe, Co, Ni)	Wolframs red salt
	$CsMCl_3.2H_2O$(M = Mn, Fe, Co)	$d^{10}:d^8$ linear/square planar halides
		$CsAuCl_3$
Direct	Square planar $3d^8$, $4d^8$, $5d^8$	Square planar $5d^{8-n}$
	$M(DMG)_2$ (M = Ni, Pd, Pt)	$A_2Pt(CN)_4X_n$
	K_2PtCl_4, $PtenCl_2$	$A_{2-n}Pt(CN)_4$
	$Pt(NH_3)_4PtCl_4$	$A_{2-n}Pt(C_2O_4)_2$
	$APt(CN)_4$	
	$Ir(CO)_2acac$	

B to B separation is much smaller parallel to the *c*-axis of the hexagonal unit cell than perpendicular to it. If the BX_6 octahedra were undistorted, and if the ionic radius of A^+ were equal to that of X^- the ratio of the nearest neighbor B-B distance along the stacks to the distance between stacks would be 2.449. In many examples, however, the ratio is larger since one can use cations such as $N(CH_3)_4^+$ which are bigger than X^-. On the other hand, with undistorted BX_6 the nearest neighbor B-B distance is $2/\sqrt{3}$ times the B-X bond length. It is clear therefore that the anions will play an important role in any exchange processes between the B ions. Whatever the pathway of the B-B interaction though. A convincing demonstration of its one-dimensionality is provided by the ratio of intra-chain and inter-chain exchange integrals in a compound such as $N(CH_3)_4MnCl_3$, which has been determined as 10^3 (2).

The other class of single valence chain compounds in which anion bridging plays a role are the hydrated ternary transition metal halides. Here the coordination of each bivalent metal ion is again octahedral, consisting of four halide ions and two *cis*-water molecules (3). The octahedra are joined into chains through approximately 180° bridges involving the *trans* halide ions.

Halide bridges between cations of differing oxidation state are mainly confined to situations in which one of the cations has a low spin d^8 configuration, and hence basically square planar coordination. The vacant sites perpendicular to the plane containing the ligands may then be occupied by halide ions which are themselves coordinated to another cation. In the most common examples of this type, the second cation has a low spin d^6 configuration and thus octahedral coordination. The halide bridges may also be the terminal groups of linearly coordinated d^{10} cations such as Au^I. In fact, a famous example of the latter is the black compound known as Wells' salt, which has the empirical formula $CsAuCl_3$. The structure, determined many years ago by Elliot and Pauling (4) contains chains of alternating linear $AuCl_2^-$ and square planar $AuCl_4^-$. However, each $AuCl_2^-$ also has four chloride ions from four $AuCl_4^-$ coordinated at right angles to its principal axis, so *in toto*, the lattice could be viewed as a distorted version of the cubic close packed perovskite.

To achieve a close enough approach between metal ions for direct interaction between them to outweigh interactions through bridging ligand groups, the most favorable situation is clearly to have complexes which are coordinatively unsaturated, so that another metal ion, or complex, can act, as it were, as a ligand. The most familiar examples of coordinative unsaturation are square planar complexes, so it is among low spin d^8 compounds that some of the most spectacular intermetallic interaction effects are found, both in single and mixed valency compounds. Also, because it is the *trans* sites in such a geometry, which are available for coordination, polymerization is bound to give linear systems. Single valence examples are found with $3d^8$, $4d^8$ and $5d^8$ configurations, both where the units are neutral molecules like $PtenCl_2$, anions like $Pd(CN)_4^{2-}$ or alternating anions and cations, as in Magnus' Green Salt. In all of these, important intermetallic interaction effects are observed, as we shall indicate later. At the present time examples of direct metal-metal interactions in mixed valence chains are only known for systems based on partial oxidation of square planar complexes in the third transition series. We and others have made repeated efforts to prepare Ni and Pd analogues of the partially oxidized Pt chain compounds, but with no success. Even attempts to dope $K_2Pt(CN)_4Br_{0.30}\cdot 3H_2O$ with $Pd(CN)_4^{2-}$ or $Ni(CN)_4^{2-}$ by co-crystallizing it in the presence of large excesses of these two ions do not lead to any detectable incorporation of $3d^8$ or $4d^8$ complex. Whether this is because the greater radial extension of the 5d orbital is needed to obtain sufficient overlap between the cations to stabilize the band structure, or whether it is connected with the smaller nd-(n+1)p separation in the third transition series we cannot say.

Following this brief and generalized survey of the main structure types of the metal chain compounds, some account can now be given of the physical properties associated with each type. In order to avoid this having too much of the character of a review, reference will mainly be made to work on chain compounds which has been carried out in Oxford in the last few years.

4. Single Valence Metal Chain Compounds

(a) Anion Bridged Compounds. An outline of the properties of single valence metal chain compounds is given in Table 2, which highlights the differences found between compounds containing anion bridges rather than directly interacting metal ions. The key observation is that none of the anion bridged compounds have any low energy excited states of metal-to-metal charge transfer type. Since, as we have noted, it is mixing of this kind of state into the ground state which leads to collective electronic behavior, it is no surprise that all known examples are insulating, with localized ground states. Ligand-to-metal charge transfer states exist in the ultraviolet, however, and insofar as these involve the bridging ligands, their mixing into the ground state provides a mechanism for superexchange, leading to magnetic ordering at low temperatures. At lot has been written about one-dimensional magnetic ordering, but a single representative example will illustrate some of the characteristics of this kind of system.

Unlike $CsCuCl_3$, which distorts from the hexagonal perovskite structure so as to provide each Cu atom with four short and two long bonds, the other first transition series ion which

Table 2. Properties of Single Valence Transition Metal Chains

	Anion Bridged	Direct Interaction
Optical		
(a) Locally excited (ligand field)	Weakly perturbed by exciton-magnon interaction (unusual temperature dependence of intensity)	Vibronic transitions strongly perturbed by UV charge transfer states
(b) Collective (d→p or charge transfer)	None at low energy Usual halide → metal transitions only	Large Davydov shifts when E∥ chain
Magnetic	Ferromagnetic or antiferromagnetic at low temperatures Weak superexchange	Diamagnetic (anisotropic?)
Transport	Insulator	Semiconductor (probably extrinsic)

customarily shows strong Jahn-Teller distortion, Cr^{II}, forms a compound $CsCrCl_3$ which, at least above 172 K, has an undistorted hexagonal perovskite structure (5). That one-dimensional spin correlations are important in this compound follows at once from the temperature variation of the magnetic susceptibility which has a broad maximum near 170 K, then decreases with the onset of antiferromagnetism. In fact though, three dimensional magnetic ordering does not take place until the Neel temperature of 16 K is passed, so between 170 and 16 K only antiferromagnetic interactions within the chains are of any importance. An even more direct measure of the one-dimensionality of the spin correlations is the dispersion of the spin-wave excitations parallel and perpendicular to the chain (6). Just as we have to write the electronic excited state wavefunctions of a crystal as Bloch functions, so as to allow the excitation an equal probability of residing on any of the constituent ions, so, in the same way, must a magnetic excitation (that is, the deviation of a spin from the direction it would have in the totally ordered lattice) be delocalized. Spin waves of differing wavelengths (or wave vectors) have different energies, which can be determined by inelastic neutron scattering, thus building up a picture of the dispersion curve experimentally. The experimental spin-wave dispersion of $CsCrCl_3$ near 4 K parallel and perpendicular to the chains is shown in Figure 1. Although the results do not cover the entire Brillouin zone, because of the small size of the crystal available to us, they do show very clearly that the dispersion perpendicular to the c-axis is essentially zero. The exchange integral between neighboring Cr ions in the basal plane is therefore negligibly small whilst within the chains curve fitting to the experimental points on the dispersion curve gives $J_1 \sim 26.7\ cm^{-1}$.

Since customarily there are no charge transfer states in the visible or near ultraviolet in this class of compound, the lowest energy excited states are ligand field in type, and can therefore be described as Frenkel, or tight-binding, excitons. Many ligand-field excited states have spin projections different from the ground state and transitions to them should consequently be electric dipole forbidden. It has been known for a number of years, however, that such transitions can be rendered allowed in antiferromagnetic compounds by using the exchange interaction to couple an exciton formed with decrease of spin (e.g., sextet to quartet) with a spin deviation among the rest of the ions in the lattice, which remain in their ground states. The most famous examples of such 'exciton-magnon' combination bands are found in manganese salts (7), so it is interesting to see how the effect manifests itself in Mn^{II} salts of the hexagonal perovskite type, which contain chains of antiferromagnetically coupled ions like that found above in $CsCrCl_3$.

In overall appearance the ligand field spectra of Mn^{II} chain compounds are much like those of other six-coordinate Mn^{II}

complexes, confirming again that the metal-metal interactions in anion bridged single valence chains are weak. Two features stand out, however. First, the oscillator strengths of the transitions are greater than in non-bridged ('outer-sphere') salts where magnetic interactions are negligible. Second, the temperature dependences of the oscillator strengths follow a rather curious pattern, quite unlike that of the usual 'coth' plot for a simple vibronically induced transition. Some experimental examples are shown in Figure 2(a) (8). In all cases the oscillator strength increases rapidly at first when the temperature is raised above 4.2 K, then passes through a broad maximum at a temperature which varies from one compound to another, but appears to be the same for all the bands in each compound. Then after dropping slightly it finally increases slowly and monotonically towards room temperature. This variation has some general resemblance to the curve of susceptibility vs. temperature for these materials (9). In fact, the results in Figure 2(a) stimulated Tanabe and Ebara to calculate the intensity, frequency and linewidth variation of the magnon sidebands in linear antiferromagnets, with the result shown in Figure 2(b) (10). It can be seen that the overall form of the variation is nicely reproduced. Tanabe's theory requires that the temperature at which the broad maximum occurs in the intensity is approximately $|J/k|S(S+1)$. For the three compounds we investigated, the values of J/k derived from the spectra are listed in Table 3, which also shows that for the two compounds for which we have susceptibility data (9,11), the level of agreement between the latter and the optically derived value of J/k is very satisfactory.

(b) Directly Interacting Metal Ions. When metal ions are brought close enough together in a chain to interact directly, without the intermediary of a bridging ligand, the effect of the neighboring ions on the electronic states of each metal ion is naturally increased. Unfortunately no examples of such chains are known in which the constituent ions have unpaired spins because, as we noted already, they are all based on square planar d^8 complexes. The single exception known to us is $Pt(NH_3)_4CuCl_4$ which, however, contains chains of alternating $Pt(NH_3)_4^{2+}$ and $CuCl_4^{2-}$ and so has a diamagnetic ion between each pair of d^9 ions. Consequently it has a Neel temperature of only 0.5°K, from which near neighbor exchange integral of about 0.25 cm^{-1} can be estimated (12).

Turning to optical properties, one finds two different types of situations in the single valence chains with directly interacting metal ions, depending on whether the lowest energy excited states of the constituent complexes are forbidden or allowed. The former are most likely to be ligand field states, which are parity forbidden from the ground state in centrosymmetric complexes like $PtCl_4^{2-}$ and may in addition be

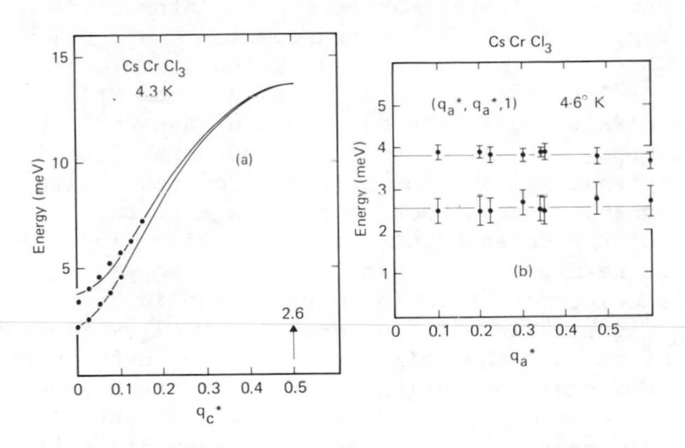

Figure 1. Dispersion of spin waves propagating (a) parallel and (b) perpendicular to the Cr chains in the linear antiferromagnet $CsCrCl_3$

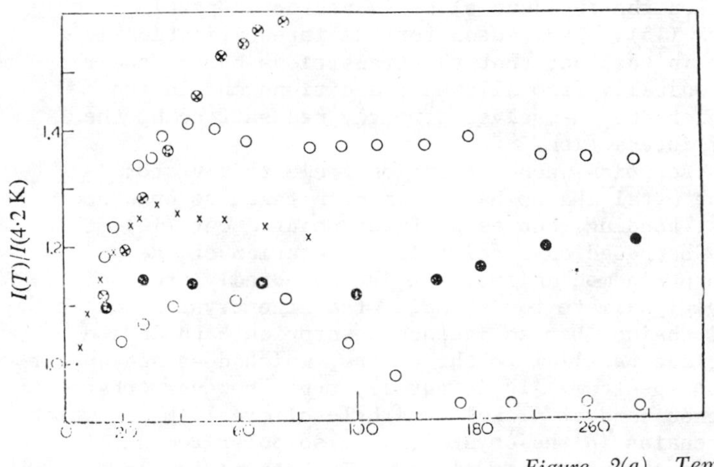

Figure 2(a). Temperature dependence of oscillator strengths of ligand field transitions in linear antiferromagnets $N(CH_3)_4$ $MnCl_3$ and $CsMnX_3 \cdot 2H_2O$ (X = Cl, Br) (8)

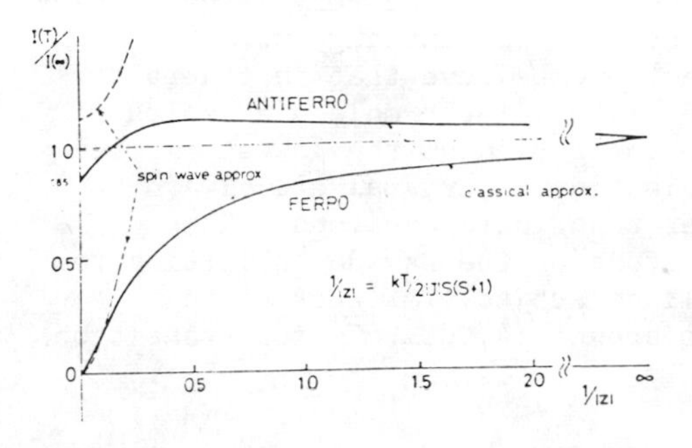

Figure 2(b). Calculated temperature variation of oscillator strengths of exciton–magnon transitions in linear antiferromagnets and ferromagnets (10)

spin-forbidden. The latter may be either 'atomic' (e.g., 5d → 6p in Pt^{II} complexes) or charge transfer.

Cases where the lowest excited states are of ligand field type are the salts K_2PtCl_4, $PtenX_2$(X=Cl,Br) and $Pt(NH_3)_4PtCl_4$, Magnus' Green Salt (MGS), although some of the earlier workers thought that the unusual color of the last example pointed to 'metal-metal bonding,' and the existence of new excited states which could not be traced back in parentage to any intramolecular transitions of the constituent anion or cation (13). At first sight it does seem peculiar that MGS should be green, since $Pt(NH_3)_4^{2+}$ is colorless, both in solution and in $Pt(NH_3)_4Cl_2$, while $PtCl_4^{2-}$ is pink in solution and in K_2PtCl_4. Nevertheless, by looking at the way the polarized crystal spectra vary along a series of MGS analogues $Pt(RNH_2)_4PtCl_4$ as the R group is made more bulky, we showed some years ago (14) that the visible bands in MGS are all traceable to ligand field transitions appearing in the spectrum of $PtCl_4^{2-}$ in K_2PtCl_4, albeit with red shifts up to 4000 cm^{-1}. Examples of these are shown in Figure 3. They are also considerably intensified compared to K_2PtCl_4, most dramatically when the incident electric vector is parallel to the metal stack (15). The reason for the intensification is assumed to lie in the fact that the transitions borrow their intensity vibronically from allowed transitions out in the ultraviolet which are themselves strongly red-shifted by the intermolecular interaction.

Intense absorption when the incident electric vector is parallel to the metal chains has often been taken as evidence for metal-metal bonding, but as we first pointed out (16), this conclusion may be based on a false interpretation of the available absorption mechanisms. It is undoubtedly true, of nickel dimethylglyoximate for example, that the crystal, which contains metal chains, has an intense absorption band in the visible, polarized parallel to the chains, which does not appear in the solution spectrum. It is equally true, however, that the first intense band of nickel N-methyl-salicycaldimine, which does not form chains in the crystal, is also polarized perpendicular to the molecular planes. In both cases the transition is most probably a $d_{z^2} \rightarrow \pi^*$ charge transfer, but the feature distinguishing them is that when the molecules are stacked plane to plane the transition dipoles are all parallel, so interactions between them are maximized and very large Davydov shifts occur. No intermolecular electron exchange need be invoked in either case. In fact, we believe that in general, when the lowest energy excited states of the molecular units are polarized perpendicular to their planes, the lowest excited states of the stacks of molecules in the crystal are always neutral Frenkel excitons rather than ionic excitons. This prejudice is based on the magnitude of the Davydov splitting to which the neutral excitons will be subject (a 'back of an envelope' calculation suggests around 14,000 cm^{-1} for transition

Table 3. Temperature Dependence of Intensity of Magnon Sidebands in Linear Antiferromagnets

	T_{max} (exp.)	J/k	J/k (susceptibility)
$CsMnCl_3.2H_2O$	20–40 K	3.3 K	3.0 K
$CsMnBr_3.2H_2O$	30–60	4.5	-
$(NMe_4)MnCl_3$	50–80	6.7	6.3

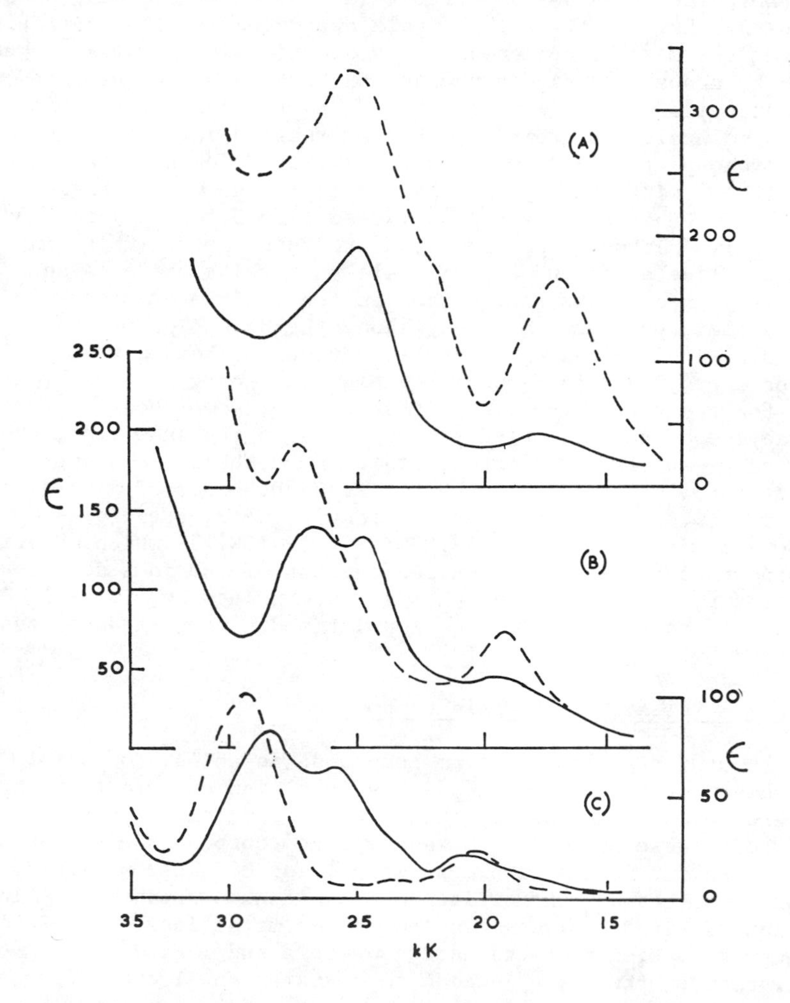

Figure 3. Polarized absorption spectra of (a) $Pt(CH_3NH_2)_4PtCl_4$, (b) $Pt(C_2H_5NH_2)_4PtCl_4$, and (c) K_2PtCl_4. Dotted lines are E c, full lines E c (15).

dipole length of 1Å in a stack of molecules separated by 3.25Å). On the other hand, if the parent transition is polarized within the molecular planes, Davydov splitting of its neutral excitons in the crystal is smaller (actually one half of the other in the point dipole approximation) and ionic excitons may become observable. An example here is the work of Martin and his colleagues on $PtenX_2$ (17,18).

One way of verifying that 'new' electronic transitions appearing in chain compounds with polarizations parallel to the stacks are actually Davydov components of transitions which take place in the ultraviolet in the isolated units is to determine how their energy varies as one changes the intermolecular spacing in the stacks. In the point dipole approximation the separation between the Davydov components is proportional to $|M|^2/R^3$, where M is the transition dipole moment and R the distance between the centers of the molecules (19). A particularly favorable set of molecules to exemplify this expression is that of the tetracyano-platinites. Many salts of this anion contain stacks of $Pt(CN)_4^{2-}$ with their planes parallel, though with changing cation the Pt-Pt spacing can be varied from 3.13Å (Mg) to 3.69Å (Rb) (20). Although the lowest excited state of $Pt(CN)_4^{2-}$ in solution lies at 35,600 cm^{-1}, in all such salts there is an extremely intense absorption band in the visible or near ultraviolet, polarized entirely along the direction of the stacks. With increasing Pt-Pt separation the band moves to higher energy, and in Figure 4 we plot its energy against $1/R^3$. Both for the tetracyano-platinites and the isomorphous tetracyanopalladites, which are included on the same plot, there is a very satisfactory correlation between these two quantities. Furthermore, extrapolating the two straight lines of Figure 4 to $R = \infty$, i.e., to the isolated molecule, we find energies of 44,800 cm^{-1} ($Pt(CN)_4^{2-}$) and 52,900 cm^{-1} ($Pt(CN)_4^{2-}$) which, bearing in mind that the aqueous solution spectra of the ions do not strictly represent those of the isolated molecules, is well within the energy range of the z-polarized allowed transitions assigned by Mason and Gray (21).

5. Mixed Valence Metal Chain Compounds

About forty elements form compounds in which, to satisfy the observed stoichiometry, one has to assign different oxidation numbers to different metal ions of the same element in the same lattice. These so-called 'mixed valence' compounds often have properties far from a simple superposition of those we would predict for each of the oxidation states taken separately. In particular, if the sites occupied by the metal ions of differing valency have similar coordination numbers and ligand environments, the energy needed to transfer an electron from one to the other, i.e., to interchange the valencies, may be very low. Indeed, if the metal ion sites in the lattice of our

Table 4. Properties of Mixed Valence Transition Metal Chains

	Anion Bridged	Direct Interaction
Optical		
(a) Locally excited (ligand field)	Only weakly perturbed when E⊥ chain	No ligand field states seen. Weak allowed transitions when E⊥ chains
(b) Collective (charge transfer or metallic)	Metal → metal charge transfer for E∥ chains	Opaque at all frequencies for E∥ chains. Plasma edge in visible
Magnetic	Diamagnetic	Diamagnetic
Transport	Semiconductor (probably intrinsic)	Metallic at 300K. Insulator at 4K
Mixed Valence Class (ESCA, Mossbauer etc.)	II (Valences trapped)	IIIB (Incipiently II??) (Metal atoms equivalent)

mixed valence compound all turn out to be crystallographically identical, then one may have a non-integral number of electrons at each site, even in the ground state. Searching out the known mixed valence compounds from all corners of the Periodic Table, and correlating their properties with the observed structures, we find that they can be divided into three broad classes (22). If the ligand fields around the ions of differing valence are very different, ionized excitons, or in chemists' language metal-to-metal charge transfer states, lie a long way above the ground state whose properties, along with those of many lower excited states, are very close to being just a superposition of the properties of the constituent ions. This category, called class I, is of little interest here. At the opposite extreme, a truly non-integral number of electrons at each metal ion site, which would be required if they are all crystallographically identical, can only be achieved in a metal if the lattice is continuous. This category we call class III. Between classes I and III lies a continuous spectrum of cases (class II) in which two types of metal ion sites can be distinguished by crystallography, but in which ionized excitons lie close enough to the ground state that they contribute to the optical spectrum in the visible or even the near infrared. Some of the ionized excitons may even have the correct symmetry to mix with the ground state, partly smudging out the formal oxidation numbers defined by counting electrons at the two metal ion sites. We will now examine the properties of the known mixed valence metal chain compounds in the light of this simple classification. Just as in the single valence chains, we distinguish the two cases of anion bridging and directly interacting metal ions.

(a) Anion Bridged Compounds. All the anion bridged Pt and Au compounds have two sets of crystallographically distinct metal ion sites, and may therefore be characterized as containing trapped valences in their ground states, (II,IV) in the former and (I,III) in the latter. X-ray photoelectron (XPS) spectroscopy should be a powerful tool for deciding whether different partial changes must be assigned to the ions or each type of site, because we might hope to see core ionizations from each type. If we do not (assuming that the reason is not simply the poor resolution of the technique), then the partial charges are equal or, said another way, the electrons are exchanging between the sites more rapidly than the time taken for the core electron to be ejected. Figure 5 shows some XPS spectra of Pt chain compounds in the 4f region (23), including the mixed valence $PtenCl_3$ (en: ethylenediamine). Crystallographically, the latter can be thought of as made up of alternating square planar $Pt^{II}enCl_2$ and octahedral $Pt^{IV}enCl_4$ molecules. Comparison between the spectra of $PtenCl_3$ and its separated constituent molecules is made more difficult by the ease with which $PtenCl_4$ and $PtenCl_3$ reduce in the X-ray beam. However, by carefully

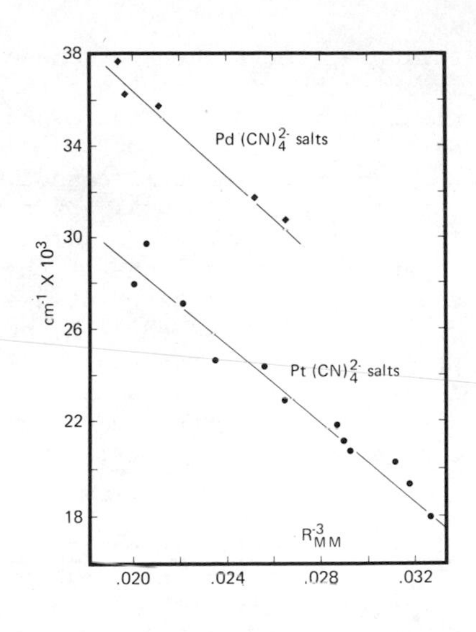

Figure 4. Variation in energy of the lowest intense transition in solid palladocyanides and platinocyanide salts with intermetallic spacing

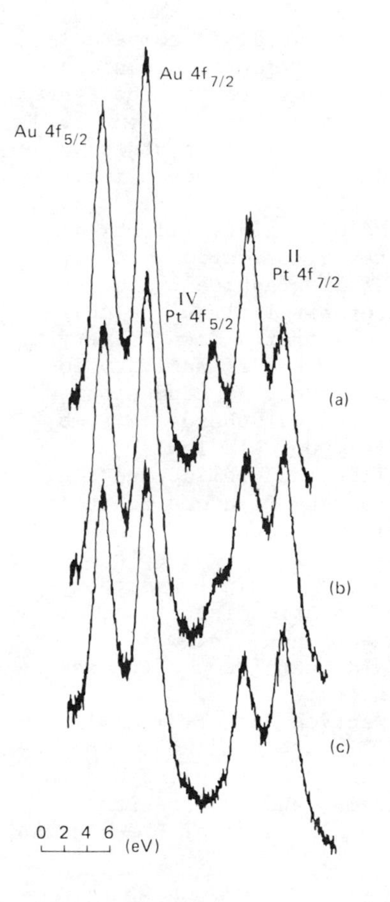

Figure 5. XPS spectra of Pt chain compounds in the Pt 4f region, (a) $PtenCl_4$, (b) $PtenCl_3$, (c) $PtenCl_2$

following the spectra of these two compounds as a function of irradiation time one finds that the $PtenCl_3$ spectrum is indeed close to a superposition of those of $PtenCl_2$ and $PtenCl_4$ (Figure 6).

However, the same cannot be said of the optical properties, for in addition to weak absorption bands, polarized perpendicular to the chains, which do identify closely with locally excited ligand from transitions of the constituents, $PtenCl_3$ and other similar salts all have intense absorption across most of the visible, polarized entirely parallel to the chains (24). This we assign as due to metal to metal charge transfer, i.e., to ionized exciton states, and the compounds belong to class II defined above. Because the valences in the ground state are trapped, though, the crystals are semiconducting rather than metallic.

(b) Directly Interacting Metal Ions. In the single valence chains when the metal ions were brought close enough together to interact directly, instead of through a bridging group, large Davydov shifts of the neutral exciton states showed that the metal-metal interaction was very much increased. A comparable effect on the ionized exciton states of a mixed valence chain might bring them close enough to the ground state for the system to find itself on the other side of a Mott transition, and to have a conducting ground state in which the electron states were truly collective. This is the background to the recent interest in compounds like $K_2Pt(CN)_4Br_{0.30}3H_2O$(KCP) (25). At first sight it has all the characteristics expected of a class III mixed valency compound: despite the non-integral oxidation number, all the Pt atoms appear to occupy crystallographically equivalent sites; with the incident electric vector parallel to the chains it is opaque throughout the visible and infrared down to very low frequencies, and has a plasma edge in the reflectivity in the visible; at room temperature it is a metallic conductor. Unfortunately, if more interestingly, however, there are some features which complicate such a simple view. At low temperature, the compound is not metallic, but semiconducting. There is also evidence of a tendency for the Pt atoms to become inequivalent as a result of long wavelength acoustic phonon instabilities. If such an instability, which showed up by inelastic neutron scattering as a Kohn anomaly in the acoustic phonon dispersion became locked in as a static distortion, it would provide enough reason for the energy gap at low temperature. A crucial experiment would therefore be to examine the crystal structure at liquid helium temperature.

We recently collected X-ray diffraction data from a single crystal of KCP at room temperature, 77° and 4.2°K (26). The arrangement of the diffractometer and cryostat required one to concentrate on one crystal axis at a time, and we chose the $(0,0,\ell)$. Within that limitation we found no sign of Bragg peaks

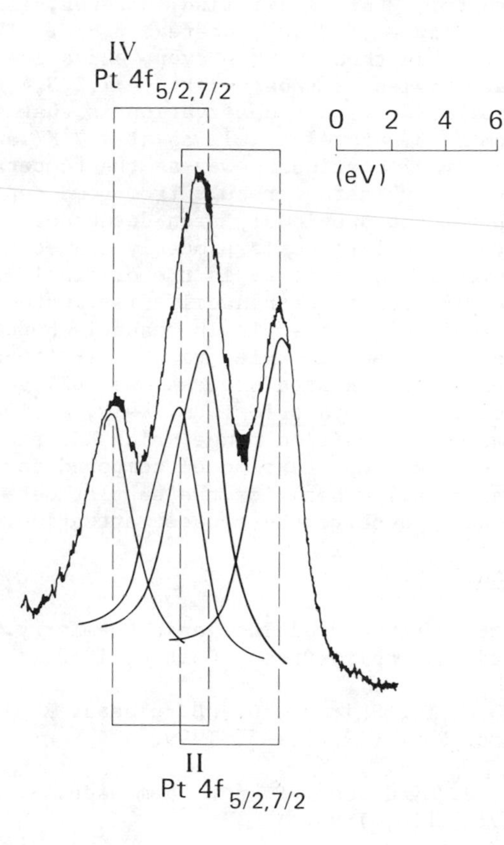

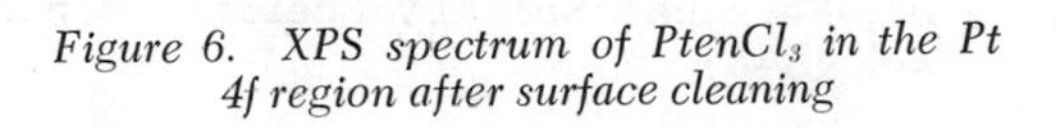
Figure 6. XPS spectrum of $PtenCl_3$ in the Pt 4f region after surface cleaning

corresponding to 'locking in' a Peierls distortion at the repeat distance (6.67 times the Pt-Pt spacing) suggested by the neutron scattering results. What we did find, however, is a set of seven peaks indexing as $(0,0,\varepsilon)$, where $\varepsilon = n/\zeta$. The peaks with n odd are weaker than those with n even, and a least squares fit to the Bragg angles of those with n = 1,2,3,4,6,8,10 yields a value for ζ of 8.97. A key observation is that the peaks are present at room temperature as well as at 4.2°K, and that they do not vary in any discontinuous way as the temperature is lowered. They must therefore result from some superlattice ordering which has not previously been detected. Our present view is that the superlattice is probably connected with ordering among the bromide ions, supposed in the original crystal structure determination to be randomly distributed on three fifths of the available sites within channels between the stacks of $Pt(CN)_4$ groups. From analogies with other 'tunnel' compounds, such as the hexagonal tungsten bronzes, we believe a random distribution of anions is *a priori* unlikely, a view reinforced by the very narrow composition range which KCP exhibits. Since it has become such an important model compound for one-dimensional metallic behavior the subtler details of its structure obviously deserve the closest attention.

Literature Cited

1. Wells, A. F., "Structural Inorganic Chemistry," 3rd ed., p. 375, Oxford University Press, Oxford, 1962.

2. Hutchings, M. T., Shirane, G., Birgeneau, R. J. and Holt, S. L., Phys. Rev. (1972), B5, 1999.

3. Jensen, S. J., Andersen, P. and Rasmussen, S. E., Acta Chem. Scand. (1962), 16, 1890.

4. Elliot, N. and Pauling, L., J. Amer. Chem. Soc. (1938), 60, 1846.

5. McPherson, A. L., Kistenmacher, T. J., Folkers, J. B. and Stucky, G. D., J. Chem. Phys. (1972), 57, 3771.

6. Hutchings, M. T., Day, P., Gregson, A. K., Leech, D. H. and Rainford, B. D., Physical Society Meeting, Manchester, England, January 1974.

7. Sell, D. D., Greene, R. L. and White, R. M., Phys. Rev. (1967), 158, 489.

8. Day, P. and Dubicki, L., J.C.S. Faraday II (1973), 69, 363.

9. Dingle, R., Lines, M. E. and Holt, S. L., Phys. Rev. (1969), 187, 643.

10. Ebara, K. and Tanabe, Y., J. Phys. Soc. Japan (1974), 36, 93.

11. Smith, T. and Friedberg, S. A., Phys. Rev. (1968), 176, 660.

12. Soos, Z. G., Huang, T. Z., Valentine, J. S. and Hughes, R. C., Phys. Rev. (1973), B8, 993.

13. Yamada, S. J. Amer. Chem. Soc. (1951), 73, 1579.

14. Day, P., Orchard, A. F., Thomson, A. J. and Williams, R. J. P., J. Chem. Phys. (1965), 42, 1973.

15. idem., ibid. (1965), 43, 3763.

16. Day, P., Inorg. Chim. Acta Rev. (1969), 3, 81.

17. Martin, D. S., Hunter, L. D., Kroening, R. F. and Coley, R. F., J. Amer. Chem. Soc. (1971), 93, 5433.

18. Kroening, R. F., Hunter, L. D., Rush, R. M., Clardy, J. C. and Martin, D. S., J. Phys. Chem. (1973), 77, 3077.

19. Craig, D. P. and Walmsley, S. H., "Exitons in Molecular Crystals," W. A. Benjamin Inc., New York, 1968.

20. Moreau-Colin, M. L., Structure and Bonding (1972), 10, 167.

21. Mason, W. R. and Gray, H. B., J. Amer. Chem. Soc. (1968), 90, 5721.

22. Robin, M. B. and Day, P., Adv. Inorg. Chem. and Radiochem. (1967), 10, 247.

23. McGilp, J. Chemistry, Part II Thesis, Oxford, 1973.

24. Yamada, S. and Tsuchida, R., Bull. Chem. Soc. Japan (1956), 29, 894.

25. See papers by K. Krogman and H. R. Zeller in this symposium.

26. Griffiths, D., Day, P. and Wedgwood, F. A., unpublished.

18

Evidence for Extended Interactions between Metal Atoms from Electronic Spectra of Crystals with Square Complexes

DON S. MARTIN, JR.

Ames Laboratory of the U.S.A.E.C., Iowa State University, Ames, Iowa 50010

Abstract

Frequently, the square-planar molecular or ionic complexes of platinum(II) or palladium(II) in crystals are aligned directly over one another in one dimensional stacks. The electronic absorption spectra of the crystals for polarized light in favorable circumstances provide information about the crystal interactions and possible intermolecular electron transfers in the solid state. For crystals with large metal-metal separations, >4.1Å, the d-d spectra correlate closely in energies and intensities with solution spectra. The spectra of crystals with $Pt_2Br_6^{2-}$ ions indicate metal-metal electron transfer between metals at 3.55Å, although the transfer may be influenced by the bromide bridges. For Magnus' green salt the spectral changes of $PtCl_4^{2-}$ are consistent with large energy shifts of Frenkel excitons by crystal effects. However, for some molecular complexes with separations of 3.4-3.5Å, electron transfers to ionic exciton states occur.

Introduction

The electronic absorption spectra have been measured recently for single crystals of a number of ionic or molecular square planar complexes of platinum(II). Frequently, such complexes stack face to face in linear arrays in crystals to provide strong anisotropies in the absorption of polarized light and in the electrical conductivities. Our concern here will be for the evidence of transfer of electrons between the platinum atoms. If such electron transfer transitions can be identified, their wave lengths provide directly the energy required for the transfer. Our work to the present has been restricted to the complexes with halide and amine ligands.

The absorption bands for the platinum complexes at room temperature are rather broad; half-widths of approximately 2,000 cm^{-1} are common at 300°K. A transition band is normally characterized by the wave length, λ, or wave number, $\bar{\nu}$, of the maximum, by the intensity, ϵ_{max}, molar absorptivity at the maximum or as the oscillator strength: $f = 4.32 \times 10^{-9} \int \epsilon \, d\bar{\nu}$ and by the temperature dependence of the intensity. For single crystals there is a limit upon the intensity imposed by the practical lower limits of crystal thickness, which has been extended down to 1-20μ. Thus, the studies are limited to bands with ε less than 1,000-2,000 $cm^{-1}M^{-1}$. These are the normally forbidden transitions. Hence, the allowed transitions are not characterized by crystal absorption spectroscopy except in their absence.

The temperature dependence of intensity is an especially powerful diagnostic property. For the d←d transitions, for example, which are symmetry forbidden, an asymmetric vibration may serve as a perturbation to give a vibronic wave function into which is mixed an asymmetric electronic wave function. Thus the transitions may obtain a non-zero transition moment and observable intensity. Since the average amplitude of the vibration, which serves as the perturbation, decreases at lower temperatures, there is normally a decrease in intensity. For a band, excited vibronicly by a vibration of wave number, $\bar{\nu}_i$, the intensity dependence (1) is given by the equation:

$$f(T) = f(0^{o}K) \coth(h\bar{\nu}_i/2kT). \tag{1}$$

However, for a band with a non-zero transition dipole, since the integrated intensity is not temperature dependent, the peak height, ϵ_{max}, will increase as the band narrows at lower temperature.

For the halide and amine-halide complexes of platinum(II) the crystal spectra have indicated the one-electron orbital ordering in Figure 1, which applies to the tetragonal symmetry D_{4h}. The lowest unoccupied orbital is the σ^* orbital based on the $d_{x^2-y^2}$ with b_{1g} symmetry. With the other d-oribtals filled, the ground state will be a $^1A_{1g}$; and d-d transitions to the singlet and triplet states of A_{2g}, E_g and B_{1g} respectively should occur, the triplet states lying below the corresponding singlets. With the high spin-orbit coupling of a heavy Pt-atom the spin designations are not exact, and the states must be described by the double symmetry group, in this case D_4'. Interactions along a stack of square planar ions will normally involve the d_{z^2} orbital, which possesses sigma character about the stacking axis. For complexes with low energy π acceptor orbitals such as cyanide and oxalate, the d_{z^2} orbital may be much higher, and indeed has been considered by Schatz and coworkers (2) to be the highest filled orbital for $Pt(CN)_4^{2-}$. Even in the halide complexes the assignment of the transition to $^1B_{1g}$ state ($\sigma^* \leftarrow d_{z^2}$) is somewhat open to question since the assigned band is surprisingly weak. Below the d orbitals in Figure 1 are shown the orbitals derived from π and σ orbitals on the ligands. Only the odd orbitals are designated for these can provide intense $Pt(\sigma^*)_g \leftarrow L_u$ transitions. A possible alternative assignment of intense transitions may be $6p_z \leftarrow 5d$, for the $6p_z$ orbital is the second lowest unfilled orbital.

Fully allowed molecular transitions in D_{4h} must be $^1A_{2u}$ (z-polarization) or ^1E_u (x,y polarization). All other transitions are seen as a consequence of vibronic and spin-orbit perturbations which mix these states into the wave functions. In a consideration of all the vibrations of an ion such as $PtBr_4^{2-}$ it is found that the $^1A_{2g}$ state is vibronicly forbidden in z-polarization but allowed in x,y. The ^1E_g and $^1B_{1g}$ transitions are allowed in the three polarizations x,y and z. Since the singlet spin functions have the symmetry, A_1', and triplets are A_2' and E_1' in the double group, D_4', the triplet state selection rules are different from those of the singlet states.

For crystalline states in which extended interactions are small, there is negligible overlap between orbitals on different complexes, and electrons are locallized. Excitations are not locallized, however, and other members of the crystal influence transition energies. In such cases the theory of Frenkel type excitons for molecular crystals can apply. The crystal ground state wave function will have the form

$$\Phi^o = \varphi_1{}^o \varphi_2{}^o \ldots\ldots \varphi_N{}^o \tag{2}$$

where φ^o is the ground state molecular wave function and the product is over the N molecules in the crystal.

A wave function for a state in which the i-th molecule of the p-th unit cell is excited is:

$$\emptyset'_{ip} = \varphi^o_{1,1} \ldots \varphi'_{i,p} \ldots \varphi^o_{h,N/h}, \tag{3}$$

where there are h molecules per unit cell.

The crystal states will constitute a band of functions:

$$\Phi'_k = (N/h)^{-\frac{1}{2}} \sum_p \exp(i\underline{k}\cdot\underline{r}_p)\ \emptyset'_{ip}, \tag{4}$$

where k is the wave vector.

However, for the absorption of light the selection rule applies: $\Phi^o \rightarrow \Phi'_0$, i.e. only the transition to k = 0 is allowed. For the case of only one molecule per unit cell, the transition energy is given by (3):

$$\bar{\nu} = \bar{\nu}^o + D' + I' \tag{5}$$

where $\bar{\nu}^o$ is the energy for the free ion.

D' is the difference in van der Waals energy between the excited molecule and the ground state molecule. This term is frequently not well characterized but is intensity independent. The I' term has received extensive investigation since such terms provide the Davydov splitting of states. I' contains a sum of interaction terms, $\sum_n I_{mn}$. Frequently, these terms are taken as the interaction of transition dipoles on the two atoms, m and n, with form.

$$I_{mn} = R_{mn}^{-3} e^2 (x'_m x'_n + y'_n y'_m - 2z'_n z'_m) \quad (6)$$

where R_{mn} is the separation of the molecules m and n, ex'_m is the x component of the transition dipole on the m-th molecule, etc.

x'_m, x'_n and y'_m, y'_n are directed along sets of parallel axes on the two molecules and the z direction is along the axis of centers of the transition dipoles. For a linear stack of molecules separated by 3.4Å and with a transition moment of 1Å in the z or stacking direction the I_{mn} terms of the two adjacent molecules in the stack provide a red shift of 12,000 cm^{-1}. A similar value for an x or y transition moment will give a shift to shorter wave length of 6,000 cm^{-1}. Hence in such stacks, rather large changes in the transition energies can occur which are proportional to the intensity, even with no electron delocalization.

K_2PtBr_4 and K_2PtCl_4: Minimum Extended Interactions

K_2PtBr_4 possesses the K_2PtCl_4 structure shown in Figure 2 with a = 7.35Å and c = 4.33Å. The Pt-Br bond length is 2.45Å. The $PtBr_4^{2-}$ ions stack directly over one another. They are separated by the K^+ ions and are at the relatively large distance of 4.33Å from one another so interactions are small. The absorption with light polarized in the c-direction provides the ionic z-polarization whereas the polarization normal to c, i.e. (a), provides the ionic x,y-polarization. This is therefore an ideal crystalline system to study, for the ion occupies a site of full D_{4h} symmetry so that the situation is not compromised by a lower site symmetry.

The crystal spectra for K_2PtBr_4 are shown in Figure 3. The bands at energies below 29,000 cm^{-1} are then the d←d transitions. The large decrease in intensity at 15°K is characteristic of vibronic excitations. The locations of the maxima in this region correspond closely to the $\bar{\nu}$'s reported for solution spectra of $PtBr_4^{2-}$ (4,5) so apparently the D' and I' terms of equation 5 match closely the solution effects. For a-polarization there is a shoulder at 300°K which is resolved as a peak with vibrational structure at 24,000 cm^{-1} in the 15°K spectrum. This band is completely absent in the c-polarization, and it can therefore be assigned unambigously as $^1A_{2g}$. The band in both polarizations at 27,000 cm^{-1} is assigned as ^1E_g from analogy to the $PtCl_4^{2-}$ spectrum where an A-term in the MCD spectrum is clearly discernable (6,7). The shoulder and peak in each polarization at ca. 17,000 and 19,000 cm^{-1} are presumably components of $^3A_{2g}$ and ^3E_g. The shoulder at 22,700 cm^{-1} with vibrational structure, which is apparent in a-polarization and also is very faint in c-polarization, is assigned as the $^3B_{1g}(\sigma^*\leftarrow d_{z^2})$. With this assignment of the

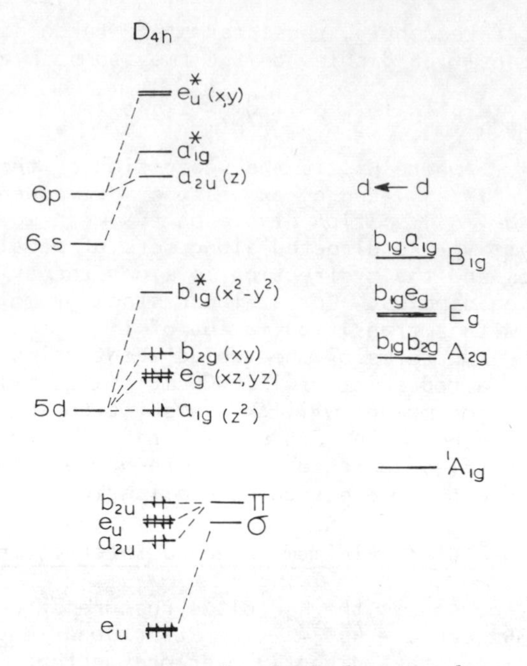

Figure 1. Molecular orbital scheme for platinum(II) complex under D_{4h} *symmetry as in* $PtCl_4^{2-}$ *or* $PtBr_4^{2-}$*. Only the* p *orbitals of the halides are included, and gerade orbitals arising from ligand* π *and* σ *orbitals are omitted. On the right are excited symmetry states arising from* d ← d *transitions from the* $^1A_{1g}$ *ground state. Both singlet and triplet states will occur for each indicated symmetry state.*

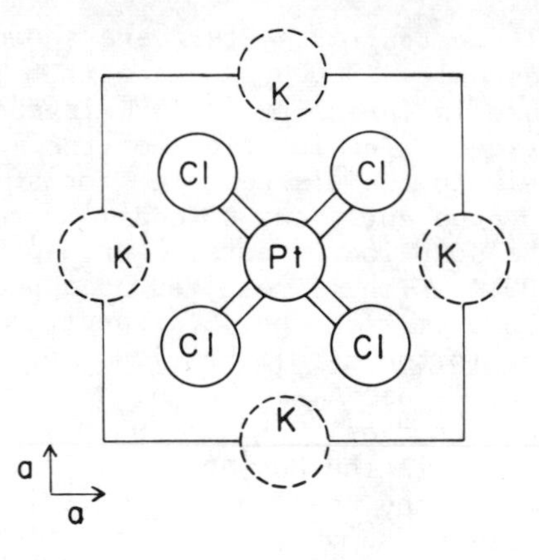

Figure 2. Unit cell for the tetragonal K_2PtCl_4 *structure. K ions are at* z = ½*; others at* z = 0*.*

triplet state the $^1B_{1g}$ state is expected to lie at some 9,000-12,000 cm^{-1} higher energy. Consequently, the transition seen at 33,800 cm^{-1} in c-polarization is assigned to this $^1B_{1g}$ transition. However, the intensity of this transition appears uncomfortably small for this assignment. It is hardly more intense then the spin-forbidden transitions despite its greater proximity to the strong bands from which intensity is borrowed, and the assignment clearly is subject to question.

There is a close correspondence between the d←d spectra of K_2PtBr_4 and K_2PtCl_4 which is clear from Figure 4 where the energies of the transitions are compared. The transition intensities are indicated by the length of lines designating the states, which for the crystals are proportional to log ϵ_{max} at 15°K. It can be seen that each d←d transition in K_2PtBr_4 falls 1,500-2,200 cm^{-1} below a corresponding transition in K_2PtCl_4. In addition, the vibrational structure is observed at 15°K on corresponding transitions. However, the intense bands of the two anions are quite different. First of all, the solution spectrum $PtBr_4^{2-}$ has a band at 31,500 cm^{-1}, ϵ - 600 $cm^{-1}M^{-1}$. Since the crystal spectra for both polarizations have a deep valley at 29,000 cm^{-1}, this band is presently assigned to a solution species other than $PtBr_4^{2-}$, probably $Pt_2Br_6^{2-}$. The solution does exhibit strong bands at 34,200 and 37,200 cm^{-1} with ϵ_{max} of 3,000 and 8,500 cm^{-1} M^{-1} respectively. The c-polarized crystal spectrum, which can be recorded to 37,000 cm^{-1}, requires that both these transitions be polarized x,y. The present assignment attributes them to singlet and triplet states of the same irreducible representation in D_4', viz. E_g', with nearly the same energy which mix, are split and share intensity. The separation of 3,000 cm^{-1} for these are consistent with the expected spin-orbit coupling for the Pt ion. The singlet state is probably the ^1E_u arising from $M(\sigma^*)\leftarrow L$. According to Jorgensen (8), the ^1E_u transition gains intensity by mixing of π and σ character in the e_u ligand orbitals. There should be a $^1A_{2u}$ state corresponding to $M(\sigma^*)\leftarrow L$ at roughly this energy. However, it arises from $\sigma^*(b_{1g})\leftarrow b_{2u}$. Since b_{2u} is pure π in character, the transition should be, according to Jorgensen, much weaker. It may lie just above 37,000 cm^{-1}. However, there can not be a strong $^1A_{2u}$ transition below 48,000 cm^{-1}. With $PtCl_4^{2-}$ the MCD studies of Schatz and coworkers (7) show that a shoulder at 43,000 cm^{-1} is ^1E_u whereas an intense peak at 46,000 cm^{-1} must be $^1A_{2u}$. The difference between $PtCl_4^{2-}$ and $PtBr_4^{2-}$ suggests that the intense transitions in $PtCl_4^{2-}$ are primarily 6p←5d. The $^1A_{2u}$ state would correspond to $6p_z\leftarrow 5d_{z^2}$.

The very weak peak at ca. 30,500 cm^{-1} seen in both polarizations deserves some comment. In c-polarization, at least, it seems that a vibronicly excited transition would be apparent in the 300°K spectrum. Even though the absorption is rising rapidly in this region, it can not be discerned at all. It appears therefore to have a small but non-zero transition dipole, and it is assigned as ^3E_u. The transition to this state is dipole-allowed

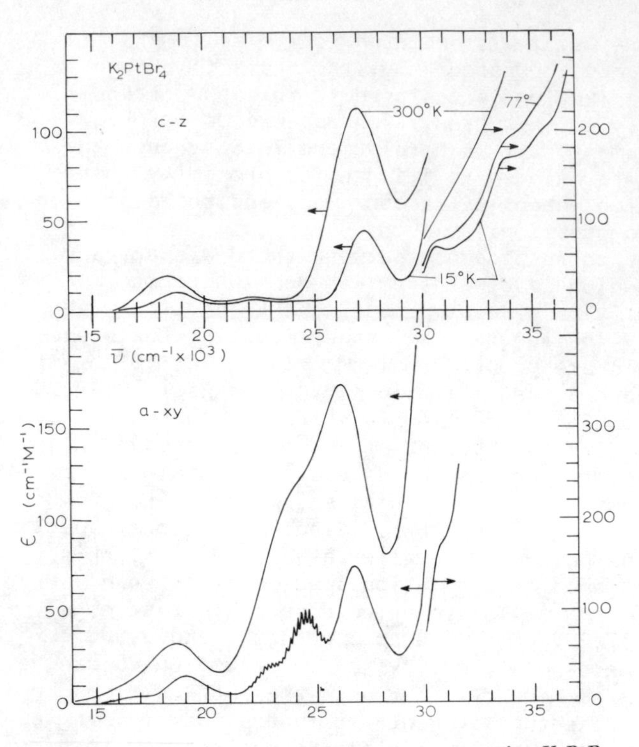

Figure 3. Polarized crystal absorption spectra for K_2PtBr_4

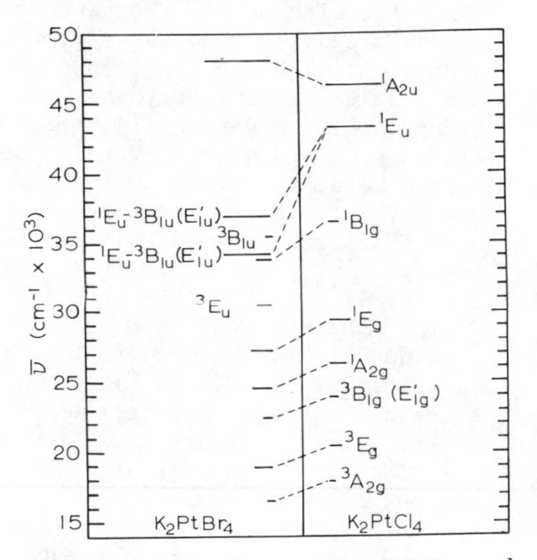

Figure 4. Excited states for K_2PtBr_4 and K_2PtCl_4. Length of the line for each state is proportional to log ϵ_{max} (15°K) for the crystal transitions and log ϵ_{max} (soln) for the intense transitions.

by virtue of the spin-orbit coupling in both polarizations.

Magnus' Green Salt. $Pt(NH_3)_4PtCl_4$.

It has long been recognized that there must be strong crystal effects on the spectra in Magnus' green salt (MGS), $Pt(NH_3)_4PtCl_4$. The component cation, $Pt(NH_3)_4^{2+}$, is colorless in aqueous solution or halide salts whereas the anion, $PtCl_4^{2-}$, is red. As the name implies the MGS is a dark green, very insoluble salt. It possesses a tetragonal crystal structure (9) with two of each ion per unit cell. There are stacks of alternating ions which are illustrated in Figure 5 with a spacing of 3.24Å between the anions and cations in the chains. The polarized crystal spectra are presented in Figure 6. The crystals are highly dichroic. A crystal which appears dark green in c-polarization is pale yellow in a-polarization. At all wave lengths the absorption is much higher in the c-polarization than in the a. The green color results from a window at about 20,000 cm^{-1}. The waves in the recording at low absorbance are due to interference by multiply reflected light (10). Although the instrumentation did not permit measurements below 17,000 cm^{-1}, it is clear that the low energy band in c-polarization is vibronic in character. At higher energies the absorption increases beyond an ϵ of 700 cm^{-1} M^{-1}, the limit of the measurements. In a-polarization the spectrum is dominated by a single peak at about 25,000 cm^{-1}, although there is a shoulder at 23,000 cm^{-1} and a weak spin-forbidden band below 17,000 cm^{-1}. The wave numbers of the three peaks in K_2PtCl_4 solution are shown by the 3 arrows in Figure 6. It is concluded that the spin-forbidden peak of $PtCl_4^{2-}$ has been red-shifted by about 4,000 cm^{-1} in MGS. The $^1A_{2g}$ transition and the ^1E_g transitions have coalesced to the single peak at about 25,000 cm^{-1}. This corresponds to a red shift of not more than 1,000 cm^{-1} for the $^1A_{2g}$ state and about 4000-5000 cm^{-1} for the ^1E_g state. The relative shift of these transitions is attributable to the D' term of equation 5. The angular dependence of the d orbitals involved is shown in Figure 7. Thus, for the transition to the $^1A_{2g}$ state, $\sigma^*(d_{x^2-y^2})\leftarrow d_{xy}$, both orbitals are concentrated in the plane of the ion and their electrons suffer similar interactions with the electron clouds of the adjacent ions. For the d_{yz} orbital, however, there is a considerable electron density out of the plane of the ion. The energy of the electron in this orbital is therefore raised by the electron-electron repulsions, and a red shift for the transition occurs.

The high absorption in the c-polarization is clarified from diffuse reflectance spectra for K_2PtCl_4 and MGS which are shown in Figure 8. It can be seen that the maximum, which correspond to a very intense transition, has been shifted from ca. 44,000 cm^{-1} in K_2PtCl_4 to 35,000 cm^{-1} in MGS. This band is apparently in the c-polarization. Such a large red shift must be largely caused by the I' term of equation 5. It would require strong

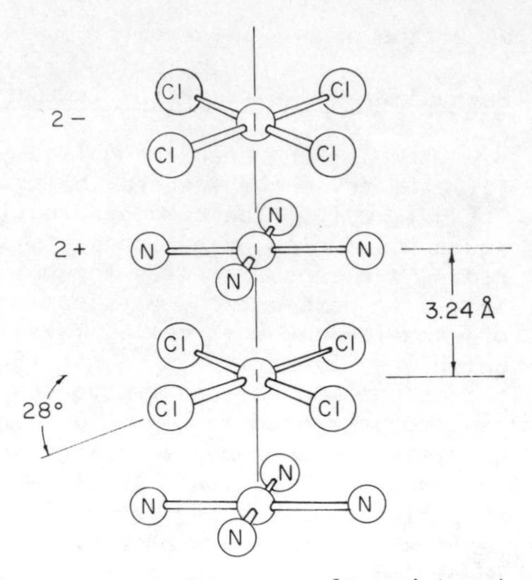

Figure 5. Alternate stacking of ions in Magnus' green salt

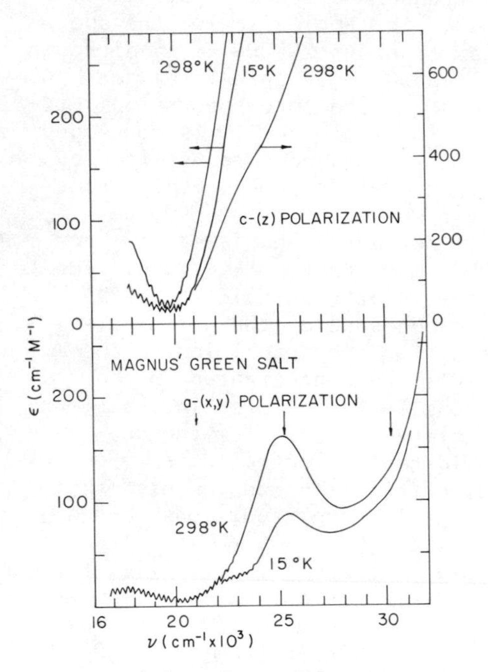

Figure 6. Polarized crystal absorption spectrum for Magnus' green salt

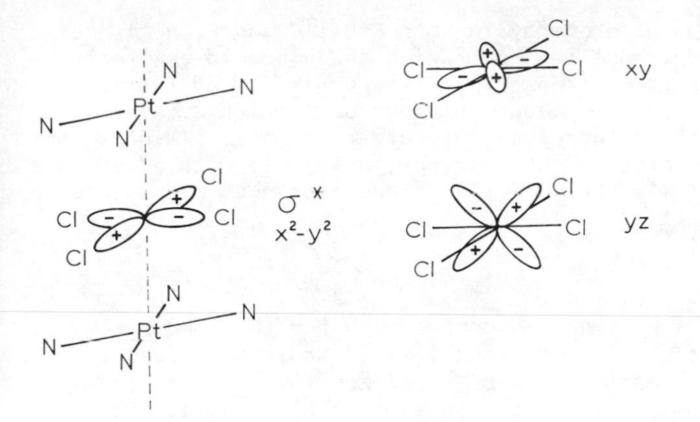

Figure 7. Angular portions of d-*orbitals involved in spectral transitions of MGS*

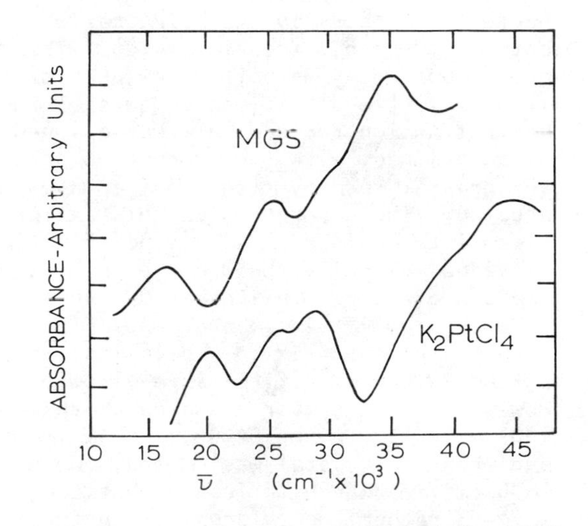

Figure 8. Diffuse reflectance spectra for MGS and K_2PtCl_4 at 300°K

transitions of comparable energy in both $PtCl_4^{2-}$ and $Pt(NH_3)_4^{2+}$. This is the allowed transition from which the vibronic excitations in c-polarization borrow intensity, and the proximity of this band enhances their intensity by about an order of magnitude.

The spectra of MGS therefore indicate that the striking color changes result from crystal interactions, primarily with adjacent neighbors, which modify the molecular transitions but without delocalization of electrons.

$Pt_2Br_6^{2-}$ Spectra

A dimer complex provides an opportunity for the demonstration of electron delocalization on a limited scale by absorption spectroscopy. The dimeric anion crystallizes in the tetraethylammonium salt which is triclinic with a micaceous cleavage for which the structure was reported by Stevenson (11). The projection of dimeric ion on the cleavage face is shown in Figure 9. There is one anion per unit cell. The platinum atoms in a dimer are separated 3.55Å, and the anions are separated 7.6Å from their nearest neighbors by the bulky cations. The light beam therefore enters dimer anion nearly end on. The molecular axes of heavy anion apparently establish the axes of the indicatrix, so there is very little directional change of these axes with wave length. One extinction for crystals was aligned, as accurately as could be determined, with the projection of the y molecular axis upon the crystal face over the entire wave length region which was studied. The absorption in this extinction direction gave a very clean intermediate axis or y-polarization. The other extinction provided primarily the short axis or x-polarization. Preliminary spectra are shown in Figure 10. The absorption in y-polarization is much higher than in the x-polarization. Thus, at room temperature there is a peak in y polarization at 23,500 cm^{-1}, very close to the 24,000 cm^{-1} first spin-allowed transition in K_2PtBr_4. However, the ϵ_{max} for the band at this $\bar{\nu}$ in K_2PtBr_4 was only ca. 100 $cm^{-1}M^{-1}$. In the $Pt_2Br_6^{2-}$ salt it was 800 $cm^{-1}/M(Pt_2Br_6^{2-})$. But when the crystal was cooled, the peak height increased to 1600 cm^{-1}/M, which indicated that the transition was dipole-allowed rather than vibronic. In the x-polarization there was not much indicated structure in the spectrum but the absorption increased as $\bar{\nu}$ increased. A peak at 36,500 cm^{-1} occurs just where the M←L charge transfer band occurs in $PtBr_4^{2-}$ solution. This may be then an x-polarization or out-of-plane M←L which apparently is very weak in $PtBr_4^{2-}$ as well. At 15°K, band maxima at 29,000 cm^{-1} and 32,000 cm^{-1} are seen as well, and there is a shoulder at 26,000 cm^{-1}.

In consideration of this spectrum each platinum atom can be considered to bring into the dimer its complement of five 5d-orbitals. Molecular orbitals for the dimer are constructed by taking the sum or difference of corresponding d-orbitals on different platinum atoms. With the choice of axes the σ^* orbitals

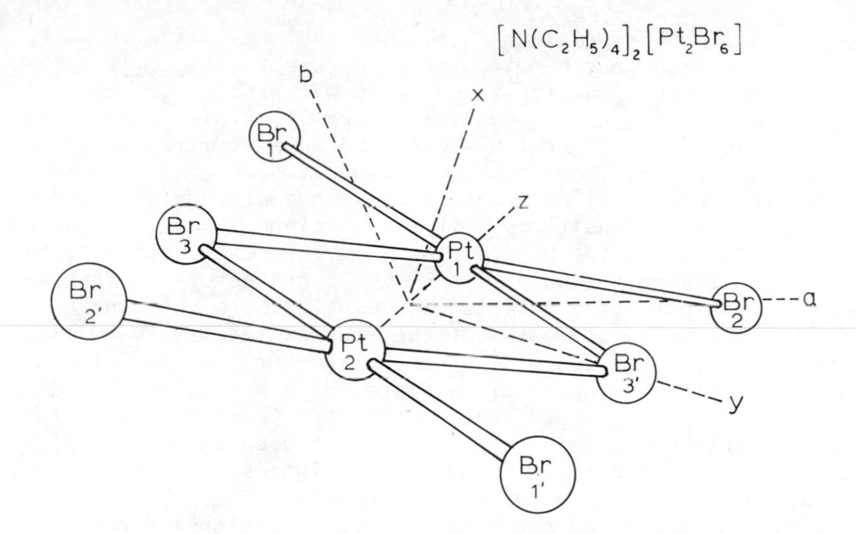

Figure 9. Projection of the dimeric ions, $Pt_2Br_6^{2-}$ upon the 001 cleavage face of $[N(C_2H_5)_4]_2[Pt_2Br_6]$

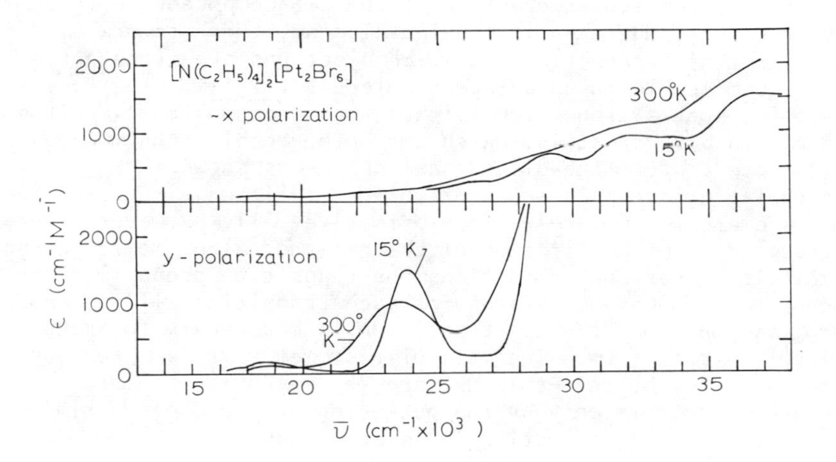

Figure 10. Polarized crystal absorption spectrum for $[N(C_2H_5)_4]_2[Pt_2Br_6]$

are formed from the platinum d_{yz}'s, and the orbitals with π-character located in the molecular plane are the $d_{y^2-z^2}$'s. The two linear combinations for these atoms with their symmetry designation under D_{2h} are shown in Figure 11. This figure also shows the LC of ligand orbitals with the same symmetry as the LC of platinum orbitals and which contribute to the σ and π bonds. Thus there are two filled orbitals and two empty orbitals and a possibility of 4 transitions. If the platinum atoms were essentially isolated and the orbitals were of pure d-character, then only a single spectral band with vibronic character is expected. However, each pair of the filled and empty M.O.'s contains a g and a u state. Therefore two of the four excited states will have u character and two will have g character. As indicated in Figure 11, both ungerade excited states have B_{2u} symmetry and should have dipole-allowed character in the y-polarization, for which the intensities will depend somewhat upon the degree of overlap of the d orbitals. However, the intensities may be modified by the bridging ligand orbital as well. The high intensity and dipole character of the band in y-polarization indicates that there is electron delocalization in the transition. The investigation of this system is continuing.

$Pt(en)Cl_2$ and $Pt(en)Br_2$. Transitions to Ionic States

For the molecular crystals of the two compounds, $Pt(en)Cl_2$ and $Pt(en)Br_2$, where en is ethylenediamine, there is now evidence for electronic transitions in which electrons are transferred from one molecule to an adjacent molecule (12,13). The excited state for such a transition is therefore a bound ionic exciton state. In both crystals, which are orthorhombic, the molecules stack face to form one-dimensional chains as shown in Figure 12. The chelating group in adjacent molecules lie on alternate sides so there are two molecules in a primative cell. However, the corresponding molecular axes of the non-equivalent molecules are parallel. Under such conditions the transition probability to one of the two Davydov states for each transition will be exactly zero, and only one crystal transition is observable for each molecular transition. The molecular symmetry is C_2. However the local symmetry at the Pt is C_{2v} and the molecular symmetry deviates from this only by the puckering of the chelate ring. It is likely that the selection rules are established by the C_{2v} symmetry. The choice of axes is shown in Figure 12. The x,y and z molecular axes are aligned with the orthorhombic a, b, and c axes, respectively. Spectroscopic quality crystals grew with well developed 100 faces so absorptions for c and b polarizations were measurable to provide the molecular z and y polarizations respectively. The symmetry of the d-orbitals are shown in Figure 13. Under the choice of axes the d_{xy} orbital is a part of the σ^*-MO. The C_{2v} symmetry splits the E_g degeneracy of D_{4h} and four distinct d←d transitions are possible. Since C_{2v} does not contain

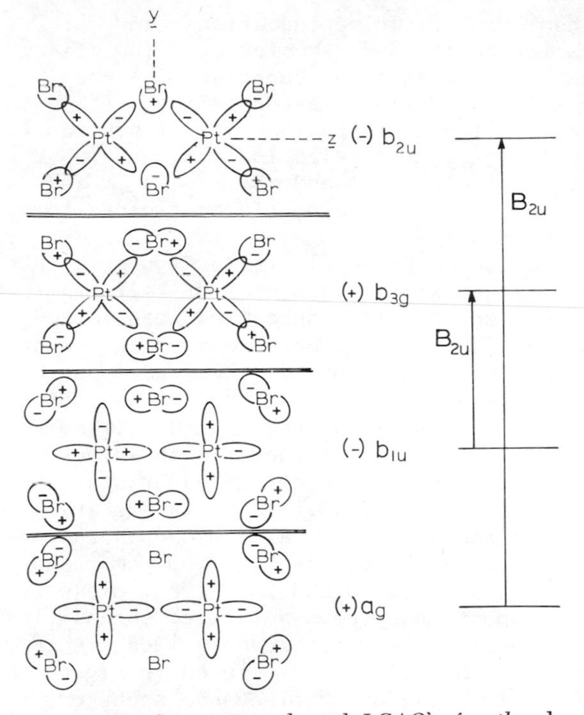

Figure 11. Symmetry adapted LCAO's for the d_{yz} *and the* $d_{y^2-z^2}$ *orbitals upon the two platinum atoms of* $Pt_2Br_6^{2-}$. *LCAO's of ligand orbitals also shown. Algebraic sign of the LC is in parenthesis; also given is the orbital symmetry under* D_{2h}. *Excited states for the two indicated* $d \leftarrow d$ *transitions are both* B_{2u}, *and both transitions are dipole allowed in y-polarization.*

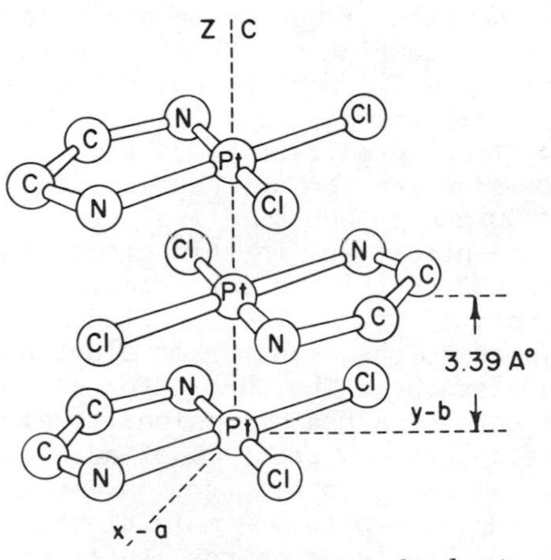

Figure 12. Stacking of the molecules in $Pt(en)Cl_2$ *crystals*

a center of symmetry, the ligand field perturbations can contribute dipole-allowed character to the d←d transitions. The states attained from spin-allowed d←d transitions together with the allowed polarization by the ligand field perturbations in the order of increasing energy as indicated by the orbital scheme of Figure 13 are: $^1B_2$, $d^*_{xy} \leftarrow d_{x^2-y^2}$, x polarization; $^1A_2$, $d^*_{xy} \leftarrow d_{yz}$, forbidden; $^1B_1$, $d^*_{xy} \leftarrow d_{xz}$, z-polarization; and $^1B_2$, $d^*_{xy} \leftarrow d_{z^2}$, x polarization. Therefore, there should be no ligand-field-allowed d←d transition with y or b-polarization.

The spectrum of a solution of $Pt(en)Cl_2$ is in Figure 14. It is quite characteristic of other solution spectra of halide-amine complexes of platinum(II). Two spin-forbidden bands can be resolved between 23,000 and 29,000 cm^{-1}. There are then two spin-allowed bands discernable. The transition at 33,000 cm^{-1} can be assigned to the $^1B_2$ and the band at 37,000 cm^{-1} to the pair of transitions $^1A_2$ and $^1B_1$ which are derived from ^1E_g (D_{4h}). Following a rather deep valley at 40,000 cm^{-1} there is an intense band at 49,000 cm^{-1}. For comparison the b-polarized crystal spectrum at 300°K is included in Figure 14. There is a sharp peak at 33,000 cm^{-1} which is nearly 3 times as intense as the one in solution at that energy. A valley occurs at 37,000 cm^{-1} and a rather definite shoulder is seen at 39,000 cm^{-1}, very close to the valley of the solution spectrum. The b-polarized intensities however, are an order of magnitude below the c-polarized spectrum. Both crystal spectra for 300°K and 15°K are included in Figure 15 together with a room temperature diffuse reflectance spectrum. The intensity features of the c-polarization and the reflectance spectra are very reminiscent of those for MGS. There is a very strong enhancement in the intensity of vibronic transitions which are derived from the red-shift of an intense band polarized in the stacking direction from an energy at least as high as 49,000 cm^{-1} in the solution down to 37,500 cm^{-1} in the crystal. This shift must be due to a considerable extent to the I' type of term in equation 5. There is a red-shift of the spin-forbidden bands as well, perhaps not quite as great as in MGS but here the stacking distance is 0.15Å greater.

The unpredicted feature of the spectrum was the dipole-allowed character of the two transitions in b-polarization at 33,000 and 39,000 cm^{-1} indicated by the temperature dependence of their intensity. In this direction the ligand field perturbation does not contribute any transition dipole to d←d transitions. Therefore, it has been concluded that these two transitions correspond to the transfer of electrons between molecules, i.e. to ionic states. The theory for electron delocalization between elements of a one dimensional chain was treated by Merrifield (14). Overlap between orbitals of adjacent atoms is required, and in general the electron from one molecule can be transferred to an unoccupied orbital, β molecules removed from the hole which results. The lower energy states are bound ionic states with a fixed separation between the electron and hole, which are however

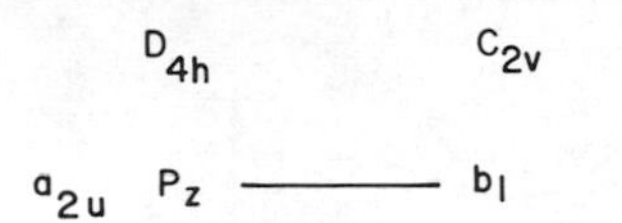

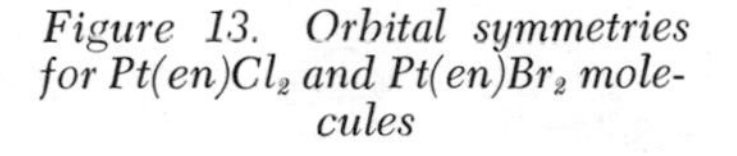
Figure 13. Orbital symmetries for $Pt(en)Cl_2$ and $Pt(en)Br_2$ molecules

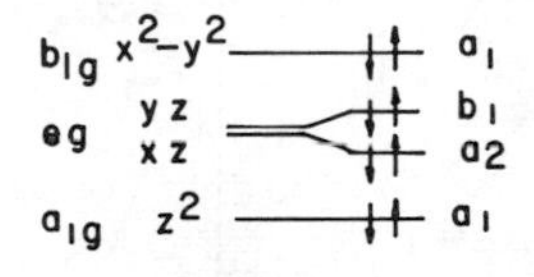

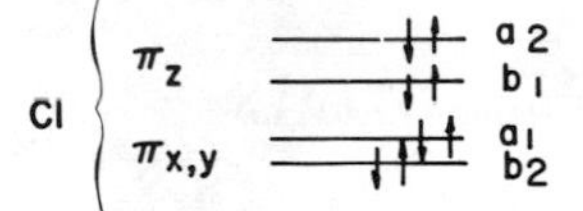

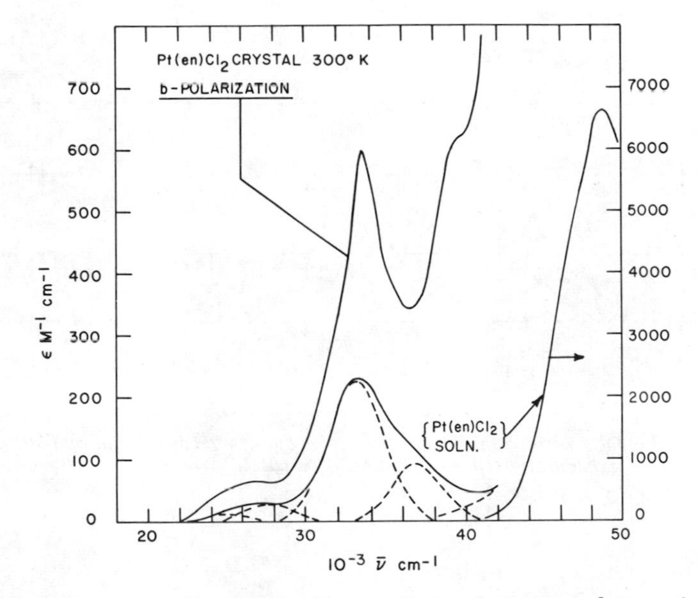

Figure 14. Absorption spectrum for an aqueous solution of $Pt(en)Cl_2$ and the b-polarized crystal absorption spectrum for $Pt(en)Cl_2$

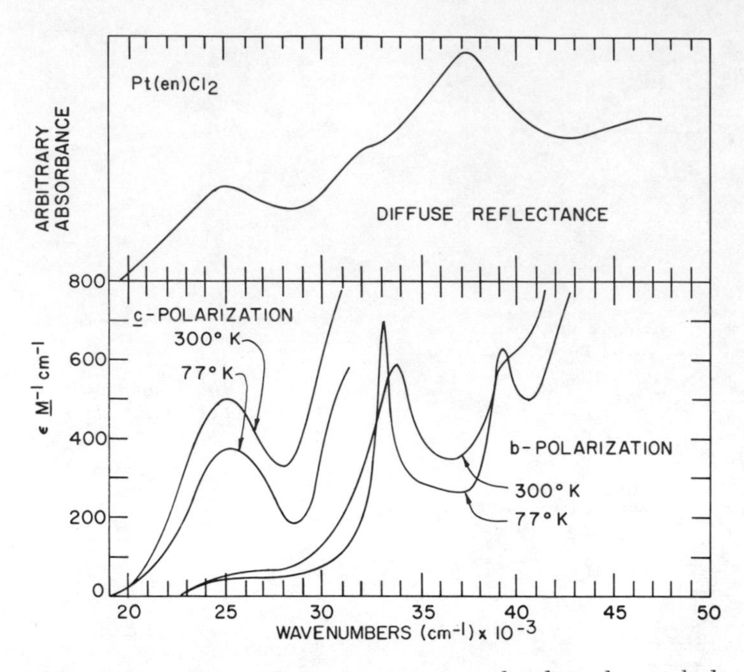

Figure 15. Diffuse reflectance spectrum and polarized crystal absorption spectrum for $Pt(en)Cl_2$

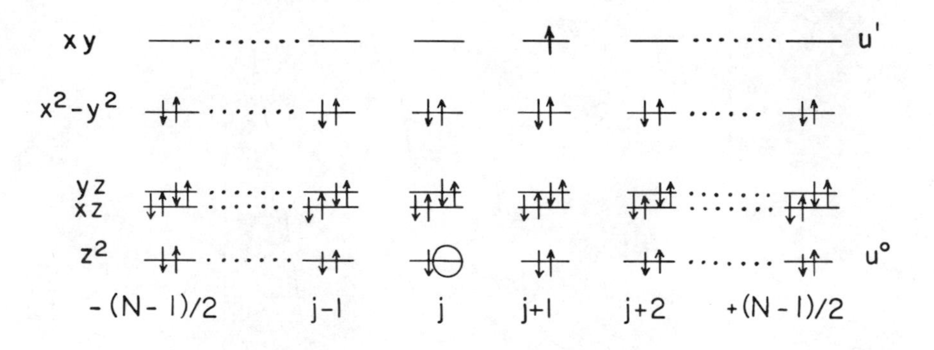

Figure 16. Schematic of an ionic excitation in a one-dimensional stack of $Pt(en)Cl_2$ *molecules:* θ_j $(\beta = +1)$ *based on* $(d_{xy})_{j+1} \leftarrow (d_{z^2})_j$

mobile in the chain. Because of the coulomb attraction of the hole and electron the lowest ionic states result when the hole and electron are on adjacent molecules. A system with such a one-molecule transfer from a d_{z^2} orbital to an adjacent d^*_{xy} orbital, $\beta = +1$, is illustrated in Figure 16. At high energies the treatment describes a conduction band for the free hole and electron. A crystal wave function for a β separation of the hole and electron and for which the one-dimensional wave vector, k, is zero (a requirement for a dipole allowed crystal transition) is:

$$\Phi'_o(\beta) = N^{-\frac{1}{2}} \Sigma_j \, \phi'_j(\beta). \tag{7}$$

Now the wave function for the transfer of the electron in the opposite direction ($-\beta$) will be degenerate with $\Phi'_o(\beta)$. The two crystal states formed from linear combination of these degenerate functions will be for $\beta = \pm 1$:

$$\Psi'_+(1) = 2^{-\frac{1}{2}}[\Phi'_o(+1) + \Phi'_o(-1)], \tag{8}$$

$$\Psi'_-(1) = 2^{-\frac{1}{2}}[\Phi'_o(+1) - \Phi'(-1)]. \tag{9}$$

The function $\Psi'_+(1)$ has the same symmetry as the Frenkel state, Φ'_o, (equation 4) and these two states may interact as is shown in Figure 17. If the Frenkel state is allowed, then both states will mix and share intensity. Unless their energy is very close, $\Psi'_+(1)$ will be much weaker than Φ'_o. However, if the Frenkel state and therefore Ψ'_+ are forbidden, then it may be possible for Ψ'_- to be allowed, since it is of different symmetry. The requirement for allowed character is that:

$$r'_{j,j-1} = -r'_{j,j+1} \neq 0, \tag{10}$$

where $r'_{j,j+1}$ is the transition moment for the transfer of an electron from the j to the j+1 member of the stack, *i.e.*

$$r'_{j,j+1} = \int u^o_j \underline{r} u'_{j+1} d\tau, \tag{11}$$

where u^o and u' are the filled and empty orbitals respectively.

Such a non-zero transition moment does require significant overlap of the orbitals. An examination of the symmetry of the d orbitals reveals that there is only one d←d electron transfer transition *viz*. $(d^*_{xy})_{j\pm1} \leftarrow (d_{xz})_j$, for which the condition of equation 10 applies for a *y* and hence a *b* polarization. Therefore this assignment has been given to the crystal transition at 33,000 cm^{-1} in the *b* polarization. It is interesting to note that it is the absorption with polarization normal to the stacking direction which provides evidence for delocalization of the electrons along the stack in the transition. However, there is the second trans-

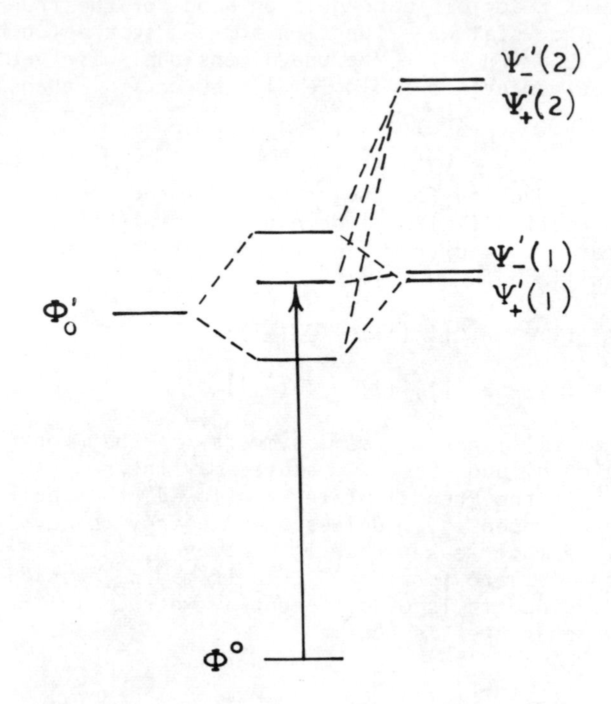

Figure 17. Energy state diagram illustrating mixing of the Frenkel Φ_0' and the Ψ_+ wave functions with an indicated transition $\Psi_-'(1) \leftarrow \Phi_0$

ition at 39,000 cm^{-1} which must be assigned. Of all possible ionic state transitions into the $\sigma^*(d_{xy})$ or into the $6p_z$ orbital, for that matter, there is only one, a $\sigma^* \leftarrow \pi$-Cl ($d^*_{xy} \leftarrow a_2$) which will provide a non-zero transition moment in accordance with equation 10. The 39,000 cm^{-1} transition has been given this assignment.

In the recent work with the $Pt(en)Br_2$ system (13), the stacking distance of the molecules is 3.50Å compared to the 3.39Å for the $Pt(en)Cl_2$. The solution spectrum for $Pt(en)Br_2$, Figure 18, exhibits similar features of that for $Pt(en)Cl_2$. However, the spin forbidden and spin allowed components have been red-shifted about 2,000 cm^{-1}; and a maximum in the intense spectrum was not reached below 50,000 cm^{-1}, where a molar absorptivity of 12,000 $cm^{-1}M^{-1}$ was attained. The crystal spectrum for $Pt(en)Br_2$ is in Figure 19. In the diffuse reflection the highest maximum is at 36,500 cm^{-1}, very close to that in $Pt(en)Cl_2$. Again, the corresponding spin-forbidden bands in c-polarization are a factor of at least 10 more intense than in b-polarization. In the b polarization a component band at 31,000 cm^{-1} has vibronic character and probably corresponds to $^1B_2(d^*_{xy} \leftarrow d_{x^2-y^2})$. The small bump at 33,500 cm^{-1} however has dramatically increased in height as the band narrows. It has therefore been attributed to the corresponding ionic state $(d_{xy})_{j\pm1} \leftarrow (d_{xz})_j$. Since normally the bands for bromide complexes are red-shifted in comparison to chloride, the occurrence of this band at nearly the same wave number as in $Pt(en)Cl_2$ is consistent with the greater separation of the molecules and the higher coulomb energy for separation of the electron and hole. The spectrum could not be extended to a high enough energy to observe a second peak as was the case for $Pt(en)Cl_2$.

In the ionic crystal state the interaction of the excited ions with the remainder of the crystal is an important energy term. Thus the electron transfer between the neutral molecules was observed whereas there was no evidence for a similar transition in MGS. In that case an electron transfer from $PtCl_4^{2-}$ to $Pt(NH_3)_4^{2+}$ would replace a pair of doubly charged ions in an ionic lattice by a pair of singly charged ions with a considerable reduction in the lattice energy. This effect therefore removes the electron transfers to higher energies than can be observed.

Summary

The absorption spectra for the systems described in this review indicate that if the metal ions in the platinum(II) complexes are separated by greater than 4.0-4.1Å, the extended interactions have negligible influence on the molecular electronic transitions. When the separations are reduced to 3.25 to 3.5Å there are observable effects. At this point the crystal effects, even in the absence of electron delocalization do provide striking changes in some of the transition energies. Finally, the identification of electron transfer transitions in $Pt_2Br_6^{2-}$ and the molecular crystals $Pt(en)Cl_2$ and $Pt(en)Br_2$ indicate the energy required, 3-4

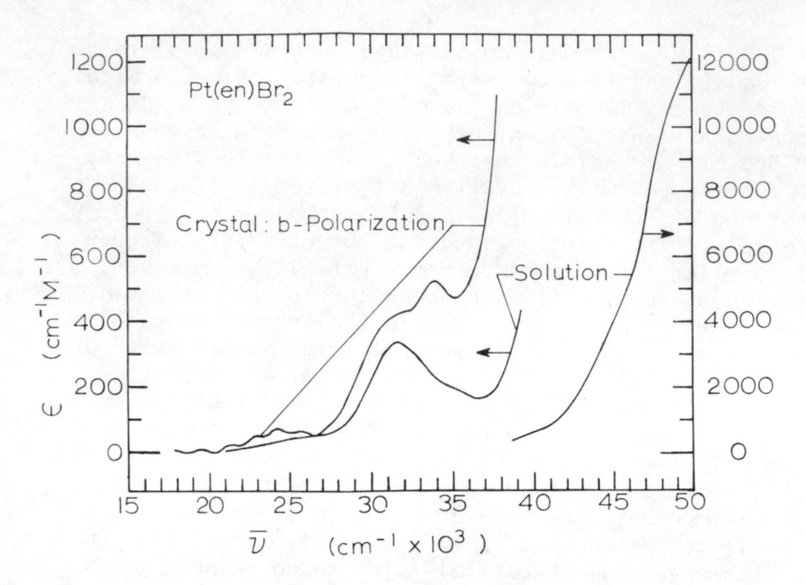

Figure 18. Absorption spectrum of an aqueous solution of Pt(en)Br$_2$ and the b*-polarized crystal absorption spectrum of Pt(en)Br$_2$*

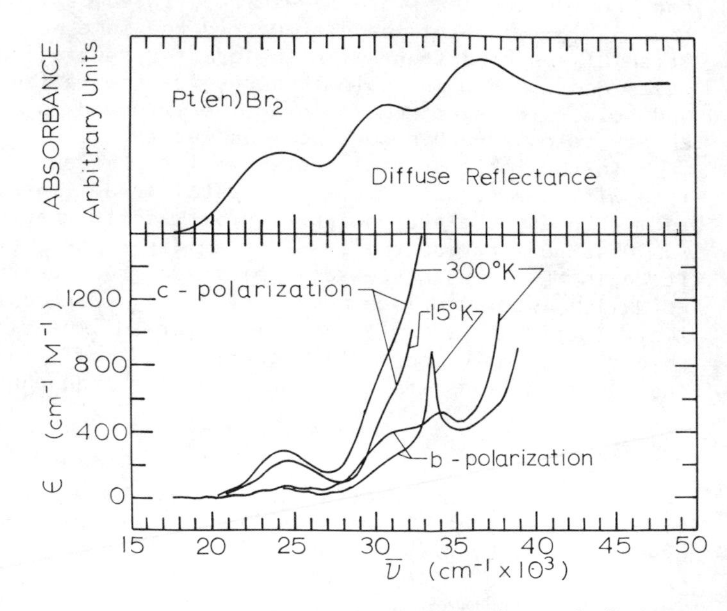

Figure 19. Diffuse reflectance and polarized crystal absorption spectra for Pt(en)Br$_2$

e.v, to provide the electron transfers between the metal ions. Such high energies for electron transfer require, of course, the characterization of the materials as insulators.

The absorption spectra which apply to weak transitions are complimented by the polarized specular reflection spectra for the intense transitions, which will be discussed by Professor Anex in the following paper.

Literature Cited

1. Ballhausen, C. J., "Introduction to Ligand Field Theory", p. 186, McGraw Hill, New York, 1962.
2. Peipho, S. P., Schatz, P. N., McCafferty, A. J., J. Amer. Chem. Soc., (1969), 91, 5994.
3. Craig, D. P., Walmsley, S. H., "Excitons in Molecular Crystals", p. 54, W. A. Benjamin Co., New York, 1968.
4. Mason, W. R. III, Gray, H. B., J. Amer. Chem. Soc., (1968), 90, 5722.
5. Jorgensen, C. K., U. S. Army Report, DA-91-506-EUC 241, (1951).
6. Martin, D. S. Jr., Foss, J. G., McCarville, M. E., Tucker, M. A., Kassman, A. J., Inorg. Chem. (1966) 5, 491.
7. McCafferty, A. J., Schatz, P. N., Stephens, P. J., J. Amer. Chem. Soc., (1968), 90, 5730.
8. Jorgensen, C. K., Private Communication.
9. Atoji, J., Richardson, J. W., Rundle, R. E., J. Amer. Chem. Soc., (1957), 79, 3017.
10. Martin, D. S. Jr., Rush, R. M., Kroening, R. F., Fanwick, P. F., Inorg. Chem. (1973), 12, 301.
11. Stephenson, N. C., Acta Cryst., (1964), 17, 587.
12. Martin, D. S. Jr., Hunter, L. D., Kroening, R., Coley, R. F., J. Amer. Chem. Soc., (1971), 93, 5433.
13. Kroening, R. F., Hunter, L. D., Rush, R. M., Clardy, J. C., Martin, D. S. Jr., J. Phys. Chem., (1973), 77, 3077.
14. Merrifield, R. E., J. Chem. Phys., (1961), 34, 1835.

19

The Nature of the Lowest-Energy Allowed Electronic Transition in Crystals of Certain d^8 Transition Metal Complexes that Possess Extended Metal Chains

BASIL G. ANEX

University of New Orleans, New Orleans, La. 70122

Abstract

The Magnus' salts and glyoximates and alkaline-earth cyanides of Pt(II), Pd(II), and Ni(II) are representatives of a group of d^8 transition-metal complexes that show striking solid-state optical properties. The most obvious manifestations of the solid-state effects involved here are either directly or indirectly related to the lowest-energy allowed electronic transition displayed by these crystals. A series of single-crystal optical studies involving primarily polarized specular reflection techniques has thus been carried out with the aim of characterizing the crucial low-energy transition in each group of compounds listed above. The results obtained for crystals containing only one metal species combined with separate work on the individual ions that constitute the Magnus' salts have allowed the identification of the single-molecule origin of the crystal transitions. Evidence is thus provided regarding the atomic orbitals involved in the crystalline excited state. The extension of these investigations to include studies of systems containing more than one kind of metal atom has allowed one to ascertain the extent to which the crystal excited state involves delocalization over more than one center. A rather comprehensive picture of the experimental situation for these systems with regard to their optical properties has thus emerged and provides a critical standard by which proposed models for their electronic structure may be evaluated.

The work reported here has been in part supported by the National Science Foundation and the National Institutes of Health.

Introduction

For a number of years our laboratory has had an interest in a rather broad group of planar d^8 systems whose crystals show absorptive behavior that is markedly and often dramatically different than that displayed by their component ions or molecules as isolated entities. The compounds studied in this work have included the glyoximates and alkaline-earth cyanides of Pt(II), Pd(II), and Ni(II) and a variety of Magnus' salts. These systems all possess crystal structures characterized by the component planar species stacking one on top of the other and the metal atoms forming parallel chains extending the length of the crystal. The interplanar spacing in the molecular stacks appears to be limited by repulsive interactions between the component molecules, since in each class of compounds one finds roughly the same lower limit to the metal-metal spacing, 3.2 - 3.3 Å, and an opening up of the stack when one introduces relatively bulky substituents into the ligands involved (or, in the case of the cyanides, introduces more bulky cations).

The optical studies we have carried out on these substances have had two broad objectives: to establish the nature of the characteristic solid-state absorption that one finds in these systems and to identify the relationship, if any, between these absorptions and the "single-molecule" (or "isolated-molecule") spectra of the component complexes. Throughout, polarized single-crystal spectroscopy has played a key role. The polarizations of the electronic absorption bands thus determined not only allowed one to relate the observed transitions to theory and to resolve overlapping bands, but also - most importantly - provided "labels" that one might use to follow a given transition as he moved from one compound to another within a given series.

One may also note that in addition to the development of evidence bearing on the questions stated above, this work has also led to the accumulation of much valuable information concerning the spectroscopy of the allowed bands of the complexes involved as such. Although ligand-field spectra of transition-metal complexes have been rather extensively studied, and polarization determinations are not uncommon for them, detailed work on the strongly-allowed bands (often loosely called "charge-transfer" bands) is less frequently carried out, and polarization assignments for them have been largely non-existent. These bands, however, contain a wealth of valuable and, as will be seen, sometimes surprising information.

The reasons for the paucity of information concerning the intense bands in these systems include the fact that they are often strongly overlapping and thus difficult to resolve, and they are sufficiently intense that polarization determinations by direct absorption become formidable undertakings which require very thin samples, very good spectrometers, or, more likely, a combination of these two factors (1). Except for the glyoximate studies, the techniques of single-crystal reflection spectroscopy provided the key in the present work for avoiding these difficulties. For this reason, in what follows, we will first briefly recapitulate the manner in which the reflection approach is applied in our laboratory. In subsequent sections we will summarize our work on the classes of compounds mentioned previously following an order that reflects the sequence in which the research was actually done.

Specular Reflection Spectroscopy

Even if one has available a spectrometer with reasonably low levels of stray light, and thus may study samples of relatively high optical density, obtaining crystal spectra for transitions that in solution studies show molecular extinction coefficients of even a few thousand may require rather exotic methods of sample preparation. Figure 1 of Reference 1 presents some extreme examples of the distortions that one can obtain if the samples employed in an absorption study are sufficiently thick that one essentially measures the stray light in his spectrometer over much of the absorbing region of the spectrum. One can have situations where distortions of the spectra due to opaqueness of the sample are much more subtle than those shown in Reference 1. Careful direct absorption studies on high absorbers thus usually involve a fairly elaborate set of self-consistent studies, such as ascertaining the optical density as a function of crystal thickness at critical wavelengths.

Although the sample preparation necessary for the study of relatively highly-absorbing crystals can sometimes be accomplished (2,3), the techniques that have been developed are not generally applicable and typically involve considerable effort. For this reason it is fortunate that specular reflection spectroscopy provides a reasonable approach to the study of such systems (1). Specular, or mirror-like, reflection from the surface of a solid (to be distinguished from diffuse reflectance, where the incident radiation actually traverses the interior of a usually-powdered sample) closely parallels the behavior of the index of refraction (n) in the vicinity of an absorption. Since n typically rises to relatively high values as one approaches an absorption from the low-energy side and then falls to relatively low values on the high-energy side, similar behavior by the reflectivity may be used to infer the existence of absorption bands. Moreover, since the strength of a reflection band is, generally speaking, dependent on the strength of the absorption, the observation of reflection bands becomes easier as direct absorption methods become difficult. The relationship between reflection and absorption also implies that if one uses a single crystal and polarized incident radiation, the reflectivity may be expected to have a polarization dependence paralleling that of the related absorption. Polarized reflection studies may thus be used in a manner completely analogous to polarized absorption studies to ascertain the polarization of the transitions involved.

Figure 1 shows the relationship that exists between the absorptive properties of a crystal, the index of refraction, and the reflection coefficient in a representative situation. The extinction coefficient, k, plotted in Figure 1 is related to the usual molar extinction coefficient through

$$\epsilon = (4\pi k)/(2.303\lambda C) \qquad (1)$$

where λ is the wavelength to which ϵ and k refer and C is the molar concentration of the absorber in the crystal.

The optical constants (n and k) shown in Figure 1 are in fact not measured quantities, but have been derived by application of the Kramers-Kronig analysis (1) to the reflection spectrum. In this procedure, one first computes through the application of the integral expression

$$\theta(\omega_i) = \frac{\omega_i}{\pi} \int_0^\infty \frac{\ln R(\omega)}{\omega^2 - \omega_i^2}\, d\omega \qquad (2)$$

the phase change that incident radiation of the frequency of interest undergoes when it is reflected from the crystal surface. In Equation 2, $\theta(\omega_i)$ is the phase change for light of circular frequency ω_i ($\omega = 2\pi\nu$, where ν is the frequency of the light in question) and $R(\omega)$ is the reflection coefficient for light of circular frequency ω. Once one knows the phase change and reflection coefficient for a given frequency, the calculation of n and k is a straight-forward matter. For k, the quantity of most direct concern here, the relevant expression is

$$k = (-2R^{\frac{1}{2}}\sin\,\theta)/(1-2R^{\frac{1}{2}}\cos\,\theta + R). \qquad (3)$$

Equations 1 - 3 allow one to obtain from reflection data detailed absorptive information that is identical to that which results from direct absorption measurements. Moreover, this analysis gives absolute intensities (molar extinction coefficients) directly, and obtaining such data from direct absorption studies entails the often difficult measurement of sample thickness on very thin specimens.

The only aspect of the Kramers-Kronig analysis that prevents its application from being totally routine is the fact that the integration of Equation 2 extends over the complete frequency range from 0 to ∞, while one is of course limited experimentally to some finite spectral region. The method of correcting for phase contributions from outside the experimental range followed in our laboratory involves placing an effective transition in the vacuum ultraviolet, the region most likely to make the most significant contribution in studies such as are dealt with here. The properties of this transition are then varied until the derived spectrum fulfills some predetermined criterion - most commonly, that ϵ should be equal to zero in regions where the crystal is known to be transparent. The effective transition thus arrived at is then used to compute the optical constants throughout the region of interest. This procedure has the effect of introducing the proper phase correction in the region where it can be verified and then using an appropriate wavelength dependency in applying this correction to the remainder of the spectrum. The application of this procedure is illustrated in Figure 2. It has in general provided very satisfactory results - especially for bands of at least moderate intensity, which make a major contribution to the phase in the vicinity of the frequencies at which they occur and thus require relatively small phase corrections, and for bands that fall well within the experimental region.

Unless noted otherwise, crystal absorption spectra discussed in the following sections of this paper have been obtained *via* the application of Kramers-Kronig analysis to the appropriate reflection spectra.

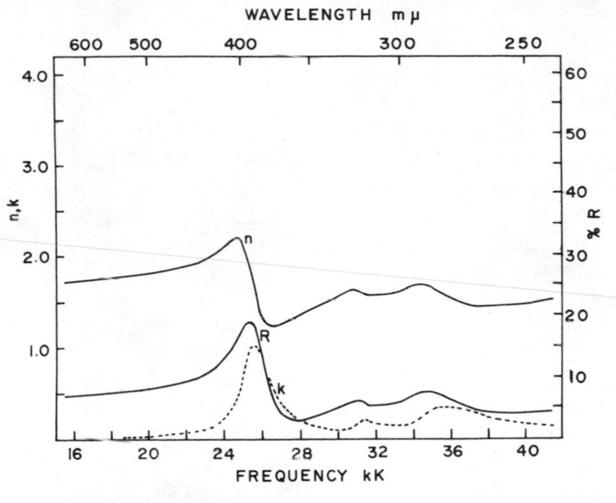

Molecular Crystals

Figure 1. Reflection coefficient (R), *index of refraction (n), and extinction coefficient* (k) *for a moderately intense absorber* (1)

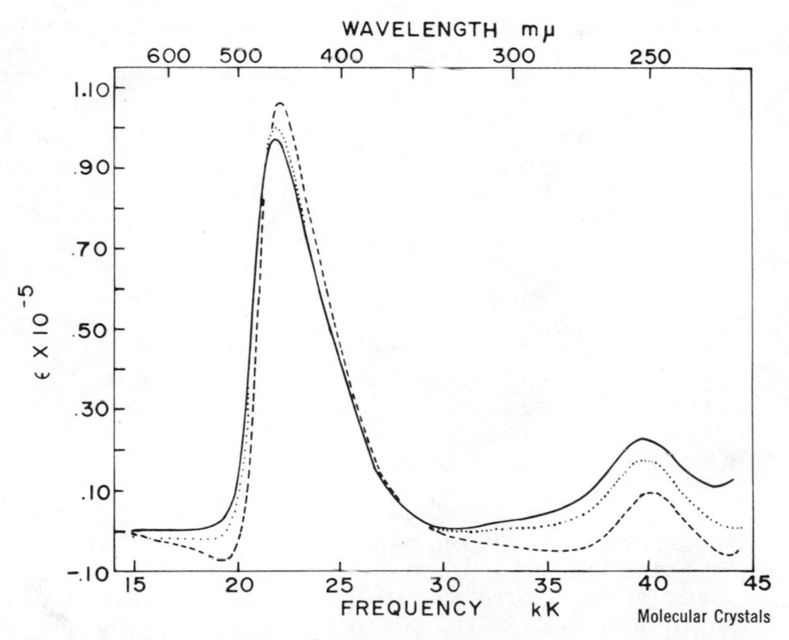

Molecular Crystals

Figure 2. Kramers-Kronig absorption spectrum in various stages of refinement: (– – –) *curve obtained by using only experimental data and straight-line high and low energy approximations from Ref.* 1; (· · · ·) *partially refined curve;* (——) *final curve obtained by varying effective ultraviolet reflectivity to obtain zero extinction coefficients in long wavelength region* (1).

The Glyoximates

Nickel dimethylglyoxime (NiDMG) itself provides a classic example of the kind of phenomena that have attracted attention to the compounds being discussed here. Those familiar with the determination of nickel as the dimethylglyoximate will recall the red flocculent precipitate that is obtained in this procedure. This material will dissolve in chloroform, however, and does so to give a pale yellow solution. The difference between the solution spectrum and that of the solid becomes even more striking when one examines a single crystal of NiDMG, which has a dark body color and displays a green metallic lustrous sheen in reflection. This sheen, when examined with a linear polarizer, is found to be polarized parallel to the needle, or stacking, axis of the crystal.

The differences between the solution and solid spectra of NiDMG are shown in a quantitative manner by Figure 3. Here the solution spectrum has been obtained in a suitable glass-forming solvent. On cooling to liquid nitrogen temperatures, a glass forms and simultaneously a finely dispersed precipitate of NiDMG appears, allowing one to compare rather precisely the solution and solid-state spectra (3). The most obvious, but not singular, new feature in the solid-state spectrum is the strong new band at 17.8 kK. It is clearly this green band, combined with other bands in the blue, that gives the red body color to NiDMG, and it is this band whose characterization we seek.

The polarization of the green band of NiDMG is displayed in Figure 4, which reports the polarized absorption spectra obtained for a thin film of NiDMG prepared by vacuum sublimation onto a quartz plate. It can be seen that this band is polarized exclusively perpendicular to the molecular planes, or parallel to the stacking axis, which is consistent with the qualitative observations on the reflectivity of the crystal noted in the remarks introducing this section and the quantitative reflection spectra reported by Anex and Krist (3).

In an effort to establish the single-molecule origin of the green band of NiDMG a series of nickel glyoximates has been studied, consisting of NiDMG itself, two forms of nickel ethylmethylglyoxime (α-NiEMG and β-NiEMG), and nickel heptoxime (NiHept). β-NiEMG and NiHept possess the stacking structure characteristic of NiDMG, while α-NiEMG, which turns out to be a crucial compound in this study, possesses a non-stacked structure. The crystal spectra of

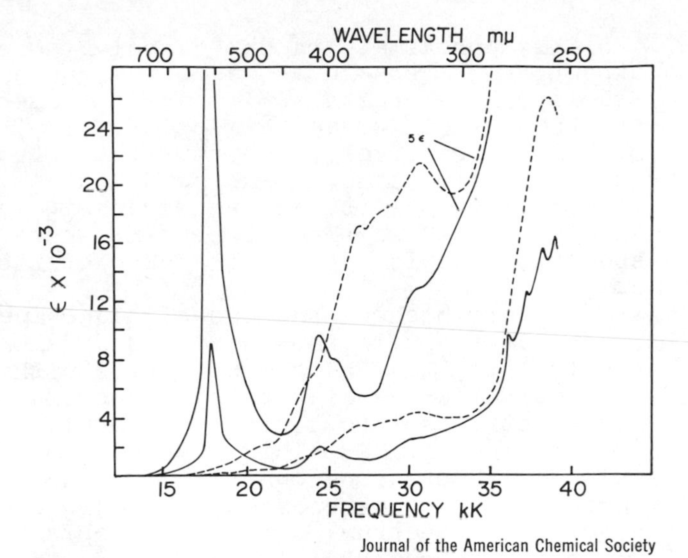

Journal of the American Chemical Society

Figure 3. Glass (——) and solution (– – –) absorption spectra of nickel dimethylglyoxime (3)

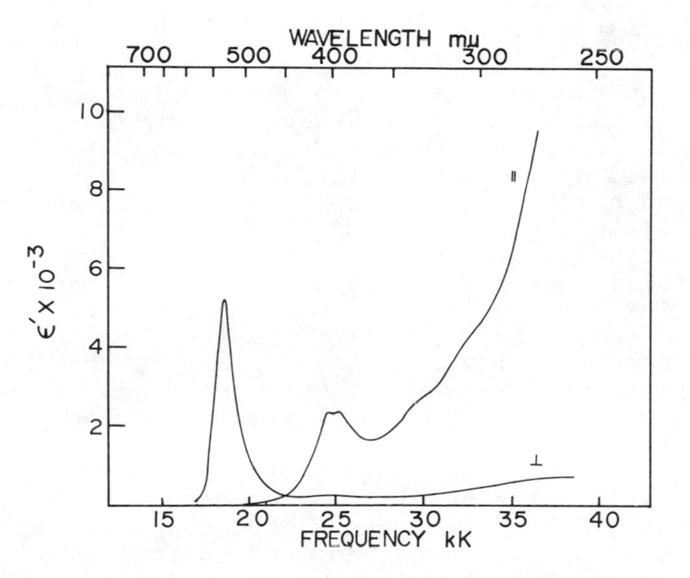

Journal of the American Chemical Society

Figure 4. Polarized absorption spectra of an oriented film of nickel dimethylglyoxime. The || curve obtained when electric vector of the incident light vibrates parallel to the molecular planes, and ⊥ obtained when it vibrates perpendicular to the molecular planes (3).

α-NiEMG approximate rather closely those you would expect from its solution spectrum and an oriented gas model for the crystal (3). The intermetallic spacings in NiDMG, β-NiEMG, and NiHept are 3.25, 3.4, and 3.60 Å, respectively, and thus this study can be viewed as an experiment in which the distances between the complexes in the stacks are successively increased until in α-NiEMG we reach the point where we obtain a spectrum that is essentially characteristic of the free complex.

Figure 5 shows the out-of-plane spectrum of each of the crystals in this series and demonstrates rather convincingly that as the complexes within the stacks are separated the characteristic solid-state band of NiDMG retreats into a single-molecule band located at 23.0 kK in α-NiEMG. This falls in the first set of solution absorptions observed for these complexes (all three complexes have essentially the same solution absorption spectrum). If one accepts Gray and Ballhausen's assignment (4) of the first allowed solution bands of the cyanides to metal-to-ligand charge-transfer transitions, one may conclude by analogy that the crucial out-of-plane band of α-NiEMG may be represented orbitally as

$$3d_{z^2}^{Ni} \longrightarrow c_1\pi_L^* + c_2 4p_z^{Ni} , \quad (4)$$

where π_L^* refers to the appropriate linear combination of anti-bonding ligand π-orbitals and c_1 and c_2 are mixing coefficients. In the Gray and Ballhausen picture, then, one would have in solution a situation where $c_1 >> c_2$ and be dealing, as noted above, with an essentially metal-to-ligand transition. Anex and Krist (3) have argued that one can then rationalize growth in intensity, red shift, and sharpening of the crystal transition relative to its solution counterpart in terms of the solid-state perturbation, whose nature was not specified, lowering the $4p_z$ orbital in energy to the point where $c_1 << c_2$. One would then be dealing with essentially a $d_{z^2} \longrightarrow p_z$ transition confined to the metal.

Explanations for the development of the green band of NiDMG that involve specific hypotheses concerning the nature of the solid-state perturbations have been proposed (5,6). These can most fruitfully be discussed after the related systems to be dealt with here have been reviewed.

The Magnus' Salts

The qualitative observation of the color changes involved in the formation of Magnus' green salt, $[Pt(NH_3)_4][PtCl_4]$, make it appear that the phenomena observed there parallel rather closely those described previously for NiDMG. One thus finds that if an aqueous K_2PtCl_4 solution, which is red, is combined with a colorless aqueous $Pt(NH_3)_4Cl_2$ solution, a green precipitate of Magnus' green salt (MGS) is obtained. A crystal color different than that of either of the components might thus lead one to assume that once again one has some "new" absorption band appearing in the visible part of the spectrum. Day and coworkers (7) and, more recently, Martin et al. (8), have measured the visible absorption spectra of MGS single crystals and shown them to be perturbed $PtCl_4^{2-}$ spectra, and not to possess any bands that could not be traced to those found in the anionic species in the K_2PtCl_4 crystal and its aqueous solution. The green color of the MGS crystal results from a red shift and intensification of the $PtCl_4^{2-}$ bands in MGS leading to a "window" in the green region of the spectrum, rather than from the development of an additional band in the visible part of the spectrum. (K_2PtCl_4 also has a crystal structure in which the complex species are stacked, but here the metal-metal separation is 4.13 Å, and one may expect that strong intercomplex interactions will not be manifested - an assumption that is born out by the reasonably close correspondence of the solution and crystal absorption for this system.)

A connection between the optical phenomena observed in MGS-type salts and those discussed previously for nickel glyoximates was to become apparent, however. Day et al. carried out a series of particularly significant experiments, analogous to those described in the preceding section on the glyoximates, in which the spacing between complexes in the MGS crystal was increased by modifying the nature of the ligands in a manner expected to affect the "single-molecule" spectra in a minimal fashion (7,9). MGS-type salts were thus prepared in which the NH_3 ligands of $Pt(NH_3)^{2+}$ were replaced by CH_3-NH_2 (Me-MGS) and CH_3-CH_2-NH_2 (Et-MGS), respectively. The MGS and Me-MGS crystals have the same metal-metal spacing (3.25 Å),while in Et-MGS this parameter has a value of 3.4 Å (5). The visible spectra of MGS and Me-MGS are very similar, and as one might expect, the Et-MGS spectra are intermediate between those of MGS and K_2PtCl_4 (7,9). Of equal importance, Day et al. also carried out diffuse reflection studies on these same systems, and found a strong correlation between metal-metal distance and the location of a strong ultraviolet transition:

Compound	Metal-Metal Distance	Position of uv Transition (Diffuse Reflectance)
MGS	3.25 Å	34.5 kK
Me-MGS	3.25	34.5
Et-MGS	3.4	40
K_2PtCl_4	4.13	42.5*

* Actual absorption to blue of this (11).

It was thus suggested that the effects observed in the visible spectrum of the Magnus' salts could be understood in terms of the visible bands' borrowing intensity from the strong metal-metal separation dependent ultraviolet band. In such a situation, the closer the uv band was to visible transitions, the greater its influence on them could be expected to be. Further, the fact that the intensity perturbations were stronger for the out-of-plane bands than the in-plane, led to the conclusion that the perturbing band must have out-of-plane polarization.

These observations and the conclusions drawn from them led directly to the study of the allowed bands in the same series of three Magnus' salts in our laboratories (10). The results were strikingly in accord with the suggestions made by Day et al., as may be seen in Figure 6, which presents the crucial portions of the out-of-plane absorption spectra obtained by Kramers-Kronig analysis of the corresponding reflection spectra. The strong ultraviolet absorption band is found to occur at 34.5, 34.4, and 39.9 kK respectively for MGS, Me-MGS and Et-MGS, in excellent agreement with the results of diffuse reflection. It was also possible to develop a quantitative correlation between the position and intensity of the strong ultraviolet band in these systems and the corresponding out-of-plane visible intensity. Strong evidence was thus provided for the proposal that interaction between the excited states responsible for the visible bands and that of the ultraviolet transition plays a substantial role in the perturbation of the visible spectra.

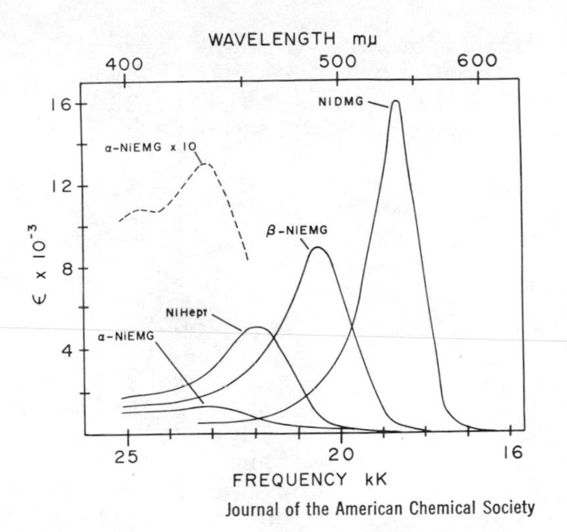

Journal of the American Chemical Society

Figure 5. Portions of out-of-plane absorption spectra for four nickel dimethylglyoximates obtained by direct absorption studies (3)

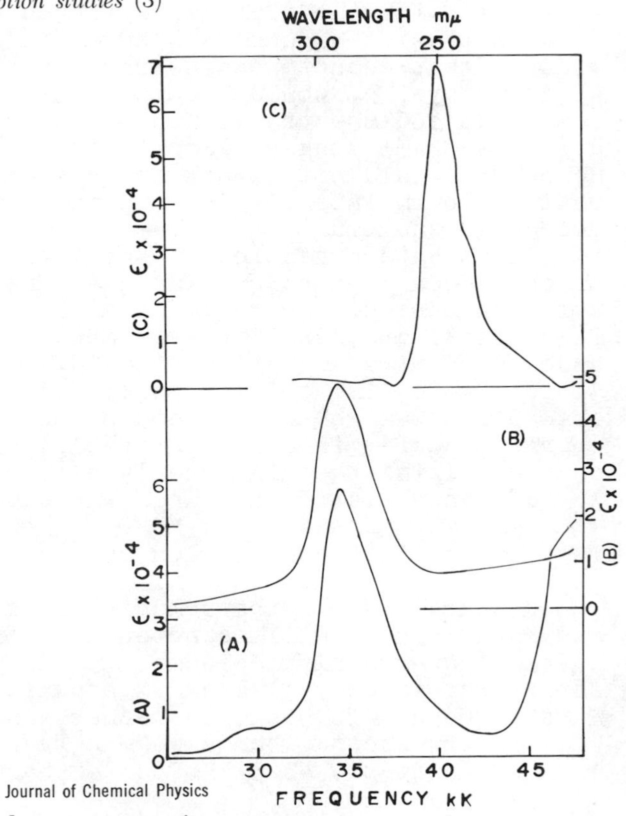

Journal of Chemical Physics

Figure 6. Kramers-Kronig absorption curves for MGS (A), Me-MGS (B), and Et-MGS (C) (10)

The bands shown in Figure 6 were thought by both Day and coworkers and our group to be correlated with the solution band of K_2PtCl_4 that peaks at 46.3 kK with a maximum molar extinction coefficient of 10,230 $cm^{-1}M^{-1}$ (11). This transition would clearly have undergone an intensification in the MGS crystal, but in estimating the magnitude of this effect one must first observe that the ϵ-values reported in Figure 6 must be divided by 3 for comparison to solution values and, secondly, note that a true comparison of intensities requires integrations over the bands. Recent studies (12) indicate that this band intensifies by a factor of approximately 2.5 in going from K_2PtCl_4 to MGS. Thus, although the behavior of the ultraviolet band in the Magnus' salts in some ways parallels the visible band of the nickel glyoximates, it differs in one crucial factor: it does not show the dramatic intensification observed for the green band of NiDMG in going from essentially infinite metal-metal separation to one of 3.25 Å. This intensity behavior is in accord with the picture developed for NiDMG, if one recalls that in Magnus' salts one is dealing with ligands that do not possess unsaturation. In Equation 4, therefore, c_1 would become zero for MGS and related compounds and the transition would be essentially d —> p at all stages of solid-state perturbation. Hence one would not expect in this case to see a dramatic drop in intensity as one moves to the "single-molecule" spectrum.

This interpretation of the Magnus' salt spectra of course carries strong implications for the 46.3 kK solution band of K_2PtCl_4: it should, if this picture is correct, at least have a component that can be associated with a platinum ion $5d_{z^2} \longrightarrow 6p_z$ promotion. Since at the time this proposal was made the 46.3 kK absorption was generally thought to be ligand-to-metal charge-transfer in character (4,11), a study aimed at characterizing the high-energy allowed bands in K_2PtCl_4 and related compounds was undertaken.

Tetrachloroplatinate(II) Ion and Related Systems

Figure 7 shows the solution spectra of K_2PtCl_4 and K_2PdCl_4. It will be noted that the 46.3 kK band of K_2PtCl_4 has roughly the same ϵ_{max} and, except for its structure, is very similar in appearance to the 35.7 kK K_2PdCl_4 band. It was thus rather generally accepted (11) that these two bands were correlated with one another, that the 46.3 kK $PtCl_4^{2-}$ band was strongly red shifted when Pd was substituted for Pt, and that the second $PdCl_4^{2-}$ band had similarly red-shifted from its vacuum-ultraviolet location in $PtCl_4^{2-}$. Moreover, the whole array of strong quartz ultraviolet bands in these ions was taken as being charge-transfer in nature.

In order to examine the validity of this picture and, more particularly, to explore the possibility of a d $\longrightarrow$ p transition being involved in the 46.3 kK $PtCl_4^{2-}$ band, the reflection spectra of K_2PdCl_4 and K_2PtCl_4 were obtained and are shown in Figures 8 and 9. It is clear that K_2PtCl_4 contains a strong high-energy out-of-plane transition, while K_2PdCl_4 displays no analogous absorption. If one makes the reasonable assumption that comparable transitions occur in $PtCl_4^{2-}$ and $PdCl_4^{2-}$, then the behavior of the out-of-plane $PtCl_4^{2-}$ band is just the opposite of that formerly thought to be the case - it shifts to higher, not lower, frequencies on moving from $PtCl_4^{2-}$ to $PdCl_4^{2-}$.

Examination of the d $\longrightarrow$ p transitions of Pt and Pt^+ and the comparable palladium species reveals a blue shift similar to that inferred for the tetrachloro ions as one moves from platinum to palladium (11), a correlation that lends additional plausability to the d $\longrightarrow$ p assignment in $PtCl_4^{2-}$. One may also note that at least two predictions flow from the foregoing discussion:

1) Since in ammonia the electrons analogous to those thought to participate in the lowest-energy ligand-to-metal charge-transfer band in $PtCl_4^{2-}$ are tied up in bonding, one would expect this band to occur at considerably higher energies in $Pt(NH_3)_4^{2+}$ than in $PtCl_4^{2-}$, while the $5d_{z^2} \longrightarrow 6p_z$ transition, being confined to the metal, should be found in roughly the same spectral region for both ions.
2) Since the strong out-of-plane transition in the solution spectrum of $PtCl_4^{2-}$ undergoes an approximately 12 kK shift in MGS, it might be expected that the d $\longrightarrow$ p transition hypothesized to occur in the vacuum ultraviolet in the free $PdCl_4^{2-}$ ion could be shifted into the quartz ultraviolet in the spectra of the palladium analog of MGS ($[Pd(NH_3)_4][PdCl_4]$).

One in fact observes a strong transition in aqueous solutions of $Pt(NH_3)_4Cl_2$ (11) and recent crystal studies (13) have shown this transition to have an out-of-plane polarization. Likewise, the palladium Magnus' salt shows a moderately strong out-of-plane absorption peaking in reflection at

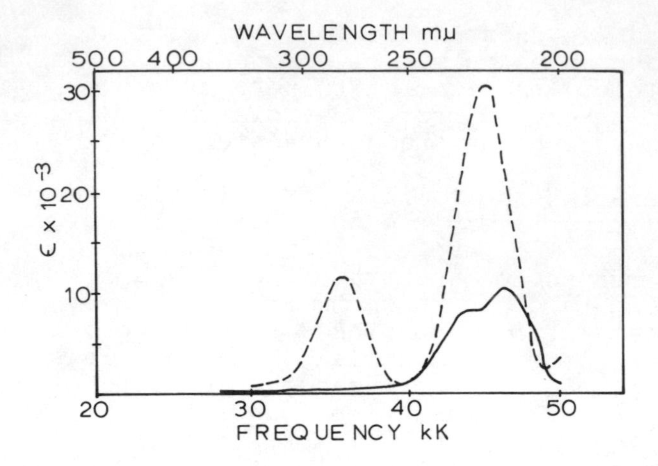

Figure 7. Solution absorption spectra (in 2M HCl) of K_2PtCl_4 (——) and K_2PdCl_4 (- - -) (data from Ref. 11)

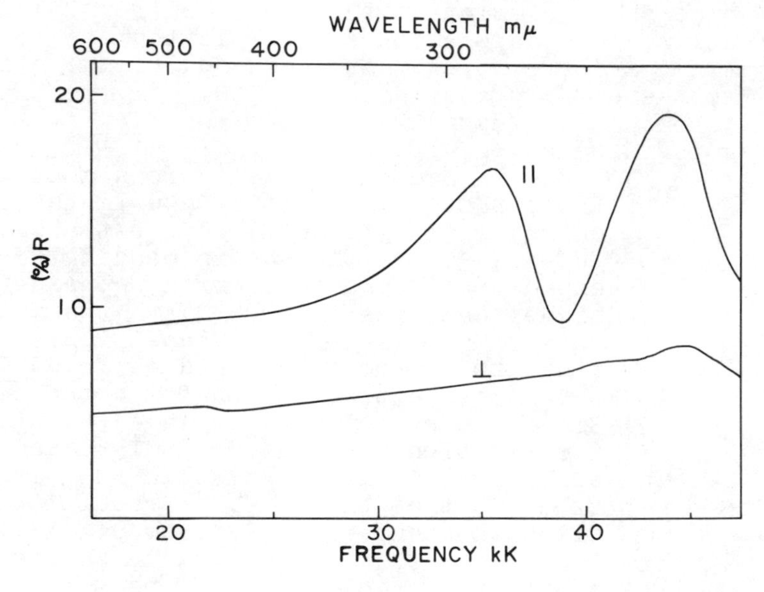

Journal of the American Chemical Society

Figure 8. Polarized single-crystal reflection spectra of K_2PdCl_4. The || curve obtained when electric vector of the incident light vibrates parallel to the molecular planes; ⊥ curve obtained when it vibrates perpendicular to the molecular planes (11).

45.8 kK (14). The findings for $Pt(NH_3)_4Cl_2$ are particularly significant relative to the $5d_{z^2} \rightarrow 6p_z$ assignment, since for $Pt(NH_3)_4^{2+}$ it is difficult to propose an alternative assignment for an out-of-plane band occurring in the 45-50 kK region.

Characterization of the Excited State

Having reached the point where one has developed a reasonably good picture of the single-molecule origin of the crucial solid-state absorption bands in these systems, the next logical step is to seek knowledge regarding the nature of the excited states associated with these bands. In particular, one would like to distinguish between the view that would characterize the transition here as an essentially intramolecular one that is highly perturbed by its environment and those that would invoke varying degrees of delocalization in at least the excited state.

As has been pointed out previously (3), mixed crystal experiments of the general type done by Banks and Barnum (15) and Basu, Cook, and Belford (16) could have an important bearing on this problem. Unfortunately, this earlier work on the glyoximates, which was carried out on colloidal suspensions and NaCl pellets, appears to have yielded spectra that were not particularly well resolved. In any case, these studies resulted in contradictory findings, in that Banks and Barnum reported a single intermediate band lying between those of the pure components in various binary mixtures of the Ni(II), Pt(II), and Pd(II) dimethylglyoximates, while Basu *et al.* reported a series of intermediate bands in such situations.

In light of the above observations, a series of mixed-crystal studies involving single-crystal measurements was undertaken in our laboratories. The results of one of these investigations was recently reported (14). In this work "mixed" Magnus' salts of the form $[Pt(NH_3)_4][PdCl_4]$ and $[Pd(NH_3)_4][PtCl_4]$ were prepared and studied *via* the techniques of specular reflection spectroscopy. The out-of-plane spectrum for $[Pd(NH_3)_4][PtCl_4]$ is shown in Figure 10, where the out-of-plane spectra of MGS and its platinum analog are also shown for comparison. It will be noted that the mixed-crystal spectrum possesses a single band with a width comparable to those of the two "pure" salts located almost midway between them. It also shows an intensity intermediate between those of the pure salts. (All three crystals have essentially the same unit-cell parameters (17). Thus, the metal-metal

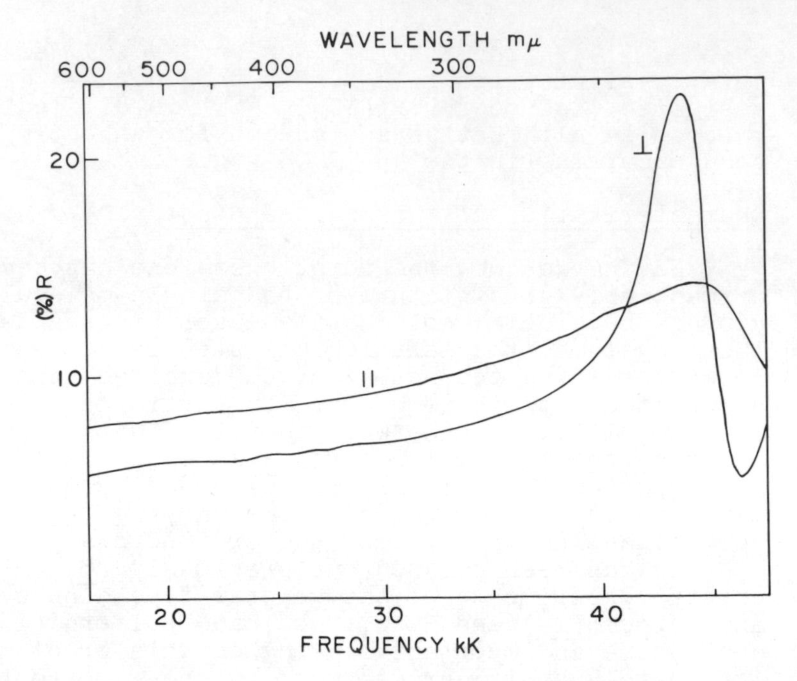

Journal of the American Chemical Society

Figure 9. Polarized single-crystal reflection spectra of K_2PtCl_4. The || and ⊥ are of same significance as in Figure 8.

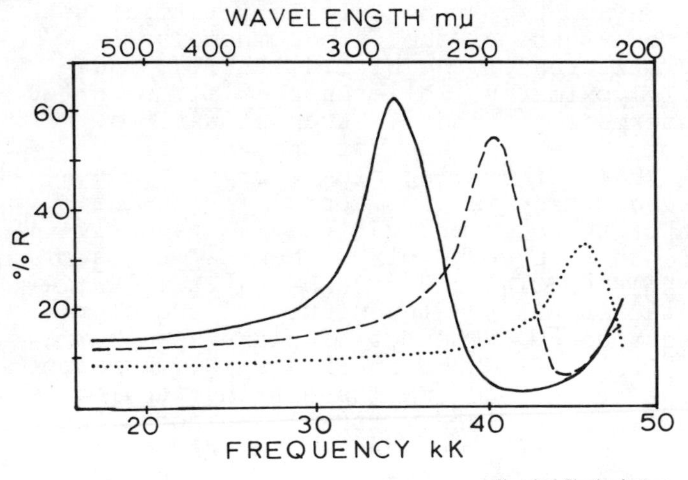

Chemical Physics Letters

Figure 10. Reflection spectra when electric vector of the incident radiation vibrates parallel to the out-of-plane direction of single crystals of $[Pt(NH_3)_4][PtCl_4]$ (——), $[Pd(NH_3)_4][PtCl_4]$ (– – –), and $[Pd(NH_3)_4][PdCl_4]$ (· · ·) (14)

spacing and molar concentrations are approximately the same for each salt.) Perhaps the most surprising fact developed in this investigation, however, is that the out-of-plane spectrum of the second mixed Magnus' salt studied, $[Pt(NH_3)_4][PdCl_4]$, corresponds closely enough to that shown in Figure 10 for $[Pd(NH_3)_4][PtCl_4]$ that it can be represented by the same curve.

These findings certainly provide telling evidence for a strongly delocalized excited state.* It is in fact as if all that matters for this transition is to have the metal ions lined up at the appropriate spacing, with the ligands, beyond this role, simply acting as spectators. As will be noted later, however, the situation has to be somewhat more complex than this.

The 34.5 kK MGS band shown in Figure 6 is the only one apparent in the out-of-plane spectrum in the quartz ultraviolet region and, in particular, no out-of-plane transitions corresponding to the $d_{z^2} \longrightarrow p_z$ bands of K_2PtCl_4 and $Pt(NH_3)_4Cl_2$ are observed in this region for MGS.** In view of the high degree of delocalization now known to exist in these systems, the most reasonable interpretation of this finding would appear to be that the 34.5 kK MGS-band has its parentage in both the cationic and anionic d ⟶ p transitions, both having been red-shifted sharply in MGS. For this reason, current practice in our laboratory is to use the total concentration of complex species present, both anionic and cationic, in computing ϵ-values for MGS-type systems. Recognition is thus made of the fact that both metal-containing species are thought to be contributing to the absorptive process in the Magnus' salts. The ϵ-values and other intensity measures - such as integrated intensities - obtained for the Magnus' salts are then directly comparable to those computed in the usual way for the component ions as "isolated" species. It will be seen that this manner of computing concentrations results in concentration values that are just twice those based on the number of formula units of the Magnus' salt in question per liter, which was the basis of the concentrations used and listed in Reference 10.

---

*The emphasis here is on the excited state, since one finds little evidence in, for instance, the d ⟶ d transitions of MGS or transitions of the NiDMG crystal other than the crucial green band to make one feel that the ground state is extensively delocalized.

**Analogous comments apply to other Magnus-type salts.

Comment should also be made at this point on the differences in the levels of reflectivity shown by the MGS curves of Figure 2 of Reference 10 and Figure 10 of the current publication. The improvement shown in the later measurements primarily reflects improved sample size and quality resulting from recently-developed techniques of sample preparation. The increased reflectivities for MGS in themselves would lead to increased ϵ-values for the Kramers-Kronig absorption spectra that result from them. This effect is, however, opposite to that resulting from the change in the basis of computing the crystal concentration noted above, and one thus obtains roughly the same ϵ_{max} for the d $\longrightarrow$ p transitions of K_2PtCl_4 and $Pt(NH_3)_4Cl_2$ and the 34.5 kK MGS band (12). An increased width of the MGS band then accounts for the approximate 2.5-fold increase in integrated intensity in MGS alluded to in an earlier section of this paper.

Another group of substances whose study is particularly informative concerning the extent of delocalization in the crucial excited state in crystals of the type being dealt with here are the barium salts of the tetracyano complexes of Ni(II), Pd(II), and Pt(II). These compounds, which form single crystals having the general formula $BaM(CN)_4 \cdot 4H_2O$, present a situation that in many respects parallels that described previously for NiDMG. Once again one has a stacking of the complexes in the solid state - with consequent "metal chain" formation - and the development of a strong, relatively low-energy band that has no obvious counterpart in the free ions (5,18,19). The "solid-state" bands here are again thought to have as their free-ion parentage relatively weak bands in the "charge-transfer" region of the solution absorption spectrum (18).

One of the interesting aspects of the solid-state studies on these compounds is the fact that one can prepare mixed crystals of them in which two different complex species appear in the same crystal. One can, for instance, grow crystals of the following type:

$$Ba[Ni(CN)_4]_x[Pt(CN)_4]_{1-x} \cdot 4H_2O^* \qquad (5)$$

*Isomorphism of these crystals with the single-anion crystals has not been ascertained by direct experiment, but is inferred from the gradual gradation of optical properties as one varies the composition of the mixed crystals.

While the mixed-Magnus' studies had the advantage of definite ionic distribution in the metal stacks (one knows that in Magnus' salts the anionic and cationic species alternate along the chain), one is limited, at least with the crystals prepared to date, to a 1:1 ratio of the two metals involved. In the present situation the advantages and disadvantages are just reversed relative to the Magnus' salts: one can vary composition, but the sequencing along the chain is no longer known.

If one prepares crystals with a composition corresponding to x in Equation 5 having a value of approximately 0.5, the spectral results are completely analogous to those observed for the mixed-Magnus case: a single band whose energy and intensity are midway between those of $BaNi(CN)_4 \cdot 4H_2O$ and $BaPt(CN)_4 \cdot 4H_2O$ (18). The unique results of this study are, however, illustrated by Figure 11, where there is plotted the position of the out-of-plane reflection band as a function of $Pt(CN)_4^{2-}$ content for a series of $BaPt(CN)_4$-$BaNi(CN)_4$ mixed crystals in which the $Pt(CN)_4^{2-}$ content is varied over rather wide limits. Regardless of the amount of $Pt(CN)_4^{2-}$ that was introduced into the $BaNi(CN)_4$ crystal, only one reflection band was found, and for even the lowest levels of concentration of one species in the presence of the other, this band was assuming "intermediate" characteristics.

The behavior shown in Figure 11, which is paralleled by other spectral characteristics of the bands in question, presents very convincing evidence for extensive delocalization in the excited state responsible for the strong out-of-plane band in these crystals. The implication here is that if, as appears to be the case, the effects of substitution became apparent after 5% "impurity" is introduced, then one would have to have averaging over a rather large number of centers in the relevant excited state. Otherwise, one might expect the development of one or more "intermediate" bands, along with a band corresponding to the major component, in the earlier stages of substitution (5,16).

General Discussion

The overall picture that emerges here is one of a strong parallelism in the optical phenomena observed for all of the closely stacked crystals discussed. Thus, in the glyoximates, the Magnus' salts, and the

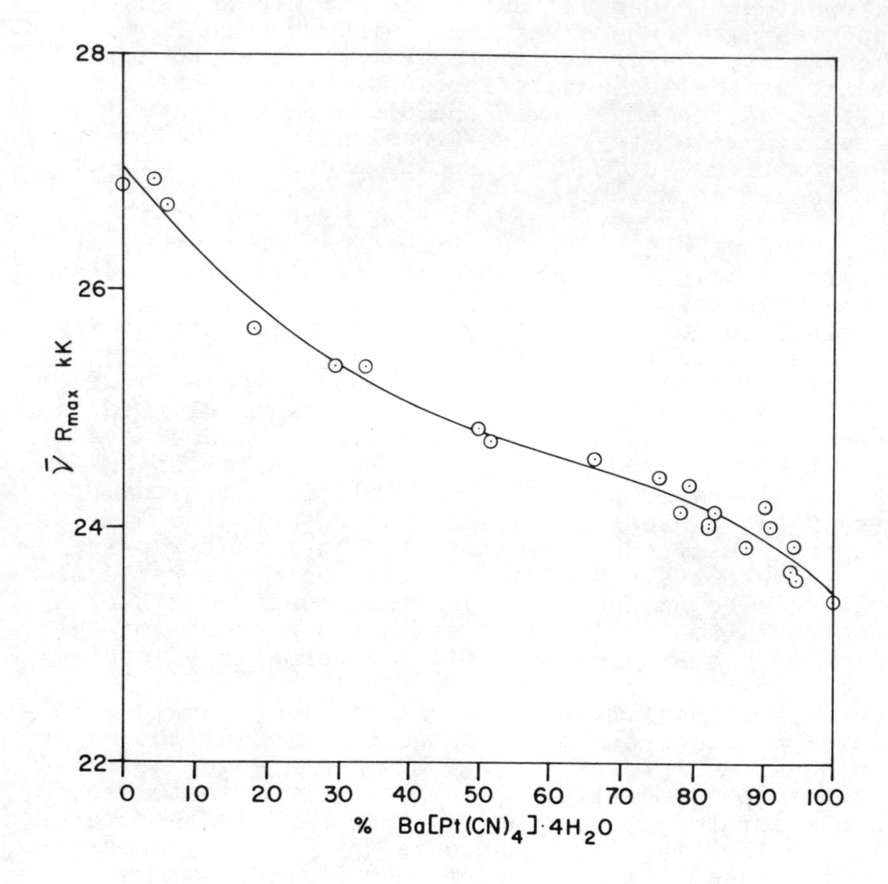

Figure 11. Variation of position of the out-of-plane reflectivity maximum as a function of $BaPt(CN)_4 \cdot 4H_2O$ content in mixed crystals of the form $Ba[Ni(CN)_4]_x[Pt(CN)_4]_{1-x} \cdot 4H_2O$.

cyanides, one finds for the metal-separation dependent out-of-plane band:

1) Similar behavior with respect to relative position as a function of the central metal ion involved (14,16,18).
2) Similar behavior with respect to increasing metal-metal separation (3,10,18).
3) Similar behavior when mixed crystals are studied (14,15,18).*

These observations provide a convincing case for the proposition that one is dealing with basically the same kind of transition in each instance. If one accepts the various arguments presented in the foregoing sections, this transition becomes identified as a highly perturbed $d_{z^2} \longrightarrow p_z$ promotion.

As was remarked in connection with the mixed Magnus' salt studies, it is tempting to view this as a situation in which one is essentially involved with a chain of interacting metal atoms and to deemphasize the role of ligands in attempting to understand the phenomena being dealt with. That this would be an oversimplification becomes clear when one recognizes, for instance, that Magnus' green salt, with a Pt-Pt distance of 3.25 Å (17), has its strong out-of-plane absorption located at 34.5 kK (10), while the comparable absorption in $BaPt(CN)_4 \cdot 4H_2O$, which possesses a Pt-Pt separation of 3.27 Å, occurs at 22.6 kK (18,19). Similarly, the strong out-of-plane absorption band for β-NiEMG occurs at 20.5 kK (3) and the analogous band in $BaNi(CN)_4 \cdot 4H_2O$ is observed at 27.2 kK (18), in spite of approximately equal metal-metal separations (3,21). Finally, one may note that the orthorhombic form of Ni(II) N-methylsalicylaldiminate, whose structure is similar to NiDMG and has a metal-metal spacing of 3.30 Å (22), shows no band analogous to the NiDMG green band throughout the visible and quartz ultraviolet (23). Thus, the nature of the ligand appears to play an important role in the energetics and even the formation of the out-of-plane band of special interest here.

The specific nature of the solid state perturbation operative in these systems is as yet not determined with any certainty. Early speculations on

---

*Preliminary studies on single-crystals of the mixed dimethylglyoximates (20) show that the behavior here parallels that found for the mixed Magnus' salts and the Pt(II) and Ni(II) cyanide mixed crystal with x equal approximately 0.5 in Equation 5.

the electronic structure of these compounds (24) invoked band formation involving the highest filled d_{z^2} and the lowest empty p_z orbitals on the metal ions making up the chains, and some of the latest work in this area continues these ideas (25). Alternate proposals (5,6,16,26) have been made in which electrostatic perturbations, excitonic interactions, and charge-transfer effects are invoked, and in some cases the d ⟶ p assignment to a greater or lesser extent challenged. The experimental information now on hand makes it unlikely that an elementary application of any of these approaches will adequately account for all the observations, although each of them can be discussed in terms of a perturbed d ⟶ p transition. For instance, band theory for MGS (25) predicts separate d_{z^2} bands and separate p_z bands for the anionic and cationic species, and thus more than one electronic transition. Even if one invokes lack of spectroscopic resolving power as the reason for the failure to observe more than one transition, such arguments will not apply to the mixed-crystal experiments. Similarly, in addition to having to invoke in a somewhat *ad hoc* manner some extremely large second-order effects to explain the intensity behavior in the glyoximates and cyanides, it does not appear that exciton theory will, at least in a straight-forward manner, account for the mixed-crystal results. Comments similar to those regarding the exciton mechanism apply to that of charge-transfer.

The foregoing remarks are not meant to imply that exciton effects, charge-transfer, and/or band formation will not figure in the final explanation of the optical properties of these systems. At this point all one can say is that none of them obviously and clearly explains the observations, and more sophisticated approaches appear to be required. Whether one or a combination of the old approaches, or a totally new one, will prove adequate remains to be seen. In any case, a sufficient amount of experimental evidence now exists to provide rather critical qualitative and quantitative tests for any proposals that are put forward.

Acknowledgment:

The author wishes to acknowledge the efforts of the many undergraduates, graduate students, and postdoctoral associates who have been in large part responsible for the account that has been given here.

A number of these workers' names appear in the references that have been cited. Others, who made contributions to the development of the equipment and techniques used in this work in the context of other types of studies, also deserve a share of the credit. Worthy of special note among this latter group are several of my early graduate students: Drs. A. V. Fratini, S. C. Neely, J. Bernstein, and C. J. Eckhardt.

Literature Cited

1. Anex, B. G., Mol. Cryst. (1966) 1, 1.
2. Stewart, R. F., Davidson, N., J. Chem. Phys. (1963) 39, 255.
3. Anex, B. G., Krist, F. K., J. Amer. Chem. Soc. (1967) 89, 6114.
4. Gray, H. B., Ballhausen, C. J., J. Amer. Chem. Soc. (1963) 85, 260.
5. Day, P., Inorg. Chim. Acta, Rev. (1969) 3, 81.
6. Ohashi, Y., Hanazaki, I., Nagakura, S., Inorg. Chem. (1970) 9, 2551.
7. Day, P., Orchard, A. F., Thomson, A. J., Williams, R. J. P., J. Chem. Phys. (1965) 42, 1973.
8. Martin, D. S., Jr., Rush, R. M., Kroening, R. F., Fanwick, P. E., Inorg. Chem. (1973) 12, 301.
9. Day, P., Orchard, A. F., Thomson, A. J., Williams, R. J. P., J. Chem. Phys. (1965) 43, 3763.
10. Anex, B. G., Ross, M. E., Hedgcock, M. W., J. Chem. Phys. (1967) 46, 1090.
11. Anex, B. G., Takeuchi, N., J. Amer. Chem. Soc., accepted for publication.
12. Anex, B. G., Foster, S. I., Fucaloro, A. F., unpublished work.
13. Anex, B. G., Foster, S. I., unpublished work.
14. Anex, B. G., Foster, S. I., Fucaloro, A. F., Chem. Phys. Lett. (1973) 18, 126.
15. Banks, C. V., Barnum, D. W., J. Amer. Chem. Soc. (1958) 80, 4767.
16. Basu, G., Cook, G. M., Belford, R. L., Inorg. Chem. (1964) 3, 1361.
17. Miller, J. R., J. Chem. Soc. (1961), 4452.
18. Anex, B. G., Musselman, R. L., unpublished work.
19. Moncuit, C., Poulet, H., J. Phys. Radium (1962) 23, 353.
20. Anex, B. G., Narain, G., unpublished work.
21. Larsen, F. K., Hazell, R. G., Rasmussen, S. E., Acta Chem. Scand. (1969) 23, 61.
22. Ferguson, J., J. Chem. Phys. (1961) 34, 611.
23. Anex, B. G., Furtado, L. T., unpublished work.
24. Rundle, R. E., J. Phys. Chem. (1957) 61, 45.
25. Interrante, L. V., Messmer, R. P., Inorg. Chem. (1971) 10, 1174.
26. Zahner, J. C., Drickamer, H. G., J. Chem. Phys. (1960) 33, 1625.

20

The Single Crystal Polarized Reflectance Spectra and Electronic Structures of $Rh(CO)_2acac$ and $Ir(CO)_2acac$

THOMAS A. DESSENT and RICHARD A. PALMER
Duke University, Durham, N.C. 27706

SALLY M. HORNER
Meredith College, Raleigh, N.C. 27602

Introduction

Among square planar d^8 complexes it might be expected that those of the lowest formal metal valence (in the absence of "partial oxidation"(1)) would exhibit the most striking evidence of extended metal-metal interaction. This prospect was borne out in 1966 by the work of Pitt and others (2), who first documented the anisotropic electrical conductivity of dicarbonyl acetylacetonato-rhodium(I) and -iridium(I).

As illustrated in Figure 1, these compounds crystallize (as do many of their analogs discussed in this Symposium) with the planar molecules stacked parallel to each other in chains and short metal-metal distances. The compounds were first prepared by Bonati and Wilkinson (3), and the crystal structure work by Bailey, et al (4), has shown the two structures to be isomorphous, with a 3.26Å M-M separation for the rhodium compound and a 3.20Å separation for the iridium compound. The space group is P1 with Z = 2. The molecular symmetry is C_{2v}, although the site symmetry is, of course, only C_1. Pitt found the dc electrical conductivity of single crystals to be 100 times greater along the M-M direction than perpendicular to it for the rhodium compound and 500 times greater for the iridium compound (2). He suggested that we might try to determine the polarized single crystal electronic spectra as an adjunct to the electrical measurements, but it was obvious at a glance, from the lustrous, metallic appearance of the crystals, that absorptivities in excess of 10^3 were involved, which would preclude consideration of single crystal absorption techniques. However, the success of Anex (5) in deriving polarized single crystal absorption spectra from specular reflectance measurements suggested that this might be a useful technique to try in this instance. We have constructed equipment for this purpose and the results of our measurements on these rhodium and iridium complexes form the basis for this paper.

Single Crystal Micro Specular Reflectance Attachment for the Cary-14R

Figure 2A shows a schematic diagram of our polarized micro specular reflectance attachment for the Cary 14R spectrophotometer. We chose to base the equipment on a high quality spectrophotometer rather than starting "from scratch", in order to take advantage of the inherent long term stability of the double beam instrument and its high resolution scanning monocromator. The attachment fits within the extended cell compartment of the Cary 14R; it is not attached to the compartment, but is placed within it and aligned by screws vertically and horizontally. Once aligned, the accessory can be taken out and replaced without further alignment. As shown in Figure 2A, the monochromatic light beam enters from the signal generator compartment through a calcite Glan-Thompson polarizer (P) in the left wall of sample compartment, passes through a quartz substrate beam splitter (BS) and is focused on the surface of the sample (S) by a 36X Cassegrain microscope objective (O). The sample is aligned by a goniometer (G) so as to reflect the beam back through the objective. The light thus impinges on the crystal at nominally normal incidence. The reflected intensity strikes the beam splitter and that which is reflected is directed by plane mirrors (M1, M2, M3) back into the normal path for the instrument and thus to the detector. Concurrently the reference beam is attenuated by neutral density filters and an iris diaphram so as to balance the beams within the range of the slide wire. In order to avoid back reflection from the central mirror of the Cassegrain optics, the objective is used in an off axis orientation, as illustrated in Figure 2B. The design of this equipment is very similar to that used previously by Anex (5) except that it is an easily interchangeable accessory to an existing instrument. It is described more completely elsewhere (6).

The Electronic Structure of Planar d^8 Complexes

For a planar C_{2v} d^8 complex the orientation of the axes given in Figure 1 leads to the highest filled d orbital being the $d_{x^2-y^2}$ (a_1 in C_{2v}) and the lowest unfilled, the d_{xy} (b_2 in C_{2v}). This orientation em-

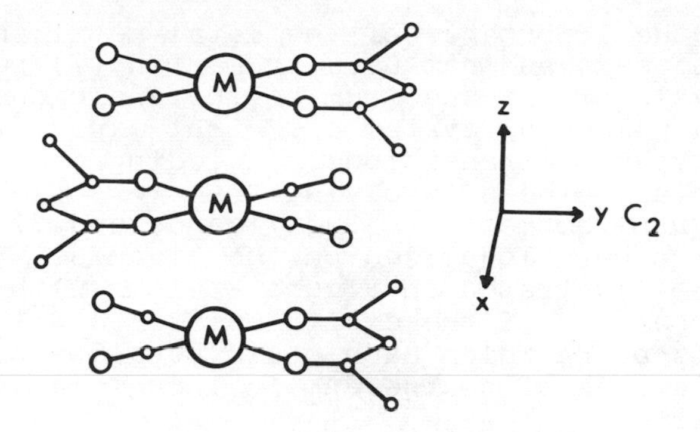

Figure 1. Molecular structure and crystal packing of $M(CO)_2acac$

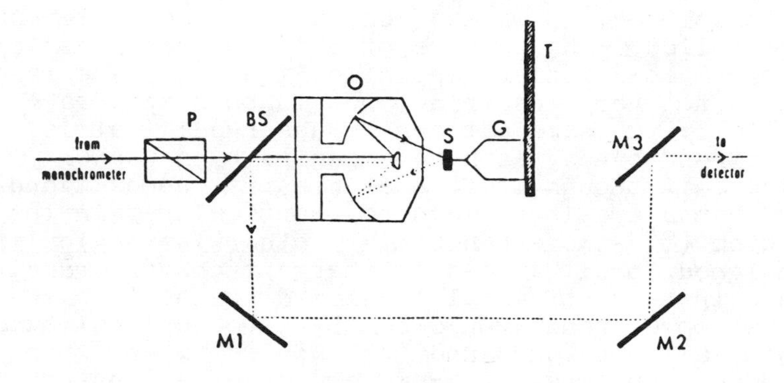

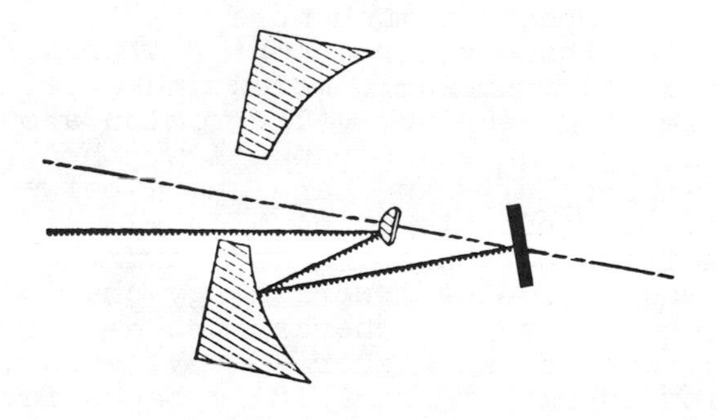

Figure 2. Single crystal microspecular reflectance attachment for the Cary 14R spectrophotometer. (A) Incident (——) and reflected (- - -) light paths. (B) Orientation of light path with respect to objective axis.

phasizes the importance of the metal-metal interaction and is the same as that used by Martin (7) recently in his analysis of the spectrum of the analogous Pt(en)Cl_2. In such a molecular system one might expect to observe several types of transitions, including ones localized on the metal, others involving transfer of electron density from the metal to the ligands and visa versa, and still others localized on the ligands. Some of these are illustrated in Figure 3 along with the symmetry properties of the excited states and the molecular axis along which each would be allowed. The ground state is of course $^1A_1$ in these diamagnetic compounds.

Results and Discussion

Solutions of $Rh(CO)_2$acac and $Ir(CO)_2$acac are only faintly colored, in marked contrast to the lusterously opaque solids. The spectra of the two compounds in chloroform solution are shown in Figure 4. The resolution is not very inspiring, but in both spectra essentially the same pattern is observed; that is, three strong bands, each accompanied by a weaker shoulder. Although it is possible that some ligand-ligand bands are involved here--particularly in the 33 kK region (8)--these bands might otherwise logically be assigned to either d-d or charge transfer transitions. In terms of metal localized transitions the three stronger bands would be the three spin-allowed, symmetry-allowed bands and the shoulders, the spin-forbidden counterparts. However, observations on Pt(II) complexes would suggest that the spin-forbidden bands should be expected at considerably lower energy than those observed here. Charge transfer, on the other hand, appears a much more likely explanation--at least for the more prominent features at 30.5 and 39.0 kK for the rhodium complex and 29.2 and 39.2 kK for the iridium complex. (The region around 33 kK likely involves an acac transition, though the difference in absorptivity and band shape between the two compounds may indicate a third prominent charge transfer transition here also.) The larger size of the iridium would predict lower energy charge transfer than in the rhodium analog, whereas the d-d transitions would be expected at higher energy. Thus, the relatively low energy of the iridium bands favors the charge transfer assignment. These tentative assignments are summarized in Table 1.

Before we completed our specular reflectance accessory we attempted to quantify the intense absorp-

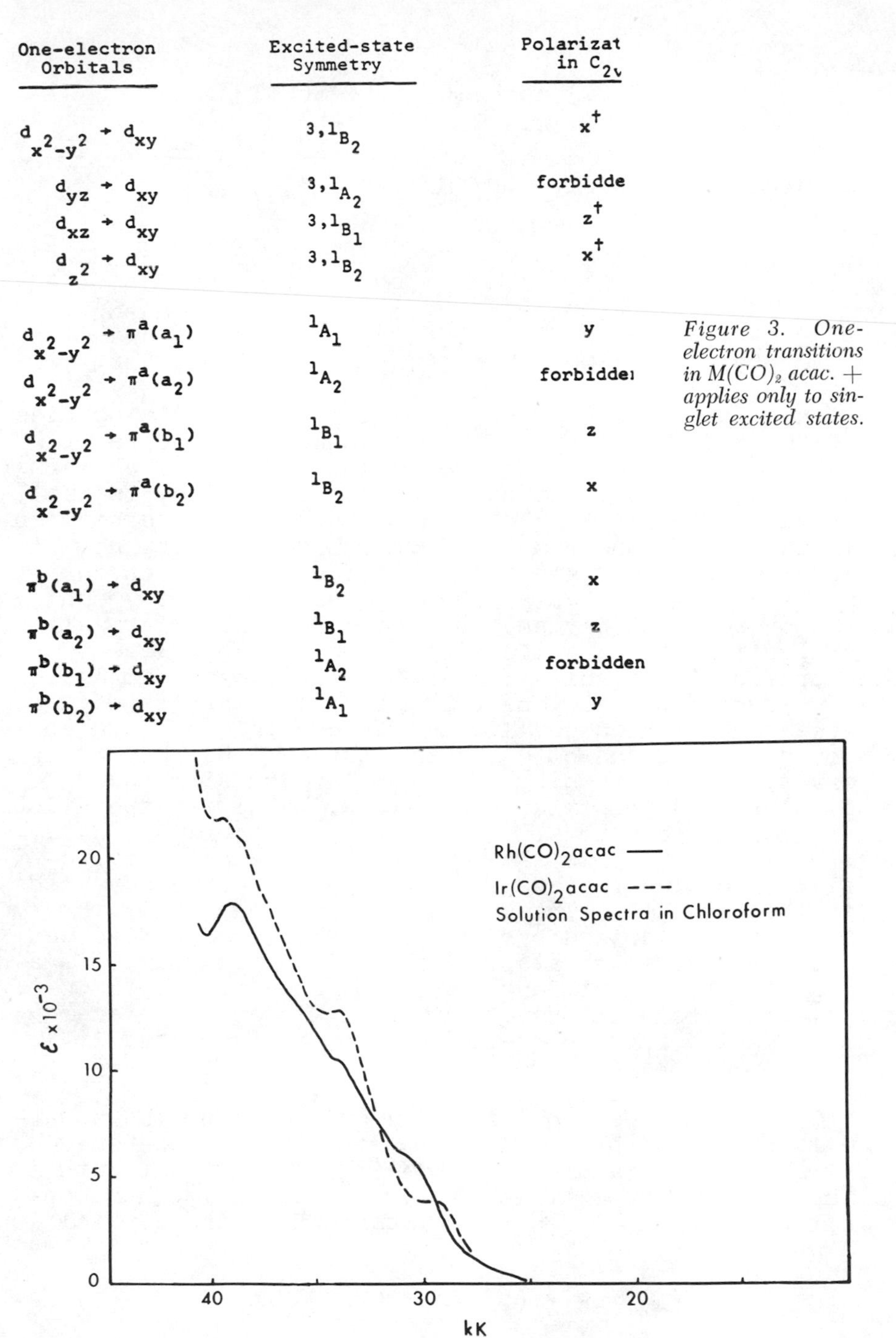

One-electron Orbitals	Excited-state Symmetry	Polarizat in C_{2v}
$d_{x^2-y^2} \rightarrow d_{xy}$	$^{3,1}B_2$	$x^{\dagger}$
$d_{yz} \rightarrow d_{xy}$	$^{3,1}A_2$	forbidde
$d_{xz} \rightarrow d_{xy}$	$^{3,1}B_1$	$z^{\dagger}$
$d_{z^2} \rightarrow d_{xy}$	$^{3,1}B_2$	$x^{\dagger}$
$d_{x^2-y^2} \rightarrow \pi^a(a_1)$	$^1A_1$	y
$d_{x^2-y^2} \rightarrow \pi^a(a_2)$	$^1A_2$	forbidde
$d_{x^2-y^2} \rightarrow \pi^a(b_1)$	$^1B_1$	z
$d_{x^2-y^2} \rightarrow \pi^a(b_2)$	$^1B_2$	x
$\pi^b(a_1) \rightarrow d_{xy}$	$^1B_2$	x
$\pi^b(a_2) \rightarrow d_{xy}$	$^1B_1$	z
$\pi^b(b_1) \rightarrow d_{xy}$	$^1A_2$	forbidden
$\pi^b(b_2) \rightarrow d_{xy}$	$^1A_1$	y

Figure 3. One-electron transitions in $M(CO)_2$ acac. † applies only to singlet excited states.

Figure 4. Solution spectra of $M(CO)_2$acac

TABLE 1

$M(CO)_2$acac SOLUTION SPECTRA

M = Rh		M = Ir		Assignment
Energy (kK)	ϵ_{max}	Energy (kK)	ϵ_{max}	
27.0	500	26.5	500	d→d
30.5	4000	29.2	3000	d→Lπ
32.0	500	31.9	200	d→d
33.8	2000	33.8	2500	d→Lπ
35.5	800	36.3	200	d→d
39.0	4500	39.2	3000	d→Lπ

tion of these compounds in the solid phase giving rise to the metallic luster, by measuring the spectra of mulls and KBr pellets. The mull spectra were of rather poor quality and, although the KBr disks did, in fact, reveal a strong visible band in both compounds between 14 and 20 kK, the results from run to run were not reporducible and gave evidence of decomposition during preparation--particular when the samples were ground in the presence of oxygen. Furthermore, the polarization properties of the solid state band or bands were of course not available from these nonoriented samples.

From material generously supplied by Pitt we were able to grow high quality single crystals of both compounds with faces of ca. 1 mm square. The rhodium crystals were grown by slow sublimation at reduced pressure and the iridium crystals, by slow evaporation of acetone solutions.

In Figure 5 is shown the polarized specular reflectance electronic spectrum of $Rh(CO)_2$acac measured in the Cary 14R. The most prominent feature is the intense z-polarized band at 19.1 kK. In Figure 6 we see the analogous data for the iridium compound. Again the most prominent feature of the spectrum is the z-polarized band in the visible. It is at lower energy than in the rhodium analog--so low in fact that it is the lowest energy band observable within the range of the GaAS detector.

The data were obtained by substracting the apparent absorption of the instrument with the reference mirror in the sample position (A_{ref}) from that with the crystal in position (A_{cryst}), with all other conditions the same. The reflectance R of the crystal is then given by

$$R_{cryst}(\omega) = R_{ref}(\omega) \times 10^{-(A_{cryst}-A_{ref})} \qquad (1)$$

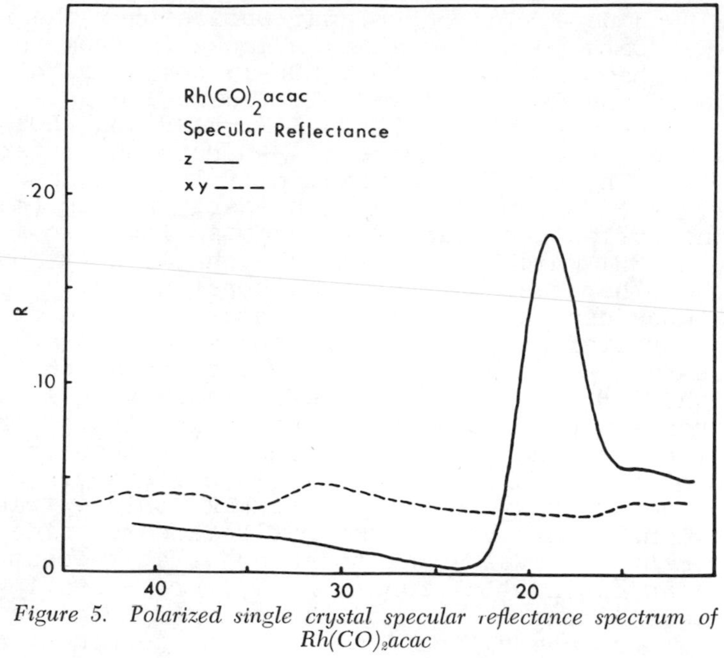

Figure 5. Polarized single crystal specular reflectance spectrum of $Rh(CO)_2acac$

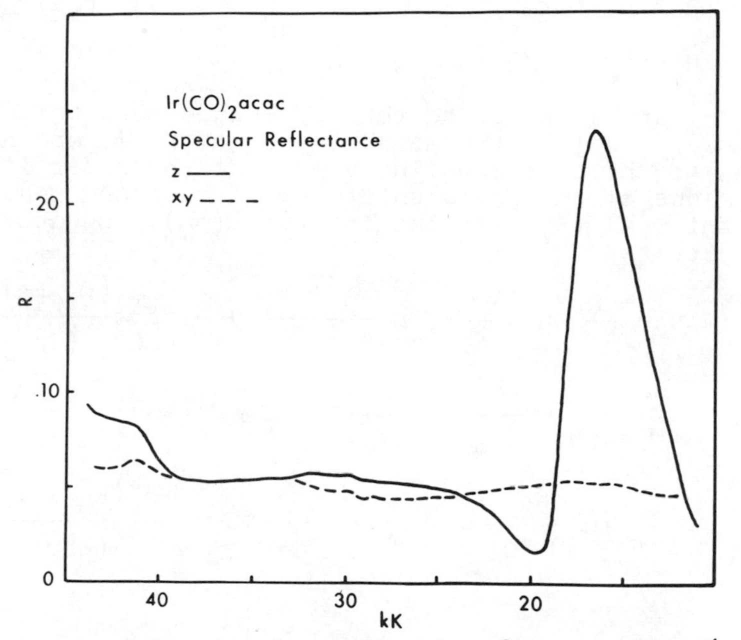

Figure 6. Polarized single crystal specular reflectance spectrum of $Ir(CO)_2acac$

$R_{ref}(\omega)$ was determined by Planar Optics over the range 5-50 kK. Data points were taken manually from the Cary recorder chart and fed into a PDP-15 computer which calculated the $R_{cryst}(\omega)$ and plotted reflectance vs. energy spectra (programs PSRC, SRI, FIXR and PLOTR). A values used were the average of 3-5 runs on different crystals. The standard deviation of the raw data was normally $\leq$ 5% except in the high energy region (<40kK).

The reflectance data were used as input to a program (KRAMER) which calculated the phase shift and from this the other necessary optical constants n, k, and ϵ. The methods used are similar to those reported by Anex (5) and by Stern (9) and others and amount to a Kramers-Kronig analysis. In such a calculation the phase shift θ depends on the reflectance at all values of ω from $-\infty$ to $+\infty$, and thus some estimation of the reflectance at energies higher and lower than the region of observation (11-40 kK) is necessary. Several procedures have been used to make this extrapolation, including setting the reflectance outside the measurement interval equal to an adjustable constant (10) or an exponential function (9), or generating a dummy band in one or both of the outlying regions (5).

We found a modification of Stern's exponential method (9) most satisfactory. In this method

$$R = R_1 \ (\omega/\omega_1)^p \qquad \omega < \omega_1 \tag{2a}$$

$$R = R_2 \ (\omega_2/\omega)^q \qquad \omega > \omega_2 \tag{2b}$$

where R_1 and ω_1 refer to the reflectance and energy at the lower limit of the measurement interval, and R_2 and ω_2 to the corresponding values at the upper limit. The values of the phase shift for ω below the measurement interval (θ_1), in the interval (θ_2), and above it (θ_3) are then

$$\theta_1(\omega_0) = \frac{1}{2\pi} \ln \frac{R(\omega_0)}{R_1} \ln \frac{\omega_0 + \omega_1}{\omega_0 - \omega_1} + \frac{p}{\pi} \sum_{n=0}^{10} \frac{(\omega_1/\omega_0)^{2n+1}}{(2n+1)^2} \tag{3a}$$

$$\theta_2(\omega_0) = \frac{\omega_0}{\pi} \int_{\omega_1}^{\omega_2} \frac{\ln R(\omega) - \ln R(\omega_0)}{\omega_0^2 - \omega^2} \, d\omega \tag{3b}$$

$$\theta_3(\omega_0) = \frac{1}{2\pi} \ln \frac{R_2}{R(\omega_0)} \ln \frac{\omega_2 - \omega_0}{\omega_2 + \omega_0} + \frac{q}{\pi} \sum_{n=0}^{10} \frac{(\omega_0/\omega_2)^{2n+1}}{(2n+1)^2} \tag{3c}$$

From equation (3) values of $\theta = \theta_1+\theta_2+\theta_3$ were calculated using values of the parameters p and q empirically chosen according to the following criteria:

(a) θ is positive on the interval $\omega_1 < \omega < \omega_2$;
(b) the calculated extinction coefficient, k, is nearly zero in the neighborhood of ω_1;
(c) the smallest possible values of p and q are used to obtain the conditions in (a) and (b).

Criterion (b) was established because it was assumed that the complexes studied are nearly transparent in the neighborhood of ω_1 (ca. 11 kK). Criterion (c) was established so that the calculated results would have the least possible dependence on the empirical parameters. No workable criterion could be established for choosing the high-frequency parameter, q. It was expected that the complexes would have rather strong absorption bands beyond the high-frequency cutoff, ω_2 (ca. 45 kK), and it was impossible to guess the magnitude of the phase change which would be imparted by these bands. Since the calculated value of θ near ω_2 could be widely varied, depending on the value q chosen, not much confidence can be placed in the magnitude of θ above 30 to 35 kK. The shape of the curve was not changed to an appreciable extent, however, by wide variation of q; therefore the general features of the calculated spectra are valid.

Once suitable values of θ were established, the refractive index, n, and the extinction coefficient, k, were calculated, and lastly, values of the molar absorptivity, ϵ, were computed from k by the following expression:

$$\epsilon(\omega) = \frac{4\pi\omega k(\omega)}{2.303M} \qquad (4)$$

where ω is the frequency in cm^{-1} and M is the concentration of the crystalline complex in moles per liter.

Figure 7 shows the derived molar absorptivity of $Rh(CO)_2acac$. Four bands are evident, the most prominent being at 19.1 kK with an ϵ of $15x10^3$, allowed, as noted before, only along the metal-metal axis (z). Of particular interest also are the two lower energy bands at 14.6 and 15.4 kK, one xy, the other z polarized. Although, as Anex (11) has pointed out, spurious bands are often observed adjacent to regions of high reflectivity, it is interesting to note in this case a band is observed in the xy polarization even though there is no adjacent region of high reflectivity in that orientation of the polarizer. The fourth band is also xy polarized and is seen at 32.5 kK.

Figure 8 shows the corresponding results for the iridium analog. In this spectrum only the strong z-polarized band at 17.6 kK seems to be well defined. This is almost certainly analogous to the 19.1 band of the rhodium crystal. Because of the low-energy limit of the detector it was not possible to cover the region where weaker bands corresponding to the 14.6 and 15.4 kK bands of the rhodium compound were observed.

In assigning these spectra, either a molecular orbital band model or an exciton model is possible. Both of these models are considered in detail in other papers in this symposium with reference to closely analogous "square planar" complexes. Our own consideration of these possibilities led us first to estimate the gap which might develop between a filled d_{z^2} molecular orbital band and an empty p_z band. The orbital energies were calculated by a modified Wolfsberg-Helmholtz scheme. In the absence of available molecular orbital overlap integrals, the overlaps were based on <u>atomic</u> orbitals. These relatively crude calculations lead to estimated band widths for the rhodium d_{z^2} and p_z bands of 6 and 50 kK, respectively, with a band gap of 26 kK. For the iridium compound the p_z band width is estimated to be 13% smaller, but the d_{z^2} band width, 50% greater, leading to an appreciably smaller band gap than in the rhodium complex. This would predict lower transition energy and greater conductivity for the iridium compound, in agreement with observation. Although the size of the estimated band gaps is reasonable compared to the 19.1 and 17.6 kK energies of the prominent bands, this model offers no explanation for the lower energy bands observed in the rhodium compound. Presuming that these bands are not artifactual, this is a serious drawback to the simple MO model.

The above considerations led us to explore the predictions of an exciton model. Martin (7) has recently presented arguments for assigning the solid state bands in $Pt(en)Cl_2$ to both Frenkel and ionic exciton transitions. Assuming the same basic dipolar model (7), one should expect in these rhodium(I) and iridium(I) complexes also that Frenkel exciton states based on z-polarized bands in the isolated molecule should be shifted <u>ca.</u> 12 kK to lower energy. The absorptivities of the prominent z-polarized bands of <u>ca.</u> $15x10^3$ are just three times those of the solution spectral transitions at 30.5 and 29.2 kK for rhodium and iridium, respectively. This is the required relationship between the intensities of the z-polarized

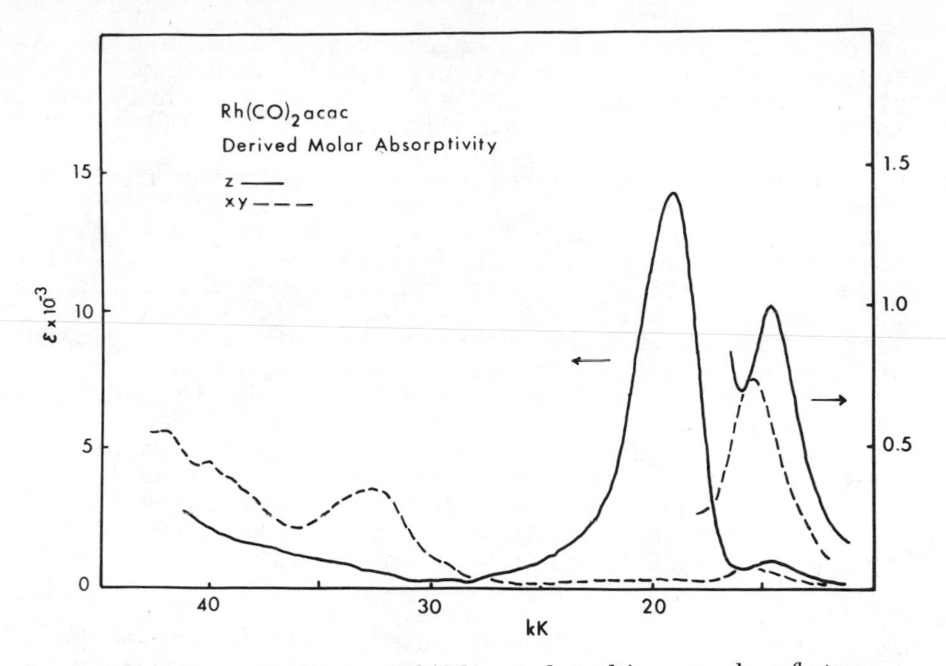

Figure 7. Molar absorptivity of $Rh(CO)_2acac$ derived from specular reflectance

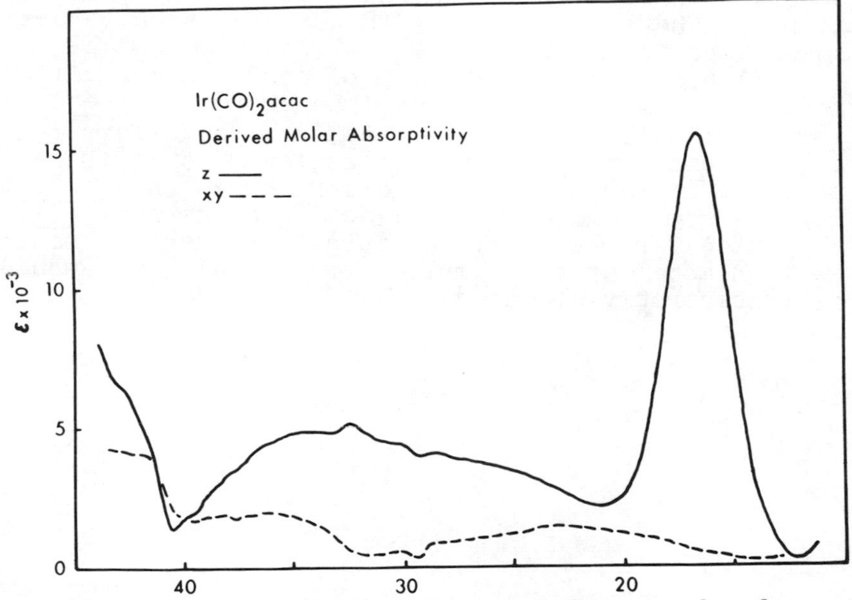

Figure 8. Molar absorptivity of $Ir(CO)_2acac$ derived from specular reflectance

spectrum and those of the solution spectrum, and supports the assignment of these bands to Frenkel excitons based on the $d_{x^2-y^2}$ (a_1) → L_π (b_1) transitions in the isolated molecules, with Davydov splittings of 11.4 and 11.6, respectively.

Turning to the weaker transitions in the rhodium crystal, we might consider first those at higher energy than the 19.1 kK band. Those at 32.5 and _ca._ 40 kK appear most likely to be unperturbed molecular transitions. The tail on the high energy side of the 19.1 kK band could conceivably be due to spin-forbidden d-d transitions which gain intensity by spin orbit coupling and excited state mixing with the exciton band.

The two bands which appear at 14.6 kK (z) and 15.4 kK (xy) are more difficult to rationalize. The z-polarized band may also be associated with a Frenkel exciton state, based perhaps on the d → L_π transition at 27.0 kK. However, since a Davydov splitting can only increase the energy of an xy-polarized transition, the band at 15.4 kK cannot be the result of a Frenkel state. On the other hand, it could be due to an _ionic_ exciton state similar to those described by Martin (_7_) for $Pt(en)Cl_2$. Such states involve excitation of an electron from a filled d_{xz} or d_{yz} orbital on one molecule to an empty d^*_{xy} orbital on an adjacent molecule. One of these excitations (d_{xz} → d^*_{xy}) is y-polarized and the other, x-polarized. If this transition is derived from the 32.0 kK d-d band in the solution spectrum, the observed shift is 16.6 kK, which is within reasonable bounds. The assignments of the crystal spectra are summarized in Table 2.

Although the possibility of confirming additional ionic exciton states as described above is appealing, and the flexibility of the exciton model offers reasonable explanations for all the features of these spectra, if further investigation should show that the lower energy bands in $Rh(CO)_2acac$ are (as Anex (_11_) has suggested) artifactual, the apparent advantage of the exciton model over the molecular orbital model would be considerably diminished.

TABLE 2

$M(CO)_2acac$ SINGLE CRYSTAL ABSORPTIVITY (DERIVED)

Energy (kK)	$\epsilon_{max}(z)$	$\epsilon_{max}(xy)$	
M = Rh			
14.6	1000	--	d→d Frenkel exciton
15.4	--	720	d→d Ionic exciton
19.1	14000	--	d→L_π Frenkel exciton
24	1000	--	d→d Spin forbidden(?)
32.5	--	3500	d→L_π
40	--	1000	d→L_π
M = Ir			
17.6	15000	--	d→L_π Frenkel exciton
22	--	1400	d→d Spin forbidden(?)
25	4000	--	?
35	5000	2000	d→L_π (?)

Literature Cited

1. Krogman, K., et. al., in paper no.   of this Symposium discuss "partially oxidized" compounds.
2. Pitt, C. G., et. al., J. Amer. Chem. Soc. (1966) 88, 4288.
3. Bonati, F. and Wilkinson, G., J. Chem. Soc. (London)(1964) 3156.
4. Bailey, N. A., et. al., Chem. Comm. (1967) 1041.
5. Anex, B. G., Mol. Cryst. (1966) 1, 1.
6. Dessent, T. A. and Palmer, R. A., to be published.
7. Martin, D. S., et. al., J. Amer. Chem. Soc. (1971) 93, 5433.
8. Cotton, F. A., Harris, C. B. and Wise, J. J., Inorg. Chem. (1967) 6, 909.
9. Stern, F., Solid State Phys. (1963) 15, 338.
10. Schatz, P. N., et. al., J. Chem. Phys. (1963) 38, 2658.
11. Anex, B. G., Ross, M. E. and Hedgcock, J. Chem. Phys. (1967) 46, 1090.

21

Directed Synthesis of Linear Chain Metal Complexes with Well-Defined Cooperative Properties

R. ADERJAN, D. BAUMANN, H. BREER, H. ENDRES, W. GITZEL, H. J. KELLER, R. LORENTZ, W. MORONI, M. MEGNAMISI-BÉLOMBÉ, D. NÖTHE, and H. H. RUPP

Anorganisch-Chemisches Institut der Universität Heidelberg, D 6900 Heidelberg 1, Im Neuenheimer Feld 7, GFR

Abstract

A simple band model deduced from the one-electron metal functions split by an electric field of the ligands can be successfully used as a guideline in the directed synthesis of linear chain transition metal complexes with strong intermolecular interactions. One-dimensional metallic behaviour can be found for instance in planar transition metal compounds with 8 d-electrons but only if small and strongly π-electron accepting equatorial ligands are used. Suitable ligands are carbonyl and isonitrile groups in connection with iridium(I) and rhodium(I) ions. I.r., ^{1}H-n.m.r., ^{195}Pt-n.m.r., e.s.r., u.v. and ^{193}Ir-Mößbauer spectra show that increasing bulkiness and increasing electron donating properties of the ligands considerably decrease the strength of the intermolecular metal interactions. A planar transition metal complex with less than 8 d-electrons per metal should be far better suited for the formation of 1d-metals. But because of the strong Lewis acid activity of these compounds the donating solvent molecules occupy the axial positions of the central metal atom and hence prevent the self-association necessary for the building up of a linear chain. Preparative procedures to overcome these problems are proposed. The structure and chemical properties of one of the obtained "mixed valence" solids which contains a linear I_3^--chain additionally is especially interesting since two non degenerate conduction spines are running closely parallel to each other through the lattice.

I. Introduction

Very recently it was shown by a.c. and d.c. conductivity measurements as well as by optical, X-ray and neutron scattering experiments that at least one of the many linear chain metal complexes - a platinum complex of stoichiometry $K_2[Pt(CN)_4]Br_{0.3} \cdot 3H_2O$ - behaves like a one-dimensional metal at room temperature (1). While the evidence for a 1d-metallic state in this "mixed valence" platinum system at room temperature is overwhelming and widely accepted, the interpretation of the phase transitions which occur at lower temperatures and which destroy the 1d-metallic state seems to be controversial. This controversy motivates the search for other 1d-metallic transition metal complexes.

In order to study the aptitude of linear metal chain compounds to form 1d-metals, first principles must be developed to direct a synthesis of the best suited molecules for building up linear 1d-metallic chains. This problem with respect to the highly conducting organic charge transfer salts of the "TCNQ" type was recently discussed by Heeger and Garito (2) and will be outlined in the following sections using planar metal complexes as lattice elements. These planned syntheses give rise to the following questions: "Which kind of molecular parameters of the individual elements govern the properties of these solids?" and "What special rules can be set up to prepare suitable compounds in a systematic way?" In order to find such rules for the synthesis of tailor-cut molecules, the electronic structure of planar transition metal complexes, which crystallize preferably in columnar stacks with direct metal contacts, has to be investigated.

II. Electronic structure of suitable metal complexes

The intermolecular contacts in the known linear chain metal compounds with direct metal-metal bonds are brought about mainly by the central metals d-electrons. Though most of the complexes have covalently bonded ligands an overlap of d-functions of the central metal ions with appropriate ligand functions can be omitted in a first approximation, to keep the model as clear as possible. A very simple and perhaps oversimplified pure crystal field approach can be used to explain the electronic structure of the solids. The central metal ions of planar transition metal complexes residing in columnar stacks are exposed to a tetragonally distorted octahedral ligand field (fig.1) exerted by the four equatorial ligands and the two adjacent metal ions in the axial positions. The degree

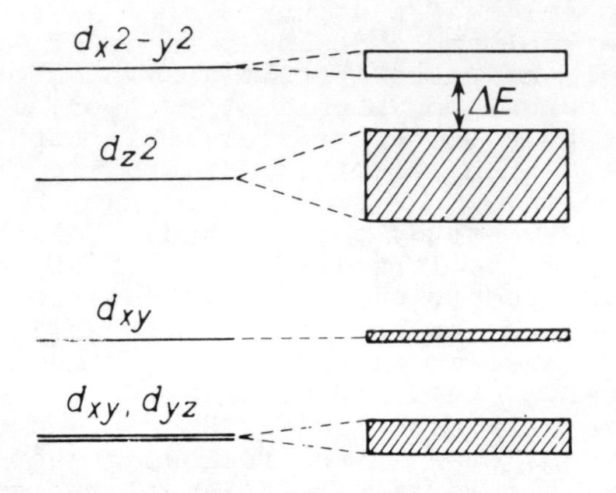

Figure 1. Energies of d*-electrons in a one-dimensional linear chain* d^8*-configurated transition metal complex*

of tetragonal distortion depends on the relative "strength" of the axial and equatorial ligand groups. Equal strength of axial and equatorial ligands results in an "octahedral" field with degenerate d_{z^2} - and $d_{x^2-y^2}$ - levels. If intermolecular interactions are included bands are obtained the width of which depends upon the degree of overlap between the one electron d-functions. The best intermolecular overlap is assumed between the d_{z^2}-functions, while the orbitals localized mainly in the complex plane are less suited for interactions and therefore give very small bands. Participation of ligand orbitals could be added easily if necessary but would not give a qualitatively different picture. This model could in contrast to earlier proposed energy level diagrams (3, 4) give a more adequate description of the electronic structure of these solids, because an appreciable mixing of metal s- and p-states seems fairly unlikely with respect to the large energy differences of about 15 eV involved (5).

What kind of consequences can be drawn from this scheme? This question is first answered in context with d^8-complexes. Most of the known linear metal chain compounds with direct metal-metal contacts contain metal ions with 8 d-electrons (4). In this case, all the bands are filled with electrons with the exception of the $d_{x^2-y^2}$-band. This model predicts if the tetragonal distortion is appreciable and the energy gap ΔE is quite large, a semiconducting behaviour and only weak intermolecular interactions. On the other hand the model could explain 1d-metallic behaviour of d^8-compounds as well. Strongly electron accepting ligands will decrease the energy separation between the d_{z^2}- and $d_{x^2-y^2}$-orbitals and the energy gap would vanish in a formally "octahedral" field. The former case is that what is actually found in most known linear chain d^8-metal compounds. But this model clearly hints at the strengthening of the predicted weak interactions in d^8-complexes with large ΔE. This goal could be achieved if it is possible to:

i) diminish the energy gap between the d_{z^2} and $d_{x^2-y^2}$-bands by using suitable electron accepting ligands,

ii) oxidize transition metal complexes crystallizing in columnar stacks, so the filled d_{z^2}-band will be depopulated and a partially occupied band results no matter how large ΔE is,

iii) synthesize linear chain metal complexes with less than 8 d-electrons per metal site.

III. Choice of Spectroscopic Methods for Identification of Co-operative Phenomena

After having set up rules for selecting suitable compounds there remains the question of how to identify the variation of intermolecular interactions in different complexes on a quantitative basis. It is necessary to find spectroscopic methods which are able to identify intermolecular interactions and to classify their strength. The methods should work on polycrystalline samples since the preparation of large single crystals is often very difficult.

In general, the usually successful spectroscopic methods: optical, e.s.r., n.m.r., e.g. can be used to identify collective electron behaviour. Especially obvious are the results of X-ray structure determination (M-M distances) and of optical methods in the visible and near infrared region. The intermolecular interactions cause an intense electronic transition, which in single crystals is linearly polarized parallel to the direction of the metal chains and often appears as a striking anisotropic metallic lustre in cases where strong interactions take place (1).

There are two methods namely n.m.r. and Mößbauer spectroscopy, which can and have been used to measure electron density at the position of metal nuclei in the chain and so should be able to identify co-operative electronic interactions. Results of these two methods are later discussed in connection with the iridium and platinum compounds. Other methods; e.s.r. and neutron scattering experiments have been used by different groups to examine the collective state in linear chain compounds.

IV. Choice of suitable transition metal compounds

What we had to do first of all was to synthesize linear chain metal compounds with moderate intermolecular interactions, to vary their molecular constitution systematically and to investigate their changing solid state properties by spectroscopic methods. It is reasonable to start with known linear chain transition metal complexes, the structures of which are proved by X-ray analysis. On these selected species a systematic variation of ligand properties by preparative chemical methods could be achieved without major structural changes. The properties of d^8-compounds were investigated firstly. The following types of complexes with linear metal chains and 8 d-electrons per central metal ion are known:

1 Tetrahalogeno metal anions of the type $[M X_4]^{n-}$ with X = Cl^-, Br^-, CN^- like $[Pt(CN)_4]^{2-}$ (3).
2 Complex-cations of the type $[M L_4]^{n+}$ with L a monodentate neutral donor like NH_3 or isonitrile. The cation $[Rh(CNR)_4]^+$ is a simple example for this group of compounds (6, 7).
3 In mixing both types of complexes stacks with alternating cations and anions are obtained (Magnus' Green Salt e.g.) (4, 8).
4 Planar dicarbonylmetal(I) species containing the metals rhodium and iridium. To this group belong the numerous derivatives of (Pentan-2,4-dionato) dicarbonyliridium(I) and rhodium(I) (9, 10, 11). A simple compound of this type is $Ir(CO)_3Cl$ (12).
5 Bis(α,ß-diondioximato)metal(II) complexes (4).
6 Bis(maleodinitriledithiolato)metal(II) compounds (13).

Other known special examples of compounds with columnar structure and d^8-central metal-ions are classified into one of the above groups.

There are only a few compounds known earlier which could be classified as "mixed valence" compounds with direct metal-metal interactions and are as follows:

The most famous examples are the "mixed valence" solids derived from the platinum(II) complex ions $[Pt(CN)_4]^{2-}$ and $[Pt(C_2O_4)_2]^{2-}$ (3).

Halogenocarbonyliridium and possibly rhodium compounds synthesized firstly by Malatesta (14).

Partially oxidized Bis(α,ß-diondioximato)metal(II) compounds containing additional linear iodine chains (15).

V. Systematic Variation of Ligand Parameters

There are several well-known examples of the variation of metal-metal interactions with the kind of ligand. The compounds K_2PtCl_4 and its derivatives may be used as a simple example of the changes which can be done on this type of complexes and what solid state consequences will be expected. The intermolecular interactions between the platinum atoms are almost zero in solid K_2PtCl_4. A large energy gap ΔE causes solid state properties to show no sign of collective electron behaviour. Therefore, the intermolecular platinum-platinum distance is quite large (4, 13 Å) (16). However, as pointed out in the described crystal field model, the intermetallic interactions can be enhanced heavily by using more electron withdrawing ligands than chlorine; especially π-acceptors (e. g. cyano-groups).

In a simple energy scheme this results in a lowering of the energy of the $d_{x^2-y^2}$-band diminishing the energy gap. Actually the Pt-Pt-distances in $[Pt(CN)_4]^{2-}$ compounds are considerably decreased and some of the solid state properties of compounds of this type clearly show the effect of anisotropic intermolecular interactions (3). Comparison of different solid $M_2[Pt(CN)_4]$ or $M'[Pt(CN)_4]$ species points out that at least one more important parameter besides the "electronic" factor has to be considered, the steric requirements of the lattice elements. This is proved by the dependance of intermolecular metal interactions on hydration sphere around the cations (3).

Furthermore the interaction can be enhanced by adding electrostatic attraction between lattice elements instead of repulsion in the purely anionic stacks of $PtCl_4^{2-}$. Magnus' Green Salt with alternating $[Pt(NH_3)_4]^{2+}$ and $PtCl_4^{2-}$ units and Pt-Pt distances of 3.25 Å shows considerably enhanced intermolecular interactions with an energy gap ΔE of about 0.7 eV (17).

But these compounds are not well suited for our special purpose. In order to find a relation between the electron accepting properties of a ligand and its sterical requirements on one side and the degree of intermolecular electron exchange on the other the ligands have to be varied systematically over a wide range but in small steps. For this reason we prepared and investigated a group of isonitrile derivatives the type $[Pt(CNR)_4][PtCl_4]$ with R = aryl and varied the ligands systematically by using different substituted isonitriles (18). As expected on the basis of the simple model, electron withdrawing groups had an effect of strengthening the metal-metal interactions and hence the co-operative properties while electron donating substituents on the ligand are weakening these interactions. This is easily shown by studying the optical properties of the compounds. Complexes with strongly electron accepting groups, decrease of ΔE, absorb strongly in the red part of the visible spectrum and are deep blue or violet in the solid state. Complexes with donating isonitrile groups are absorbing in the blue part and appear yellow for this reason. By using bulky isonitrile ligands the interactions between the metal ions are decreased as well but this effect is minor compared to the "electronic" effect. Complexes of this type, with bulky and donating ligands, look yellow generally while the ones with small accepting ligands look blue, show a typical copper-like metallic lustre which is linearly polarized parallel to the needle axis of the crystals (18).

The same systematic change in solid state properties with the variation in equatorial ligands can be observed in planar chelatodicarbonylrhodium(I) and iridium(I) compounds (group 4) some of which cristallize in columnar stacks (9). The anisotropic physical properties of these solids resulting from intermolecular interactions, which were the subject of intense investigations several years ago (19), can be varied by chemical synthesis. Complexes of this type, e. g. of stoichiometry $Ir(CO)_2(L-L)$, can be obtained easily by substituting one carbonyl group and a chloride ligand in $Ir(CO)_3Cl$. The synthesis proceeds especially well if the tricarbonylchloroiridium(I) is reacted with the "Triton B" (triphenylbenzylammonium hydroxide) salt of the ligand (10).

Some conclusions upon the intermolecular interactions can be drawn by looking at their optical absorption spectra. The interactions are diminished with increasing size of the ligands and by substituting donating groups and the absorption is shifted to shorter wave lengths. The electric conductivity is decreased drastically in compounds with large organic dye bases as ligands(20).In some cases with very large ligands even modification changes can be observed, which indicate that the interactions are not very strong in these solids and that the lattice structure is determined by the large organic ligand molecules. The conclusion that one of the modifications crystallizes without essential intermolecular interactions is drawn from the differences in the optical spectra of these systems. The optical absorption spectra of the two modifications are quite different. One of them shows a broad band in the "red" which is totally absent in the second modification, which no longer has signs of collective electron behaviour (11). We are in the process of completing the X-ray structure determination of the different modifications.

The Mößbauer effect is a more sensitive spectroscopic probe for the identification of intermolecular interactions and the more accurate classification of their strength. It works quite well with iridium (^{193}Ir) and as shown by our results, the isomer shifts as well as the quadrupole splittings depend strongly on the strength of the intermolecular interactions along the metal chains(21, 22). This is summarized in figure 2 which shows that a very consistent list of shifts ist obtained. Negative isomer shifts, relative to iridium metal as the standard, are found in solids

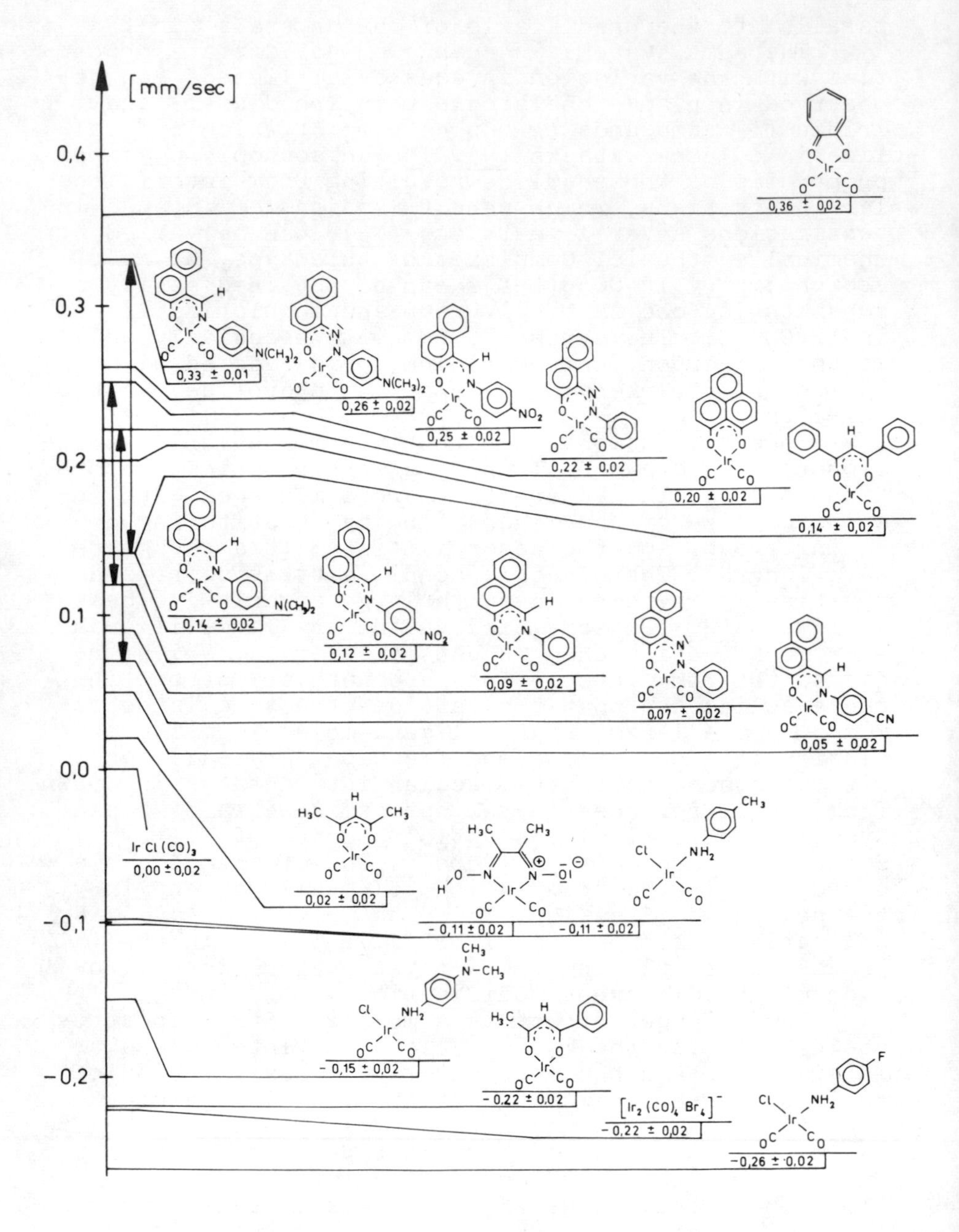

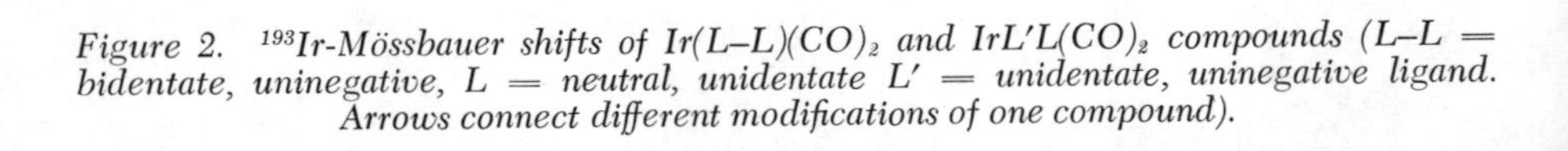

Figure 2. [193]Ir-Mössbauer shifts of Ir(L–L)(CO)$_2$ and IrL′L(CO)$_2$ compounds (L–L = bidentate, uninegative, L = neutral, unidentate L′ = unidentate, uninegative ligand. Arrows connect different modifications of one compound).

which give clear evidence to strong intermolecular interactions in the solid state. The compounds with only weak metal-metal interactions show positive shifts and quite large quadrupole splittings.

Especially interesting are the data of these complexes which crystallize in two modifications, one of which only contains columnar stacks. As expected, the isomer shifts of the modification without any sign of co-operative electron behaviour are larger than those of the deeply colored form crystallizing in columnar stacks. The quadrupole splittings are smaller for the compounds with strong intermolecular interactions growing larger with decreasing influence of the axial ligands as predicted by the simple crystal field model. By comparing the Mößbauer results with those of other spectroscopic methods, it seems that the Mößbauer effect is a well suited method in the case of iridium complexes to classify intermolecular interactions in linear chain metal complexes.

A similar relation between solid state properties on one side and sterical and electronic ligand parameters on the other hand can be found in the "dioximato" complex series. Numerous complexes of d^8 ions with these ligands were described in the past (4, 23). Especially strong interactions in these solids are found in the Bis(1,2-benzoquinonedioximato)metal(II) species the ligand of which is relatively small and is capable of delocalizing metal d-electrons (24).

These examples prove that even d^8-ions in linear metal chain complexes are capable of strong interactions giving rise to co-operative electron behaviour as pointed out by investigations of their optical spectra. Especially those candidates which have only small and strongly electron accepting ligands are capable of very strong interactions and show collective electron behaviour in one dimension such as high conductivity and long wavelength absorptions. Therefore, it seems possible to synthesize 1d-metals even with "d^8-ions" if appropriate ligands are used. $Ir(CO)_3Cl$ which considered in our opinion to be a pure d^8-compound, is an impressive example that the energy gap ΔE can be reduced very much by using electron accepting groups and that the d.c.-conductivity can be raised by using small ligands to almost metalic behaviour.

Nevertheless the intermetallic interactions in all of these d^8 transition metal complexes are not strong enough to overcome the sterical problems and a d.c.-conductivity corresponding to a metallic state has not been achieved. Especially, if using large and polarizeable organic dye molecules which are necessary in the search for a "Little model compound" (25, 26) the mobility of charge carries is decreased considerably. Therefore we tried to synthesize new "mixed valence" solids by oxidizing d^8-compounds.

VI. Preparation of "Mixed Valence" Solids

The results of oxidation reactions on d^8 linear chain metal complexes were studied thoroughly for the tetrakis(cyano)platinum(II)complexes and some oxalato derivatives. The first investigations were carried out more than 100 years ago and the results of all of the work are well known and are summarised (3).

One important question remains to be answered: "Why does the oxidation of other square planar platinum(II) compounds, crystallizing in columnar stacks, not proceed to "mixed valence" one-dimensional metals?" This can be answered by investigating the association reactions in solution, which lead to linear chains. This "piling up"-reaction is essentially a donor-acceptor association between a square planar platinum(II) donor and a solvated platinum(IV) acceptor. The reaction is prevented if solvent molecules occupy the "axial positions" in the strongly accepting platinum(IV) moiety, which cannot be removed by donating platinum(II) species. A chain is obtained if the axially bound solvent molecules can be substituted by the donating platinum(II) species, that means, if the following reactions proceed to associates:

$$[Cl \cdot Pt^{IV}(CN)_4 \cdot H_2O]^- + [H_2O \cdot Pt^{(II)}(CN)_4 \cdot H_2O]^{2-} \rightleftharpoons$$
$$[H_2O \cdot Pt^{(II)}(CN)_4 \cdot Pt^{IV}(CN)_4 \cdot H_2O]^{2-} + H_2O + Cl^-;$$
$$[H_2O\cdot Pt^{(II)}(CN)_4\cdot Pt^{IV}(CN)_4\cdot H_2O]^{2-} + [H_2O\cdot Pt(CN)_4\cdot H_2O]^{2-} \rightleftharpoons$$
$$[H_2O\cdot Pt^{II}(CN)_4\cdot Pt^{IV}(CN)_4\cdot Pt^{II}(CN)_4\cdot H_2O]^{4-} + 2H_2O;$$

This type of reaction does take place in aqueous solution in the case of $[Pt(CN)_4]^{2-}$ and $[Pt(C_2O_4)_2]^{2-}$ ions but is prevented in the same solvent when using complexes of the type $PtCl_4{}^{2-}$ or $[Pt(NH_3)_4]^{2+}$, because either the reduced form of the oxidizing agent (mostly halogenide ions) or the solvent molecules are bonded very covalently onto the axial positions of the platinum(IV) acceptor. If this assumption is correct, the oxidation should proceed to "mixed valence" compounds only if the reaction can be carried out in non-donating oxidizing solvents. The medium could be half concentrated sulfuric acid. The red K_2PtCl_4, the yellow $Pt(NH_2\cdot CH_2\cdot CH_2\cdot NH_2)Cl_2$ and the green Magnus' Salt can be oxidized to deeply coloured compounds with a remarkable red metallic lustre in half concentrated sulfuric acid. Chemical and physical evidence clearly proves that these compounds are mixed valence solids. Since the reaction is a heterogenous one and for this reason only powders could be obtained (27) this procedure is not appropriate in the synthesis of 1d-metallic single crystals. We are still looking for other better suited solvents and oxidizing agents.

On the other hand we have continued our search for "mixed valence" solids with central metals other than platinum. In order to obtain partially oxidized solids only co-ordination compounds, the ligands of which should be stable against oxidation by halogens or similar substances could be used as starting materials. As shown by many investigations dioximato ligands are chemically very stable and give metal columnar stacks. "Mixed valence" solids isolated by oxidation of Bis(1,2-diphenylglyoximato)-nickel(II) and palladium(II) have been known for a long time but were not recognized as strongly interacting metal chains. In fact the crystals contain stacks of pancaked metal complexes (with an average oxidation number of 2.33 for the central metal ion) and linear chains of I_3^- units parallel to these stacks. A structure very similar to the one found in the blue starch-iodine adducts was proposed. The essential difference is that both stacks could show collective electron behaviour. This is very interesting with respect to the prediction that mobile electrons on two parallel non-degenerate conduction spines could interact with each other in a manner proposed by Little for an excitonic high-temperature superconductor (28). A similar solid could be obtained by oxidizing bis(1,2-benzoquinonedioximate) palladium(II) and nickel(II) in dichlorobenzene to mixed valence species which differ in their solid state properties considerably from their unoxidized parent compounds. The results of X-ray analysis so far suggest that linear metal and iodine chains run parallel through the lattice of tetragonal symmetry as shown schematically in figure 3.

Furthermore we have used another well-known type of reaction to come to mixed valence solids: the so-called oxidative addition reactions. This type of reaction could possibly be used for piling up linear metal chains by adding a tetrakis(arylisonitrile) rhodium(III) species to a tetrakis(arylisonitrile) rhodium(I) compound in the following manner:

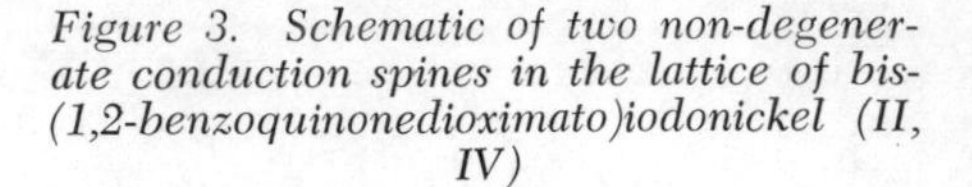

Figure 3. Schematic of two non-degenerate conduction spines in the lattice of bis-(1,2-benzoquinonedioximato)iodonickel (II, IV)

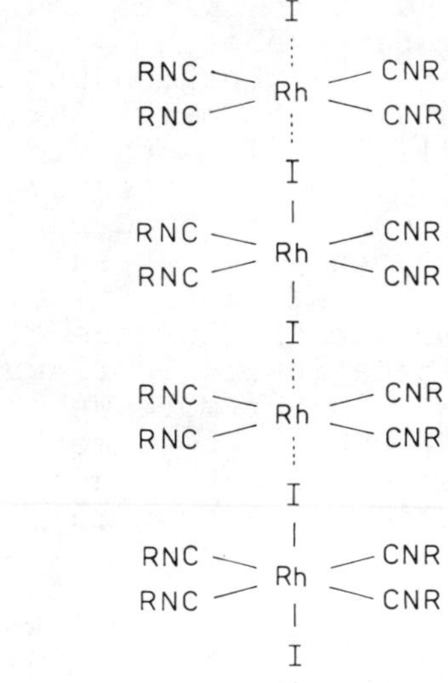

Figure 4. Structure proposed for the "mixed valence" solid diiodotetrakis(2,6-dimethylphenylisonitrile)rhodium(I, III).

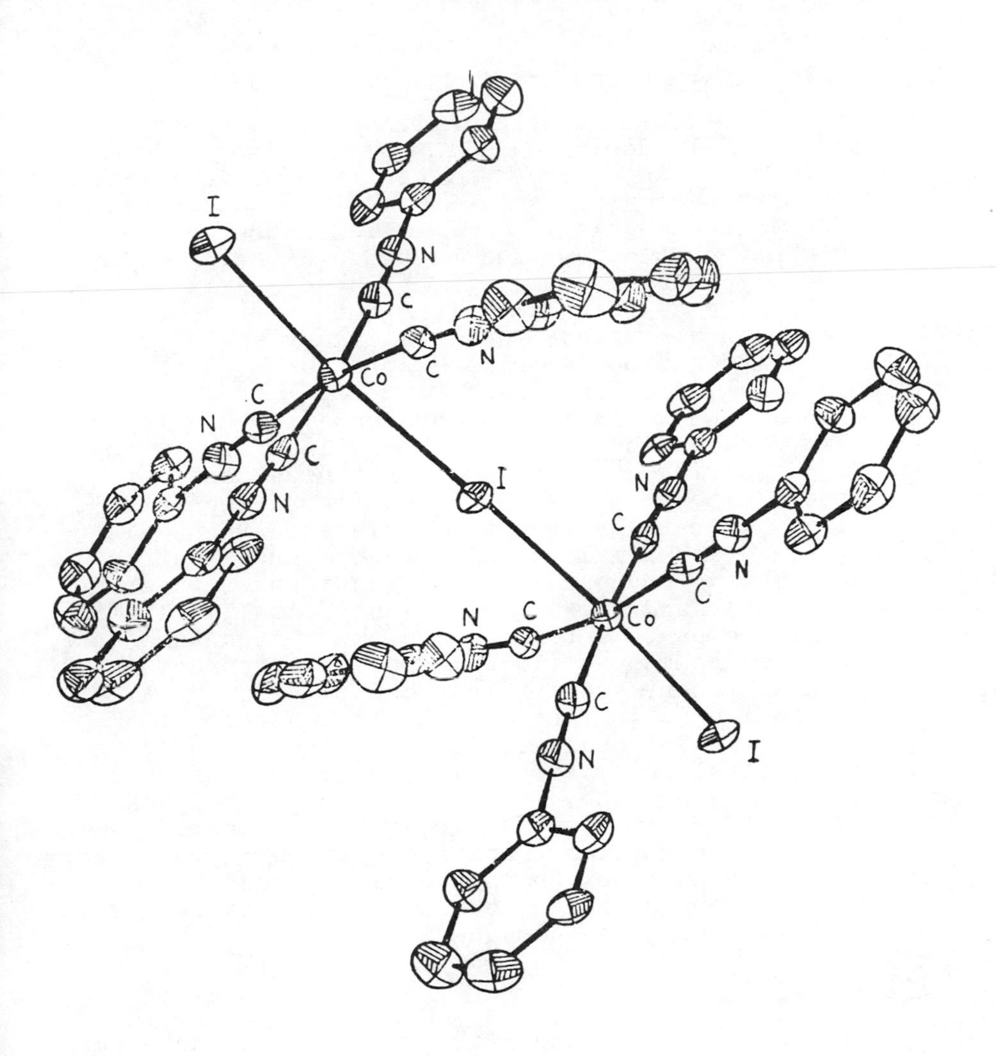

Figure 5. Molecular structure of the μ-iodobis[tetrakis(phenylisonitrile)cobalt(II)]⁺ cation in diiodotetrakis(phenylisonitrile)cobalt(II)

$$2[Rh(CNR)_4]^+ + [Rh(CNR)_4I_2]^+ \rightleftharpoons$$
$$[I\text{-}Rh(CNR)_4\text{-}Rh(CNR)_4\text{-}I]^{2+} + [Rh(CNR)_4]^+$$
$$\rightleftharpoons [I\text{-}Rh(CNR)_4\text{-}Rh(CNR)_4\text{-}Rh(CNR)_4\text{-}I]^{3+}....$$

By mixing the 2,6-dimethylphenylisonitrile rhodium(I) and rhodium(III) derivatives in chloroform a sudden change of color is observed turning from a yellow-red to dark blue. From this solution a violet solid with a striking linearly polarized metallic lustre can be obtained. Whether the compound contains directly bound metal-metal chains or an iodine bridged M - I - M - I lattice can be decided only after the X-ray determination has been completed. The preliminary spectroscopic data indicate an iodide-bridged structure (7) as indicated schematically in figure 4.

Finally we would like to present just one result which we obtained on our search for linear chain compounds containing d^7-metal ions. A typical and very stable ion with this electronic configuration is cobalt(II). The earlier report of Malatesta that diiodotetrakis(arylisonitrile)cobalt(II) compounds can be isolated in two crystalline modifications of which one is diamagnetic and shows a remarkable metallic lustre (29) prompted us to determine the exact structure of one of these compounds (30). The molecular structure of the tetrakis(phenylisonitrile)cobalt(II) is shown in figure 5 which is in fact μ-iodo-bis[iodotetrakis (phenylisonitrile)cobalt(II)]iodide. The linearly polarized optical absorption of the solid corresponds to a transition along the I - Co - I - Co - I heavy atom chains all of which run parallel to each other.

Though a pure linear chain d^7-compound would be ideal for our purposes the synthesis of such solids seems to be very difficult because of the strong Lewis acid activity of these compounds.

Acknowledgements

This work was supported by Deutsche Forschungsgemeinschaft and Fonds der Chemischen Industrie.

"Literature Cited,"

1 Zeller, H.R., Adv. Solid State Phys., (1973), Vol. XIII
2 Garito, A.F., Heeger, A.J., (preprint)
3 Krogmann, K., Angew. Chem., (1969), 81, 10
4 Thomas, T.W., Underhill, A.E., Chem. Soc. Rev., (1972), 1, 99
5 Gitzel, W., Keller, H.J., Lorentz, R., Rupp, H.H., Z.Naturforsch., (1973), 28b, 161
6 Dart, J.W., Lloyd, M.K., Mason, R., McCleverty,J.A., J.C.S. Dalton, (1973), 2039
7 Baumann, D., (1974),Dissertation, Univ.Heidelberg
8 Mehran, F., Scott, B.A., Phys. Rev. Lett., (1973), 31, 99
9 Bailey, N.A., Coates, E.U., Robertson, G.B., Bonati, F., Ugo, R., Chem. Commun., (1967), 1041
10 Aderjan, R., Breer, H., Keller, H.J., Rupp, H.H., Z.Naturforsch., (1973), 28b, 164
11 Aderjan, R., Keller, H.J., Z.Naturforsch., (1973), 28b, 500
12 Hieber, W., Lagally, H., Mayr, A., Z.Anorg. Allg. Chem., (1941), 246, 138
13 Benson, R.E., U.S. Patent, (1966), 3-255-195
14 Malatesta, L., Canziani,F., J. Inorg. Nucl. Chem., (1961), 19, 81
15 Keller, H.J., Seibold, K., J. Amer. Chem. Soc., (1971), 93, 1309
16 Dickinson, G.E., J. Amer. Chem. Soc., (1922), 44, 2404
17 Fritz, H.P., Keller, H.J., Z.Naturforsch., (1965), 20b, 1145
18 Lorentz, R., (1974), Dissertation, Univ.Heidelberg
19 Monteith, K.L., Ballard, L.F., Pitts, C.G., Klein, B.K., Slifkin, L.M., Collman, J.P., Solid State Comm., (1968), 6, 301
20 Aderjan, R., Keller, H.J., Rupp, H.H., (submitted for publication)
21 Aderjan, R., (1973), Dissertation, Univ.Heidelberg
22 Aderjan, R., Keller, H.J., Rupp, H.H., Wagner, F., Wagner, U., (to be published)
23 Banks, C.V., Barnum, D.W., J. Amer. Chem. Soc., (1958), 80, 3579
24 Endres, H., Keller, H.J., Megnamisi-Bélombé, M., Moroni, W., Nöthe, D., Inorg. Nucl. Chem. Lett., (1974), 10, 467
25 Little, W.A., Phys. Rev., (1964), A 135, 1416
26 Jagubskii, E.B., Khidekel, M.L., Russ. Chem. Rev., (1972), 41, 1011

27 Gitzel, W., Keller, H.J., Rupp, H.H., Seibold, K., Z.Naturforsch., (1972), 27b, 365
28 Little, W.A., (private communication)
29 Malatesta, L., Isocyanide Complexes of Metals, (1969), J. Wiley, London
30 Baumann, D., Endres, H., Keller, H.J., Weiss, J., J.C.S. Chem. Comm., (1973), 853

22

Magnus Green Salt Solid Solutions Containing Mixed-Valence Platinum Chains: An Approach to 1-D Metals

B. A. SCOTT, R. MEHRAN, and B. D. SILVERMAN

IBM T. J. Watson Research Center, Yorktown Heights, N.Y. 10598

M. A. RATNER

New York University, New York, N.Y. 10003

Introduction

The class of inorganic solids composed of the partially oxidized platinum chain salts are of special interest to participants in this Symposium, as they represent perhaps the most extended type of interaction realizable in the 1-D systems: that leading to free carriers and metallic-like conductivity. Such systems are typified by the compound $K_2Pt(CN)_4Br_{0.3}\cdot 3H_2O$(KCP). In terms of the known formal oxidation states of platinum (+2,+4) the stoichiometry of KCP can be written $K_2[Pt(II)_{0.85}Pt(IV)_{0.15}(CN)_4]Br_{0.3}\cdot 3H_2O$. However, it is clear from Professor Krogmann's crystal structure[1] for this compound that all Pt-sites are equivalent and separated by the extremely short intermetallic distance of 2.88 Å along the [001] stacking axis. Thus, strong metal-metal interaction involving the $5d_{z^2}$ orbitals, which are the highest lying occupied states[2], along the c-axis permits the system to attain a fractional, or partially oxidized valence of +2.3. The x-ray structure shows that the charge compensating Br^- ions are located in the interstices between the $[Pt(CN)_4]$ stacks. The interstices also contain sites for the alkali ions and water molecules.

Table I, taken in part from recent reviews[3,4], shows some of the considerable number of known partially oxidized platinum chain salts. All of these salts are composed of tetracyano- or bisoxalato-platinum complexes. In the structures of these compounds[3] the common feature is the presence of $[Pt(CN)_4]$ or $Pt(Ox)_2$ stacks comprised of closely spaced Pt ions with oxidation numbers between 2.26 and 2.40. Charge compensation occurs through the presence of halide ions in the structure, or vacancies on the alkali-cation sites. For KCP the oxidation number of 2.3 corresponds to 1.7 electrons per platinum; i.e., ≈ 5/6 filling of the $5d_{z^2}$ band. This condition leads to metallic-like electrical conductivity near room temperature. The electrical properties of KCP, as well as the results of a large number of solid state measurements, are reviewed by Dr. Zeller in this Symposium volume and will not be discussed here. Rather, our primary concern will be the phase diagram of KCP and related systems shown in Table I. Why do these compounds occur with stoichiometries corresponding to 1.6-1.74 $5d_{z^2}$ electrons/platinum? Is there a special stability in the degree of band filling associated with these electron concentrations? Answers to these questions are important to solid state chemists, providing the rules necessary to design high conductivity into new 1-D metal chain systems.

In discussing our approach to understanding the stoichiometry of the 1-D metallic-type partially oxidized metal chain complexes, it is useful to consider the oversimplified "binary" phase diagram of the aqueous $K_2Pt(II)(CN)_4$-$K_2Pt(IV)(CN)_4Br_2$ system at room temperature, as shown in Fig. 1. Here the divalent platinum complex is denoted by II and the tetravalent complex designated IV. KCP occurs at a mole fraction, x=0.15 in this system, corresponding to $n(d_{z^2})=1.7$ and the oxidation state 2.3. For $x < 0.15$ compound II and KCP crystallize out of solution, whereas for $x > 0.15$ KCP and IV crystallize on evaporation of the solvent. All of the known partially oxidized platinum salts can be represented on a phase diagram such as Fig. 1. For example, the bisoxalato salts of Table I can be designated on a phase diagram in which II = $(cation)_2Pt(II)(C_2O_4)_2$ and IV = "$Pt(IV)(C_2O_4)_2$", i.e., a pseudo-compound of tetravalent platinum. The specific composition at which the partially oxidized compound appears on the diagram ranges between x = 0.13 to 0.20. We term the exact composition of the phase for a given system "x_{magic}", because there appears to be a specific minimum, or "magic" electron concentration at which the 1-D metallic phase is stable. The reasons for this are not yet clear, but theories have been proposed[3] and others will be discussed in this as well as subsequent papers in the Symposium.

In the present paper we wish to indicate an alternative approach to the 1-D metallic state which attempts to reach x_{magic} in a way which we believe provides some clues to the stoichiometry of the partially oxidized platinum chain compounds. This approach is illustrated in Fig. 2, depicting a hypothetical phase diagram between a platinum(II) and a platinum(IV) compound. We require that II be an insulating phase with a stacking of Pt(II) ions at a separation $\lesssim 3.3$ Å, and further that the structure of II contain vacant interstitial sites for the introduction of a

TABLE I

Partially Oxidized Chain Compounds

Complex	Oxidation Number of Platinum	Pt-Pt Distance	x	$n(d_z2)$
$K_2[Pt(CN)_4]Cl_{0.32}\cdot 3H_2O$	+2.32	2.880 Å	0.16	1.68
$K_2[Pt(CN)_4]Br_{0.30}\cdot 3H_2O$	+2.30	2.887	0.15	1.70
$Mg[Pt(CN)_4]Cl_{0.28}\cdot 7H_2O$	+2.28	2.985	0.14	1.72
$K_{1.74}[Pt(CN)_4]\cdot 1.8H_2O$	+2.26	2.96	0.13	1.74
$H_{1.60}[Pt(C_2O_4)_2]\cdot 2H_2O$	+2.40	2.80	0.20	1.60
$Li_{1.64}[Pt(C_2O_4)_2]\cdot 6H_2O$	+2.36	2.81	0.18	1.64
$K_{1.62}[Pt(C_2O_4)_2]\cdot xH_2O$	+2.38	2.81	0.19	1.62
$(NH_4)_{1.64}[Pt(C_2O_4)_2]\cdot H_2O$	+2.36	2.82	0.18	1.64
$Ba_{0.84}[Pt(C_2O_4)_2]\cdot 4H_2O$	+2.32	2.85	0.16	1.68

$n(d_z2)$ = number of $5d_z2$ electrons per platinum

x = mole fraction composition parameter in pseudo-binary Pt(II)-Pt(IV) system. (See text)

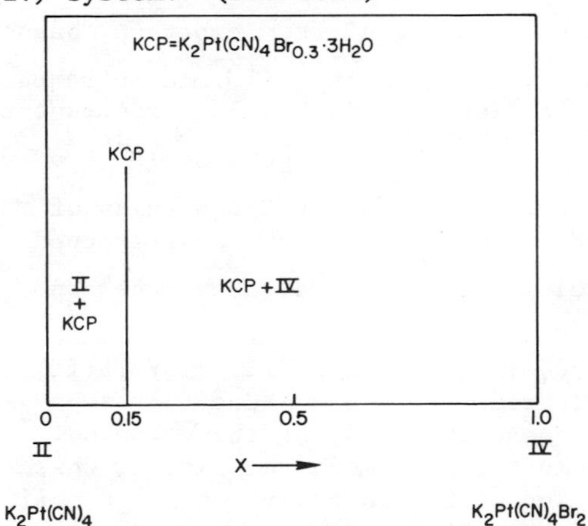

Figure 1. Oversimplified "binary" phase diagram between II = $K_2Pt(CN)_4$ and IV = $K_2Pt(CN)_4Br_2$ in water at room temperature

charge compensating ion into the lattice. Thus, we wish to create a solid solution system through the introduction of "Pt(IV)" complexes into the stacks[5] with concomitant charge compensation in the interstitial sites. By analogy with the inorganic semiconductor silicon doped with aluminum, we might expect II to transform as follows with increasing x : II(insulator) → p-type semiconductor → degenerate semiconductor ≃ 1-D metal. We contend that the final transformation will occur (possibly abruptly) at $x = x_{magic}$. For the moment we neglect the influence of disorder, the very likely possibility of one or more miscibility gaps ($\delta \neq 0$ in Fig. 2) and other perturbations which may turn off the metallic conduction (e.g. Peirels distortion), and confine ourselves instead to the first order problem: What are the solid state factors stabilizing x_{magic}? Note also that we have assumed, for generality, a finite homogeneity range, $\gamma \neq 0$, for the 1-D metal phase in Fig. 2. Fig. 2 would transform into the special case of Fig. 1 if $\gamma = 0$ and $x_{magic} = \delta = 0.15$.

In the present study we have chosen II as Magnus' green salt (MGS), $Pt(NH_3)_4PtCl_4$, and IV = $K_2Pt(CN)_4Y_2$ (Y = Cl,Br) and K_2PtCl_6. The tetragonal Magnus salt structure is shown in Fig. 3. It is characterized by the stacking of square co-planar $Pt(NH_3)_4^{++}$ cations and $PtCl_4^{=}$ anions at a 3.24 Å separation along the crystallographic c-axis, and the relatively long interchain Pt-Pt distance of 6.35 Å along the <110> direction.[6] Previous studies suggested that the electrical and optical properties[7-9] of MGS could be described by Miller's model[7] in which filled $5d_{z^2}$ metal bands are separated by $\approx$ 0.6 eV from empty $6p_z$ bands. Recent band calculations[10], conductivity[11,12] and infrared measurements[13] suggest that the $5d_{z^2}$-$6p_z$ gap is ∿ 4.5 eV and that the measured conductivities,[8,9,11,12] which range from 10^{-6} to 10^{-2} Ω^{-1}-cm^{-1}, may be impurity dominated. In our EPR studies of MGS[14] we have suggested that the signal is due to a self-trapped $5d_{z^2}$ hole whose energy level may lie ∿ 0.6 eV above the top of the $5d_{z^2}$ valence bands. Such states are formally equivalent to those introduced by the substitution of Pt(III) for Pt(II) in the Pt(II) lattice and are introduced into MGS due to the presence of Pt(IV) complexes during the preparation of the compound.[14] This situation is depicted in the band model of Fig. 4, which may be considered our starting point and equivalent to a solid solution with a very small value[14] of x ($\simeq 10^{-4}$) in the phase diagram of Fig. 2. In the present work we have found it possible to increase x to fairly large values and report the effect of such

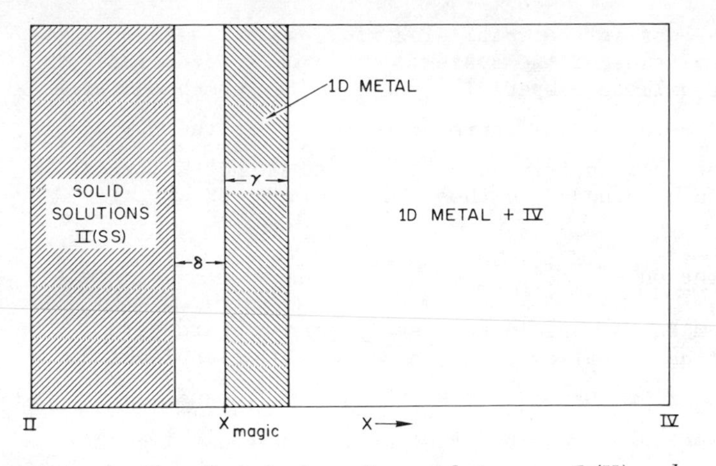

Figure 2. Hypothetical phase diagram between a Pt(II) and Pt(IV) compound illustrating the approach taken here for achieving the 1-D metallic state. Parameter δ is width of two-phase region, and γ is assumed homogeneity range for the 1-D metallic phase. In principle, many intermediate phases all separated by small immiscibility gaps could exist in such a system.

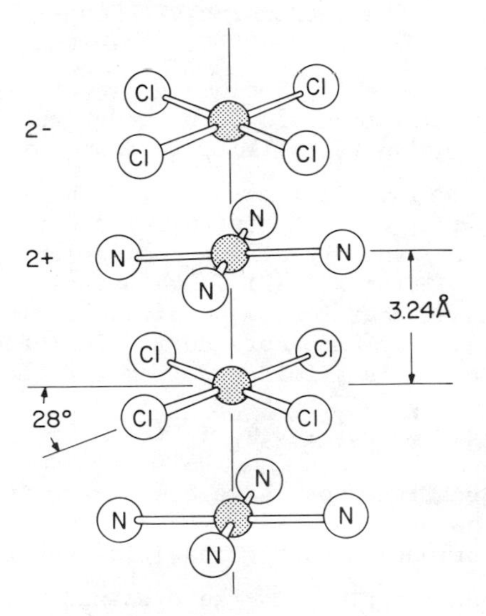

Figure 3. Stacking of $Pt(NH_3)_4^{2+}$ and $PtCl_4^{2-}$ ions in Magnus' green salt (6)

systematic variations in the total electron concentration on the physical properties of the system. It is our contention, then, that there will be a special $x = x_{magic}$ for which MGS will behave as a 1-D "metal". The approach to x_{magic} in the MGS system, and its implications for the entire class of partially oxidized metal chain compounds, is thus the main subject of this work.

Materials Preparation

All platinum salts used in this study were prepared from 99.99% pure platinum metal by standard conversion techniques as described by Brauer[15]. Samples of K_2PtCl_6, K_2PtCl_4 and $K_2Pt(CN)_4 \cdot 3H_2O$ were also obtained from Strem Chemicals (Danvers, Mass.) for purposes of comparison. The results reported herein were found to be independent of materials source.

The Magnus salt solid solutions (MGS(SS)) were prepared according to the following reaction scheme

$$(1)\quad (a)Pt(II)(NH_3)_4^{++} + (b)Pt(II)Cl_4^{=} + (c)Pt(IV)L_4Y_2^{=} \rightarrow MGS(SS),$$

where the tetrammine platinum(II) complex was present as the Cl^- or ClO_4^- salt, K_2PtCl_4 was the anion source, and the molar ratio (a):(b) = 1. Successful incorporation of higher valence Pt into MGS was found to occur under the following conditions: (i) dissolution of $Pt(IV)L_4Y_2^{=}$ dopant into the $PtCl_4^{=}$ solution prior to reaction, (ii) maintaining the solution concentrations in the range 0.01-0.1 M with $(c) \leq 0.5(a) = 0.5(b)$ and finally, (iii) rapidly mixing the cation and doped-anion solutions. The resulting MGS(SS) is metastable as it is not the expected phase under equilibrium conditions. This is because a complex redox equilibrium occurs during MGS precipitation. For example, if $PtCl_4^{=}$ is eliminated from reaction (1) and L = Y = Cl, we find

$$(2)\quad Pt^{++}(NH_3)_4 + PtCl_6^{=} \rightarrow [Pt(NH_3)_4Cl_2]PtCl_4$$

Such reactions have been extensively analyzed in the past where the platinum (II)-(IV) redox pair are both present as anions (or cations), and the resulting substitution reactions are known to occur rapidly in the presence of halide ion.[16] For this reason we have carried out reaction (1) in the presence as well as abscence of excess Cl^-, but this had no discernible influence on the degree of solid solution formation. The point to be made, however, is that a redox reaction between the cation and platinum(IV) "dopant" is also occuring during MGS(SS) preparation under the conditions of our experiments. If such experiments are carried out under conditions required for the growth of single crystals-for example, by slow diffusion of the components as in a gel[17]-the redox reactions have gone to completion prior to crystal growth and "doping" proceeds under equilibrium conditions, yielding only small concentration ($x \approx 10^{-4}$) of platinum(III) centers. For these reasons we are convinced that the initial formation of halide bridged chains of the type $Pt(II)...Y-Pt(IV)L_4-Y...Pt(II)$, as in the Pt(II)-catalyzed Pt(IV) substitution reaction[16], is essential to the doping of MGS with high concentrations of Pt(III) centers.

Experimental Results

Provided that the above conditions for the preparation of MGS(SS) are fulfilled, the reaction products from (1) are a series of single phase solid solutions ranging in color from deep green for the undoped (x = 0) Magnus salt to blue for the most partially oxidized (SS) phase. The materials were then characterized by several techniques including chemical analysis, electron spin resonance, x-ray diffraction, IR spectroscopy and electrical conductivity measurements.

Solid Solution Stoichiometry. Using Guinier x-ray diffraction techniques it was established that the products of reaction (1) were single phase materials having the structure of Magnus' green salt. This was done by searching for secondary phases on x-ray film which had been overexposed to the primary MGS(SS) diffraction pattern. The sensitivity of this technique was found to be considerably better than 1 mole% of second phase contamination as determined by analyzing standard mixtures of MGS with other Pt(II) and Pt(IV) complexes.

Chemical analysis of the precipitated MGS(SS) powders for the case of $K_2Pt(IV)(CN)_4Br_2$ dopant (oxidant) permits direct measurement of the degree of partial oxidation since both carbon and bromine determinations can be compared. This is shown in Fig. 5, where an essentially 1:1 correlation of incorporated platinum (based on the carbon analysis) to incorporated bromine is observed. This suggests that Pt is formally present as Pt(III), since this would require compensation by just one Br^-. Complete chemical analysis of the products, moreover, indicates that the MGS is $PtCl_4^=$ deficient. Similar results were obtained for $K_2Pt(CN)_4Cl_2$ oxidation of MGS, resulting in the following formulation for the solid solutions:

$$(3) \qquad Pt(II)(NH_3)_4(Pt(II)Cl_4)_{1-q}(Pt(III)(CN)_4)_qY_q,$$

where X = Br^-,Cl^-. Similarly, for doping (oxidation) with K_2PtCl_6

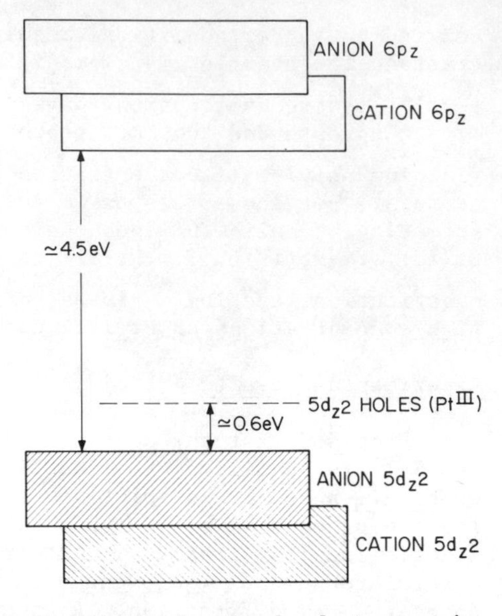

Figure 4. Schematic band structure for extrinsic MGS

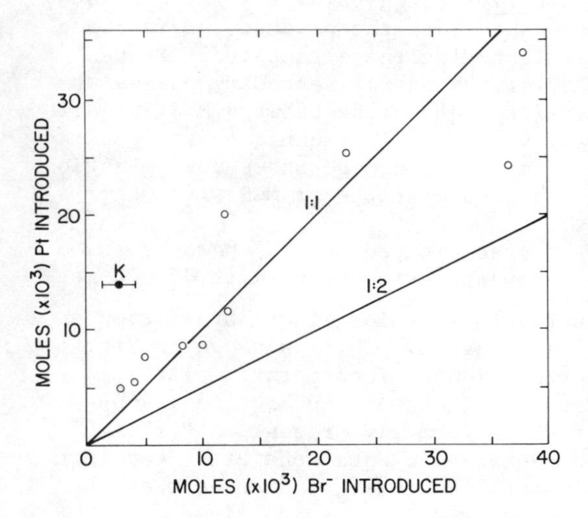

Figure 5. Number of moles of Pt vs. *Br introduced into MGS by solid solution formation (and partial oxidation) with $K_2Pt(CN)_4Br_2$. Potassium concentration is 3 ± 2 mmoles for all samples shown. The two straight lines are the results expected for Pt:Br = 1:1 or 1:2.*

$$(4)\qquad Pt(II)(NH_3)_4(Pt(II)Cl_4)_{1-q}(Pt(III)Cl_4)_qCl_q.$$

The maximum values of q obtained for the various systems is shown in Table II. It should be noted that the parameter q as used in the formulae for MGS(SS) systems is a measure of the degree of substitution of MGS anionic sites by Pt(III). It is of course related, but not equivalent to, the composition parameter x of Figs. 1 and 2. For comparison, the partially oxidized Pt(+2.3) state occurs at x = 0.15 in Figs. 1 and 2 and would be equivalent to q = 0.6 in Formulae (3) and (4) if an "averaged" oxidation state were appropriate for the MGS solid solutions. This point will be considered again in the subsequent discussion.

If the top of the $5d_{z^2}$ valence bands are mostly of anionic ($Pt(II)Cl_4^=$) character, as shown in Fig. 4, then the $5d_{z^2}$ hole is located on the $PtCl_4$ groups in both (3) and (4); i.e., (3) becomes

$$(5)\quad Pt(II)(NH_3)_4(Pt(II)Cl_4)_{1-2q}(Pt(II)(CN)_4)_q(Pt(III)Cl_4)_qY_q.$$

E.P.R. powder data for solid solutions of the type (4) gave the following g-tensor parameters: $g_{||} = 1.94$ and $g_{\perp} = 2.50$. These values are identical to those measured[14] for MGS(SS) in which $K_2Pt(CN)_4Cl_2$ or $K_2Pt(CN)_4Br_2$ is the oxidant; Thus, EPR suggests that the Pt(III) site is independent of oxidizing agent. Whether the $5d_{z^2}$ hole resides primarily on ($PtCl_4$) or $Pt(NH_3)_4$ groups depends on the relative energies of the cation and anion bands, of course, and we choose the former site to be consistent with the assignments in Miller's original model.[7]

Our formulation of Magnus salt solid solution systems places charge compensating halide ions in interstitial sites in the structure. Our main evidence for this consists of infrared data and x-ray diffraction measurements of the lattice constants of MGS(SS). The far infrared spectrum of the $K_2Pt(CN)_4Br_2$-oxidized MGS product with q = 0.2 shows no evidence for a Pt-Br stretch band. Such a band would be expected for substitution along the chain, or for charge compensation through replacement of one of the ligands within the Pt square coordination plane (e.g., replacement of NH_3). The changes in infrared pattern above 700 cm^{-1} with increasing q values are shown in Fig. 6. These spectra were taken for constant solid concentration in the KBr matrix, and show only the introduction of one new band due to the $C \equiv N$ stretching vibration near 2200 cm^{-1}. This is consistent with the replacement of $PtCl_4^=$ anions in the MGS structure with $Pt(CN)_4^=$ as determined by chemical analysis.

Unit cell data as a function of q for the $K_2Pt(CN)_4Br_2$ oxidized solid solutions, shown in Fig. 7, indicates expansion of

TABLE II

Solid Solution Limits

Dopant (Oxidizing Agent)	q*
K_2PtCl_6	0.03
$K_2Pt(CN)_4Cl_2$	0.08
$K_2Pt(CN)_4Br_2$	0.20

*See Eqs. (4), (5).

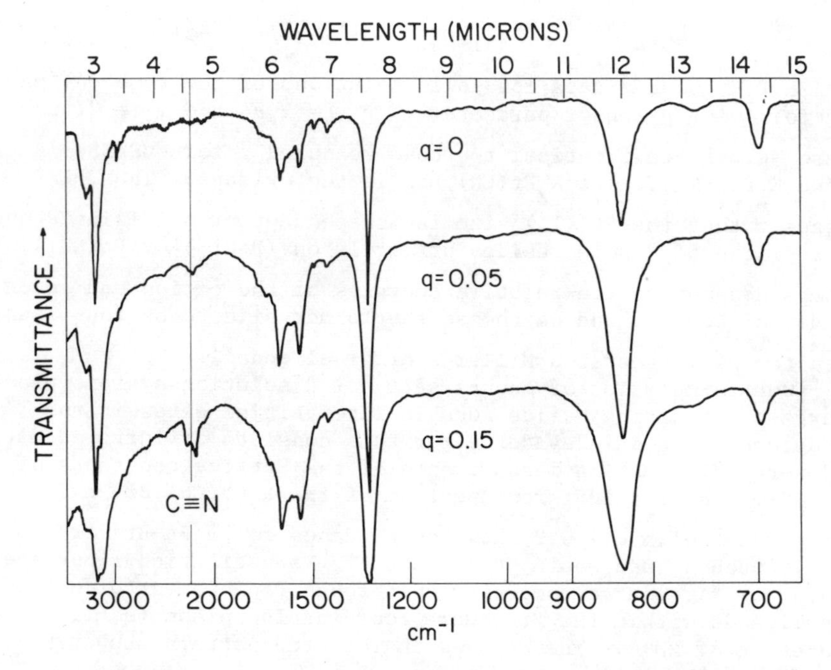

Figure 6. Infrared spectra for MGS(SS) formed by partial oxidation with $K_2Pt(CN)_4Br_2$. Composition parameter q is defined in Equation 5.

the MGS cell along the tetragonal a_o-axis and contraction along the c_o stacking direction. Expansion transverse to the linear chains is plausible for Br^- interstitials combined with the introduction of $Pt(CN)_4$ groups in anionic lattice sites, whereas contraction along c_o is consistent with the attendant introduction of the smaller Pt(III) ions in the chains. In this connection it is relevant that all the partially oxidized Pt salts shown in Table I have very short Pt-Pt spacings. Thus, our observation of a decrease in c_o for MGS(SS) is entirely expected with partial oxidation (increasing q) of this phase. However, unlike the metallic-like salts of Table I, we expect two different types of platinum sites along the chains: a majority consisting of the Pt(II) hosts, and the remainder Pt(III) sites. In fact, the expected disorder along the chain as a consequence of this arrangement is clearly observed in the x-ray diffraction as a loss and broadening of (00ℓ) reflections with increasing q.

Electrical and Optical Properties. The formation of mixed valence solid solutions through the introduction of Pt(III) has profound effects on the properties of MGS. Measurements of the d.c. electrical conductivity of powder compactions shown in Table III show a four order of magnitude change in σ as the composition was changed from ostensibly pure MGS (undoped, $q = 0$) to a solid solution containing the maximum degree of substitution, $q = 0.20$. Moreover, the hole ionization energies, E_a, determined from the activation energies for electrical conduction become smaller with increasing q. Although the measurements were performed on compacted powders, these general trends are clearly evident and quite similar to the behavior of doped three-dimensional semi-conductors as the impurity concentrations are increased.

The optical spectrum in the near infrared region also reveals some interesting features. These measurements were carried out on samples dispersed in KBr pellets. Fig. 8 depicts the spectrum for two different "pure" ($q = 0$) MGS samples. Sample A was measured in two separate pressings of the same pellet to determine the variation in optical density to be expected due to variations in sample preparatory technique. The spectra were measured with no reference in the standard compartment of the spectrometer. We ascribe the absorption rise in the 1-2.4μ region of the spectrum mostly to scattering, as there do not appear to be any distinct bands in this range. On the other hand, Fig. 9 shows the absorption in the MGS(SS) system, prepared with $K_2Pt(CN)_4Br_2$ oxidant, for the range of compositions $q = 0.05$ to 0.2. The new band appearing in this frequency region is intense, and directly proportional to the concentration of Pt(III) centers in the sample as determined by chemical analysis based on carbon

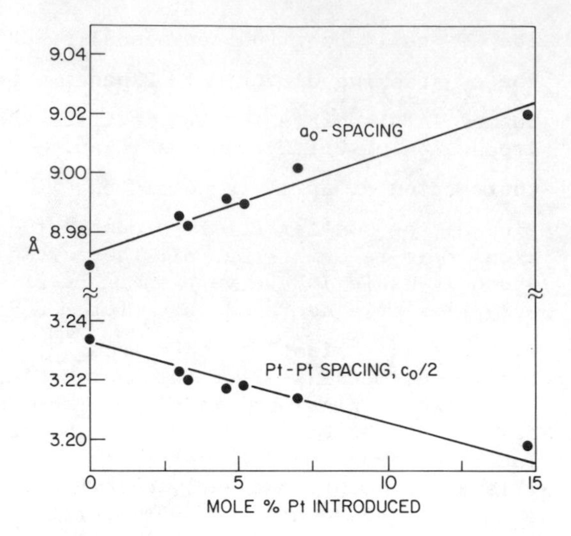

Figure 7. Lattice constants vs. *mole % platinum introduced for MGS(SS) prepared by $K_2Pt(CN)_4Br_2$. Constants obtained by least-squares refinement of Guinier camera powder data.*

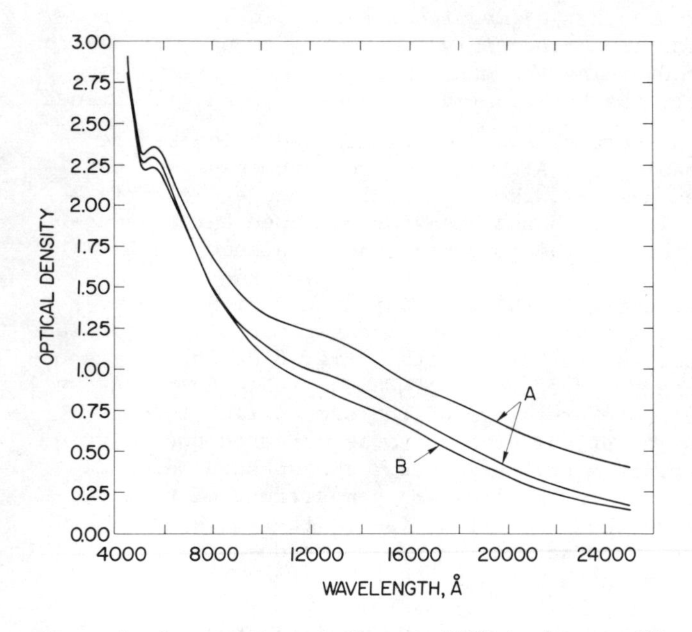

Figure 8. Spectra of nominally pure MGS in the near IR. Curves A correspond to spectra obtained in two separate pressings of the same pellet; curve B obtained from a separate MGS preparation made up to the same concentration in pellet as A.

(CN) content. Excluding the q = 0.05 samples, whose chemical analysis is least reliable, we find an absorption coefficient $\alpha = (6.0 \pm 0.3) \times 10^3$ l/mol·cm, assuming the absorption is entirely polarized along the c-axis of the crystallites (i.e., $\alpha_{\parallel} >> \alpha_{\perp}$).

In addition to the intense absorption peaking near 1.3μ, another important feature of Fig. 9 is the gradual shift of absorption edge to lower energies with increasing q. This is suggested by the general increase in optical density at 2.4μ as q increases. Although we can not obtain an accurate absorption edge from the pellet measurements, this qualitative trend is consistent with the lowering of activation energy for electrical conduction with q as shown in Table III.

It is relevant to point out that early measurements of the IR spectrum of ostensibly "pure" MGS also reported absorption in the near IR region shown in Fig. 9. Moreover, Collman et. al.[8] reported photocurrent in MGS with excitation between 0.74-2.5μ. Other measurements, including those on much purer MGS samples[13], cast doubt on the presence of this absorption,[18] but it is clear that samples prepared in the presence of Pt(IV) complexes definitely contain the broad IR peak. We have in fact found the identical IR band in MGS(SS) samples produced by $K_2Pt(CN)_4Cl_2$ and K_2PtCl_6 oxidation. However, we interpret the absorption as due to transitions from the $5d_{z^2}$ valence band to the $5d_{z^2}$ bound states, rather than to an interband transition.[7,8]

Discussion

It is worthwhile to review some of the properties of the mixed valence MGS solid solutions before examining their relationship to the class of partially oxidized platinum salts of Table I such as KCP. Previously, we found from EPR measurements[14] that "extrinsic" MGS crystals usually contain small concentrations of bound $5d_{z^2}$ holes (Pt(III)) lying ∿ 0.6 eV above the $5d_{z^2}$ orbitals. The Pt(III) holes whose occurence is due to the presence of Pt(IV) complexes during sample preparation, have a fairly wide extent-over at least several metal chain lattice sites[14]. In these studies we have been able to significantly increase the concentration of Pt(III) centers by partial oxidation with Pt(IV) complexes, resulting in as many as one out of ten Pt(II) chain sites replaced by Pt(III) for the case of oxidation with $K_2Pt(CN)_4Br_2$ (Table II, q = 0.2). Concomitant with the increasing Pt(III) concentration is an increase in d.c. conductivity and the appearance of a strong IR transition due to $5d_{z^2}$ band to

TABLE III

Electrical Data on Powder Compactions

$((IV) = K_2Pt(CN)_4Br_2)$

q	σ_{RT}, $(ohm\text{-}cm)^{-1}$	E_a, eV
0.0	3.7×10^{-7}	0.66
0.12	2.5×10^{-4}	0.27
0.20	6.3×10^{-3}	0.15

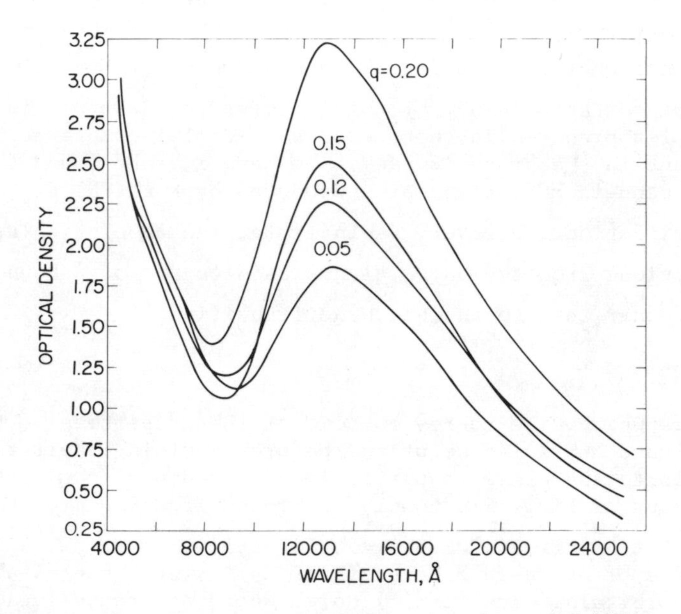

Figure 9. Near IR spectra for MGS(SS) prepared from $K_2Pt(CN)_4$-Br_2. Composition parameter q defined in Equation 5. Similar results were obtained for partial oxidation of MGS with $K_2Pt(CN)_4$-Cl_2 and K_2PtCl_6.

$5d_{z^2}$ hole excitation. Activation energies obtained from the d.c. conductivity measurements indicate a decrease in E_a, the hole ionization energy. In what follows we will consider the potential the MGS system has for exhibiting a transition to a "1-D metal" at dopant levels higher than we have been presently able to achieve and further implications concerning the stabilization of known 1-D metals such as KCP at special stoichiometries.

By considering the low hole concentration limit, one can more clearly understand just what might be responsible for a transition to metallic behavior at higher hole concentration. If we envision a full band (the MGS limit) and introduce a few local carriers by substitution, then the carrier, in this case a d_{z^2} hole, will not be freely mobile, but will instead be rather tightly bound to its site. This binding may arise simply from the coulomb attraction between the Pt(III) hole and the charge compensating anion and result in the formation of an excitonic or quasibound particle-hole state rather than in a free carrier.[19] This coulomb particle-hole attraction can however be reduced in several ways. As the concentration of dopant and density of carriers increases, the single carrier potential resulting from the charge compensators will tend to be screened. This mechanism, analogous to the Mott mechanism in higher dimensional isotropic conductors, should lead to a semiconductor-metal transition. Also, with an increase in doping, the density of charge compensators which an individual d_{z^2} hole sees will become effectively constant, i.e. independent of the site on which the hole finds itself. This will occur since the overlapping coulomb tails from the impurity or compensator distribution will average out the one electron (hole) potential fluctuations more effectively at high concentrations. For the case of the KCP limit, there are ∿ 0.3 halide ions for each Pt. We suggest that for KCP it is this stoichiometric ratio, corresponding to the composition $x_{magic} \gtrsim 0.15$ in our phase diagram (Fig. 1) at which the local exciton state goes over to the metallic state. The fact that this ratio occurs considerably below a Pt(III)/Pt(II) ratio of unity comes about because the holes, while localized, are in extended eigenstates which reach several Pt sites along the chain. Such interpretation can be inferred from the EPR spectrum[2,14] as well as the lattice regularity[1,3]. This indicates that the hole-anion interaction reduced by hole-hole interactions and smoothed due to increased halide concentration could result, at sufficiently high dopant level, in a single hole wave function with a coherence length sufficiently large to result in metallic-like behavior.

It is interesting to note that following Mott's original argument, the dopant density $x_{magic} \gtrsim 0.15$ is in near agreement with the original Mott criterion for the transition. Following Mott's original argument, based on Thomas-Fermi screening of the carriers, the semiconductor → metal transition for impurity conduction in a three dimensional system should occur approximately when $n_S^{1/3}\ a_H \sim .25$. As Mott and Twose point out,[21] a sharp break in the conductivity of both n-type and p-type germanium occurs in fact for $n_S^{1/3}\ a_H \sim .3$. In these expressions, n_S is the carrier density and a_H is the effective site radius. If we take a_H as one-half of the Pt-Pt spacing along the chain, then $n_S^{1/3}\ a_H \sim .3$ at the KCP stoichiometry; thus our suggestion of the role of the critical stoichiometry x_{magic} is in accord with the original Mott criterion for the transition. The dopant density, however, calculated in this manner must be taken with some reservation. First, the short distances that we are concerned with, namely the halide ion-Pt site separation, does not simply justify the use of a macroscopic dielectric constant. Second, the extreme anisotropy of the system as well as the fact that the halide ions are located off the Pt chain will introduce significant modifications into any screening calculation. It is however interesting (if not fortuitous) that the value one obtains from the Mott type estimate is so close to what is observed.

At the maximum solid solution limit found in our MGS system, $q = 0.2$, the holes still appear to be bound so that transport is an activated process. It is therefore evident that at this composition, which is equivalent to $x = q/4 = 0.05$ in the terminology of the phase diagrams of Figures 1 and 2, the doping is still not high enough to precipitate an electronic phase transition into the 1-D metallic state, i.e. x is still less than x_{magic} for this system. With increasing x we can expect such a transition to occur for MGS. KCP and the remaining partially oxidized metallic phases of Table I have crystallized already at this stoichiometry and as can be appreciated from the ranges $x_{magic} \gtrsim 0.13$-0.20 in this table, the precise value of x_{magic} will be very system dependent.

Regardless of the specific electronic model required to understand the stoichiometry of the 1-D metallic Pt salts, we may ask why KCP (and similar compounds of Table I) do not show a range of solid solutions from $x = 0$ to x_{magic}. That is, why the difference between MGS, which is following the phase diagram of Fig. 2, with solid solutions between $x = 0$ to $x = 0.05 < x_{magic}$, and the metallic partially oxidized systems of Fig. 1? Taking KCP as an example, there appear to be two main reasons for the difference: the rather large metal-metal distance (3.50 Å) in the end member $K_2Pt(CN)_4 \cdot 3H_2O$, and the apparent lack of a large enough interstitial site for the substitution of a charge compensating halide ion.[22] This last requirement may not be a necessary condition for mixed valence solid solution formation, since alkali cation vacancies can fulfill the latter role. However, a relatively short Pt-Pt distance, as found in MGS, may be a necessary condition for the stabilization of the trivalent Pt(III) state, since EPR indicates that even at very low concentrations these states extend over several lattice sites. Lattice energy may also play an important role. There is considerable evidence from organic systems[23] that separate stacking ("segregated" stacks) of cations and anions does not provide high Madelung energy. Thus, segregated stack-type end members (II) such as $K_2Pt(CN)_4 \cdot 3H_2O$ may be considerably destabilized by (SS) formation, whereas the MGS-type of crystals are not, due to the extremely high lattice energies of alternate cation and anion stacking.[24] Thus, for these reasons the <u>two phase</u> system of Fig. 1 is more stable in the cyanoplatinates than the solid solutions of Fig. 2, and the partially oxidized form occurs only when x_{magic} is reached.

Since we have yet to completely transform MGS into a 1-D metal, it is relevant to inquire whether it is possible to complete the process of continuously introducing $5d_{z^2}$ holes, with charge compensation, into the lattice until x_{magic} is reached. Our experimental work suggests that solution redox reactions are the limiting feature of the substitution process, and <u>not</u> crystal chemical factors. For example, the solid solution region is larger in MGS partially oxidized with $K_2Pt(CN)_4Br_2$ (q = 0.20) than with $K_2Pt(CN)_4Cl_2$ (q = 0.08). This is the reverse of the extent of (SS) expected on the basis of crystal chemical factors, since Cl^- is smaller than Br^-. However, the greater rate of $K_2Pt(CN)_4Br_2$ vs. $K_2Pt(CN)_4Cl_2$ reduction by $Pt(NH_3)_4^{++}$ suggests that the solution redox reaction is the important feature to be controlled if more extensive solid solution is to be attained.

There are several other compounds in which mixed valence solid solution formation may be possible. Among these are dichloro-(ethylenediamine) platinum(II), $Pt(en)Cl_2$, in which neutral $Pt(en)Cl_2$ units stack with an intermetallic separation of 3.39 Å.[25] The alkaline earth cyanoplatinates form a more interesting group, since many of them crystallize[3,4] with Pt(II)-Pt(II) separations below 3.3 Å. In particular, $MgPt(CN)_4 \cdot 7H_2O$ exhibits at Pt-Pt distance of 3.16 Å and the structure appears closely related to that of the 1-D metallic $MgPt(CN)_4Cl_{.28} \cdot 7H_2O$, according to Krogmann.[26] Thus, there exists the strong possibility of extensive solid solution in this system between x = 0 and

x_{magic} = 0.14 (δ = 0, Fig. 2). The Pt site equivalence present in these structures also makes them attractive, since the inequivalence of cation and anion sites in MGS (hence strongest mixing between alternate identical sites) may ultimately be the factor limiting the attainment of x_{magic} in that system.

Acknowledgements

The authors wish to thank Drs. F. Kaufman and J. B. Torrance for preliminary IR data and many useful discussions. We are grateful to Mr. A. R. Taranko for the optical data, and Mr. F. Dacol for help with the electrical measurements. We are also indebted to Mr. B. Olson and Mr. B. L. Gilbert for their tireless efforts in working out the chemical analytical procedures.

Literature Cited

1. Krogmann, K. and Hansen, H. D., Zeits. anorg. allg. Chem., (1968), 358, 67.
2. Mehran, F. and Scott, B. A., Phys. Rev. Lett., (1973), 31, 1347.
3. Krogmann, K., Angew. Chem., Int. Ed., (1969), 8, 35.
4. Thomas, T. W. and Underhill, A. E., Chem. Soc. Revs., (1972), 1, 99.
5. Such a solid solution should ideally occur with the elimination of the trans ligands of the Pt(IV) complex, as will be discussed subsequently.
6. Atoji, M., Richardson, J.W. and Rundle, R. E., J.A.C.S., (1957), 79, 3017.
7. Miller, J. R., J. Chem. Soc., London 1965, 713; Day, P., Orchard, A. F., Thomson, A. J. and Williams, R. J. P., J. Chem. Phys., (1965), 42, 1973.
8. Collman, J. P., Ballard, L. F., Monteith, L. K., Pitt, C. G. and Slifkin, L., in International Symposium on Decomposition of Organometallic Compounds to Refractory Ceramics, Metals, and Metal Alloys, edited by K. S. Mazdiyasni (Univ. of Dayton Research Institute, Dayton, Ohio, 1968), pp. 269-283.
9. Gomm, P. S., Thomas, T. W. and Underhill, A. E., J. Chem. Soc. A 1971, 2154.
10. Interrante, L. V. and Messmer, R. P., Inorg. Chem., (1971), 10, 1174.
11. Interrante, L. V., J. Chem. Soc., Chem. Commun., (1972), 6, 302.
12. Rao, C. N. R. and Bhat, S. N., Inorg. Nucl. Chem. Lett., (1969), 5, 531.
13. Fishman, E. and Interrante, L. V., Inorg. Chem., (1972), 11, 1722.
14. Mehran, F and Scott, B. A., Phys. Rev. Lett., (1973), 31, 99.

15. Brauer, G., Handbuch der Präparaten Anorganishen Chemie, Vol. 2, Ferdinand Enke Verlag, Stuttgart, 1962.
16. Mason, W. Roy, Coor. Chem. Revs., (1972), 7, 241.
17. Interrante, L. V., and Bundy, F. P., Inorg. Chem., (1971), 10, 1169.
18. Day, P., Orchard, A. F., Thomson, A. J. and Williams, A.J.P., J. Chem. Phys., (1965), 43, 3764.
19. If the hole becomes localized on the Pt-chain in the vicinity of an off-chain charge compensator, a non-axially symmetric EPR signal would be expected, contrary to that observed.[14] However, our approach is qualitative and so we choose the present picture of a quasi-bound particle-hole state rather than the self-trapped hole model[14] in our discussion.
20. For a review of the metal-insulator arguments see Mott, N.F., Revs. Mod. Physics, (1968), 40, 677.
21. Mott, N.F. and Twose, W. D., Adv. Physics, (1961), 10, 107.
22. Monfort, F., Bull. Soc. Roy. Sci. Liége, (1942), 11, 567.
23. Metzger, R. M., J. Chem. Phys., (1972), 57, 1876, 2218.
24. Miller, J. R., J. Chem. Soc., (1961), 4552.
25. Martin, Don S, Hunter, LeRoy D., Kroening, R. and Coley, Ronald F., J. Am. Chem. Soc., (1971), 93, 5433.
26. Krogmann, K. and Ringwald, G., Zeits. Naturforschg., (1968), 23b, 1112.

23

Studies on Some 1-D Metal Group VIII Complexes

K. KROGMANN

Institut für Anorganische Chemie

H. P. GESERICH

Institut für angewandte Physik, Universität Karlsruhe, D 7500 Karlsruhe 1, Germany

The abbreviation "TCNQ compounds" stands for a group of organic substances with strongly anistropic behavior in conductivity, reflectivity, and other properties. A similar abbreviation for an inorganic group of compounds with similar properties is still lacking besides "KCP" which stands for only one compound. I am tempted to propose TCMP (= tetra coordinated metal plane) for this group of square planar complex compounds. In both cases, the letters give a short hand notation of the essential structural units, and their similarity reflects the similarity of their properties.

There are two types of TCMP compounds, the stoichiometric ones with central atoms in integral oxidation states, and the partially oxidized (po-TCMP) compounds. The latter are known to behave like 1-D metals, whereas the former are more or less semiconducting. Of the po-TCMP, three structures have been studied to some detail : $K_2 [Pt(CN)_4] X_{0.3} \cdot 3 H_2O$ (1). $Mg_{0.82} [Pt(C_2O_4)_2] \cdot 6 H_2O$ (2), $Ir(CO)_{2.9}Cl_{1.1}$ (3). They all contain columns of TCMP units stacked in a staggered manner to minimize interligand repulsion. The partial oxidation makes the complex columns less negative than they were without oxidation. In the first structure (also named KCP), this partial positive charge per complex is allowed for by incorporating the necessary amount of halide ions (Cl^- or Br^-) at sites apart from the complexes, and occupied up to 64 percent (1). The second structure attains electroneutrality by a defect of cations, which way seems to be favoured by all dioxalatoplatinates and some Ir compounds (2). The third structure (3) is supposed to contain mainly neutral units without atoms or ions linking the different columns. The crystals of all these materials tend to grow as thin needles in one direction only, making it difficult to obtain crystals big enough for physical mearurements, which is especially true for the compound last mentioned with VAN DER WAALS forces only acting between the chains.

We supposed Ir(+I) to be a promising central atom for po-TCMP compounds because it has less effective charge than

Pt(+II) in its $5d^8$ configuration, and we found a number of salts containing the $Ir(CO_2)Cl_2$ unit, which are of the 1-D metal type (4). A list of compounds is given in table I, showing no great variation in distances except in the last case, though there are rather large differences in the degree of partial oxidation (DPG). The results are based on preliminary single crystal (H^+, K^+) or reliable GUINIER powder data. Some of these salts have been

Table I.
Columnar structures with Ir (4)

compound	oxidation number	distance (Å)
$Ir(CO)_{2.93}Cl$	+1.07	2.85
$(H_3O)_{0.38}Ir(CO)_2Cl_2$	+1.62	2.86
$K_{0.58}$ "	+1.42	2.86
$Cs_{0.50}$ "	+1.50	2.86
$(N(me)_4)_{0.55}$ "	+1.45	2.86
$(As(ph)_4)_{0.62}$ "	+1.38	2.86
Li_x "	?	2.86
Mg_x "	?	2.86
Ba_x "	?	2.86
$Ir(CO)_2Cl_2$ $Ir(CO)_2$acetat	+1.50	2.78

previously prepared either by high pressure CO treatment (5) or formic acid refluxing (6, 7) of K_2IrCl_6 (or similar compounds), but analyses reported and structures derived from them did not lead to the correct interpretation. It is interesting to note, that the $Ir(CO)_2Cl_2$ salts are similar to the dioxalatoplatinates in their cation defect structures as well as in the tendency of the free acid to from strongly absorbing blue coloured solutions, which may contain polymers like the Pt acid (8).

Dioxalatoplatinates of the 1-D type are able to crystallize in many different phases with no detectable difference in composition. We reported three Mg salt modifications (2), but for potassium, the situation is more complicated, as there are at least five different powder patterns. Recent single crystal growth and x-ray studies revealed, that one of the powder diagrams we named "Phase A" can be obtained from three crystal structures differing in their superstructures, which do not show up in the powder photographs. We can imagine several reasons for superstructures. One already mentioned is the staggered stacking of ligands leading to a ligand superstructure. There may also be a lattice period connected with the composition or the degree of partial oxidation (DPO), which may be called a PEIERLS superstructure according to the theory of 1-D metals (10). Such a lattice distortion was shown for KCP (11,12) to behave like a soft vibrational mode instead of a static superstructure. The evaluation of the diffuse scattering effects lead to shifts of about 0.01 Å in the Pt positions along the metal chain. A reciprocal lattice vector due to a PEIERLS distortion period $c_P^{\times}$

should be related to the reciprocal metal sublattice vector $c_M^{\times}$ by

$$c_P^{\times} = c_M^{\times} \cdot (DPO)/2 \quad .$$

The corresponding superstructure period c_P is then given by

$$c_P = c_M \cdot 2/(DPO) \quad ,$$

which is no simple integer multiple of c_M in most cases. Neutron and diffuse x-ray scattering proved such an "aperiodic" but dynamic superstructure for KCP (11,12).

On the other hand, the superlattices we observed (table II) are static, as we see BRAGG diffraction spots, and "aperiodic", but with still different periods, at least in the first two phases A_α and A_β .

Table II.

Superstructure periods in $K_{1.64}Pt(C_2O_4)_2 \cdot 4\ H_2O$, Phase "A"

substructure	(c_M)	superstructure ligand (c_L)	superstructure shifted metal (c_S)
triclinic A_α	2.84 Å	4.00 × c_M	6.63 × c_M
triclinic A_β	2.84 Å	4.00	9.35
monoclinic A_γ	2.84 Å	6.00	-

The PEIERLS period may be calculated from the composition : $c_P = 5.56 \times c_M$ for all phase A modifications, which does not coincide with any of the observed superstructures.

Preliminary results show that the displacements of the heavy atoms from the ideal positions are transverse instead of longitudinal as for KCP. The amplitude of the "shift wave" is about 0.2 Å (Fig. 1). There is also the case of a helical "wave" in the Ir compound last mentioned in table I, where the radius of the helix is 0.7 Å, and the period is 9.6 × c_M.

Many TCMP compounds show a metallic luster, which leads to the assumption of a 1-Dmetal system. Polarized specular reflectance spectra clearly distinguish between 1-D metal and semiconductor systems (po- and no-TCMP), as the comparison of KCP and $Ir(CO)_2$ acac shows. The latter is a no-TCMP compound with Ir(+1). In Fig. 2, the KCP strongly reflects for photon energies less than 2.0 eV even in the FIR, if the light is polarized parallel to the chain direction. The Ir(+1) complex with completely filled valence shell displays a steep maximum in reflectivity up to 60 % at 2.0 eV also, but quite normal values of about 10 % are found in the IR region. The substance is, there fore, no 1-D metal, although it is of similar appearance as KCP (13).

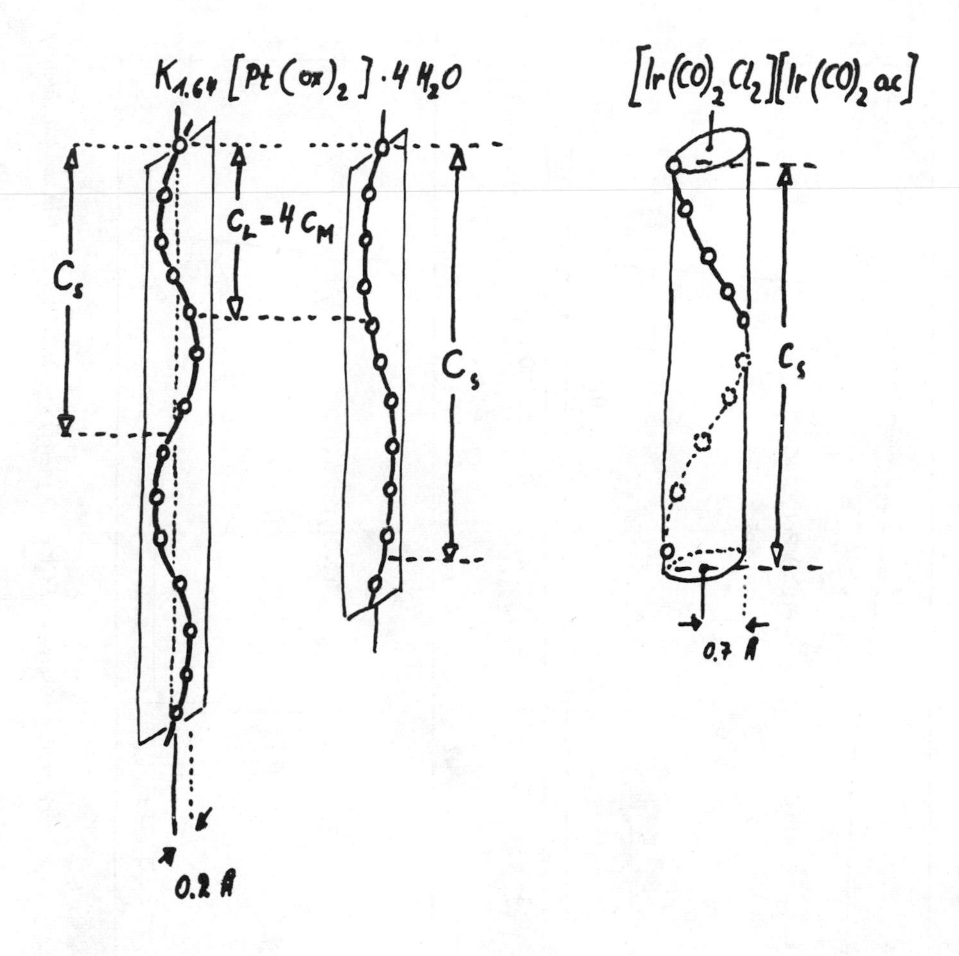

Figure 1. Shifts of heavy atoms from ideal positions in some 1-D metal phases

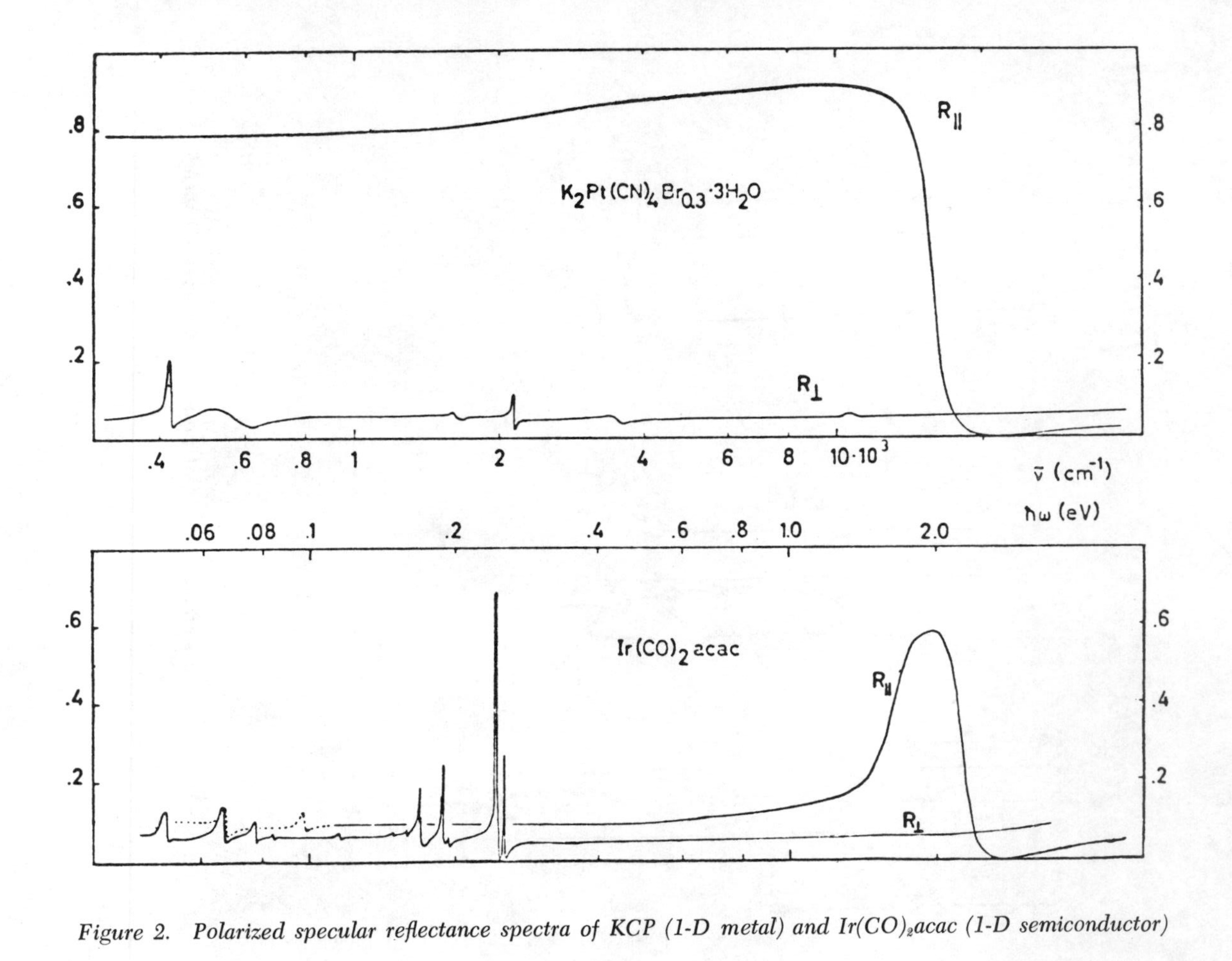

Figure 2. Polarized specular reflectance spectra of KCP (1-D metal) and $Ir(CO)_2acac$ (1-D semiconductor)

Literature Cited

1. Krogmann, K., and Hausen, H.D., Z.Anorg.Allg.Chem. (1968) 358, 67-81.
2. Krogmann, K., Z.Anorg.Allg.Chem. (1968) 358, 97-110.
3. Krogmann, K., Binder, W., and Hausen, H.D., Angew.Chem. Internat.Edit. (1968) 7, 812.
4. Zielke, H.J., Dissertation Karlsruhe 1973.
5. Malatesta, L., and Canziani, F., J.Inorg.Nucl.Chem. (1961) 18, 81.
6. Chernaev, I.I., and Novozhenyuk, Z.M., Russ.J.Inorg.Chem. (1966) 11, 1004.
7. Cleare, M.J., and Griffith, W.P., J.Chem.Soc. A (1970) 2788.
8. Krogmann, K., and Dodel, P., Chem.Berichte (1966) 99, 3408-3418.
9. Krogmann, K., Angew.Chem.Internat.Edit. (1969) 8, 35-42.
10. Peierls, R.E., "Quantum Theory of Solids", p. 108, Clarendon Press, Oxford (1955).
11. Comes, R., Lambert, M., Launois, H., and Zeller, H.R., Physical Review B (1973) 8, 571.
12. Renker, B., Rietschel, H., Pintschovius, L., Gläser, W., Brüesch, P., Kuse, D., and Rice, M.J., Phys.Rev.Lett. (1973) 30, 1144.
13. Wagner, H., Dissertation Karlsruhe 1974.

24

Some Comments on the Electronic Structure of Krogmann Salts and the Stability of Pt 2.3†

AARON N. BLOCH* and R. BRUCE WEISMAN†

The Johns Hopkins University, Baltimore, Md. 21218

I. Introduction

Salts of the divalent tetracyanoplatinate or dioxaloplatinate ions typically display crystal structures in which the planar complex anions are stacked to form linear chains, separated by comparatively wide channels containing counterions and water molecules (1). From solutions of the complex which have been partially oxidized, crystals can be grown having similar structures but considerably reduced metal-metal distances (1) and a formal platinum valence (1) close to 2.3+. Nominally, these are "mixed-valence" compounds, but the term is probably misleading inasmuch as X-ray (1), infrared (2,3), and photoelectron (4) spectroscopic studies show the oxidation states of all the metal atoms to be equal. Rather, there is ample evidence (1,5) for the formation of a well-defined quasi-one-dimensional band structure, including a partially filled conduction band of appreciable width. Hence, while their divalent parent compounds are electrical insulators (1), the partially oxidized "Krogmann salts" display large electrical (5,6) and optical (7) conductivities along the chain axis, and recently have excited widespread interest as prototypes for the study of one-dimensional conductors.

From this point of view, our understanding of the physics of these materials is by now well developed (5,8-10). Their theoretical chemistry, however, has received less attention, and several fundamental questions remain outstanding. (a) Why do these unique structures occur in only two chemical forms (1), the filled-band (Pt2+) insulators and a series of conductors whose conduction-band occupations fall in a narrow range near five-sixths filling (Pt 2.3+)? (b) Why do the intermetallic spacings (1) vary among the former group of compounds over the enormous range 3.09-3.60Å, while those in the latter group are much shorter and restricted to the range

---

*Alfred P. Sloan Foundation Fellow

†Present address: Department of Chemistry, University of Chicago, Chicago, Illinois 60637

2.88-2.98Å for the cyanides and 2.80-2.85Å for the oxalates? (c) Finally, why do cyanide and oxalate, alone among common ligands to platinum, form compounds of this class?

The final disposition of such questions must of course await detailed calculations of the electronic structure of the materials. Pending these, however, enough experimental information is available to guide a preliminary inquiry into the basic principles involved. Such a contribution has already been made by Krogmann (1), who proposed a plausible answer to (a) and (b) based on simple one-dimensional band-structure arguments. In this article we consider these arguments in more detail. We find, surprisingly, that while they probably do answer question (b), they do not resolve (a). We suggest instead that the fractional valence is a covalent molecular effect, and represents the optimum value for stability of the complex anion itself. A tentative response to (c) follows directly.

The paper is divided into seven parts. In Section II our approach is formulated within the framework of current empirical evidence. We conclude that the arguments of Krogmann (1) are to be cast in terms of the contribution of a single five-sixths-filled one-dimensional conduction band to the internal energy of the crystal. As an illustration, this calculation is performed in Section III using a simple tight-binding model. We find that the band-structure energy gained from oxidation of the Krogmann-salt structure to Pt 2.3+ is much too small to provide a realistic answer to question (a). Section IV inquires whether this conclusion is altered significantly by improving upon the simple tight-binding model. Appealing once more to empirical evidence, we find that neither the tight-binding limit nor the free-electron model recently proposed by Zeller (10) is fully consistent with experiment. A simple intermediate treatment corrects this deficiency, but still does not account for the stability of the five-sixths-filled band. We therefore suggest in Section V that the effect must reside in the covalent internal energy of the complex anion, and briefly discuss the chemical physics of a system in which the valence can be treated as a continuous variational parameter. We note that the most stable oxidation state of a molecule need not be an integer, and that in practice it can be attained only in unique molecular crystal structures, such as those of the Krogmann salts, in which the effective total charge of the molecule is not quantized. The existence of two stable oxidation states with the same general crystal structure has electrochemical implications as well, and Section VI presents a preliminary experimental account of the electrochemical oxidation of a single crystal of $K_2Pt(CN)_4 \cdot 2H_2O$ to a solid-state galvanic cell. In Section VII we summarize our conclusions.

II. General Remarks

We shall find it useful at the outset to review briefly the salient physics of the materials. The Krogmann salts represent a particularly interesting class of model systems in which the roles of instabilities (11), fluctuations (9), and disorder (8,12) in the physics of the one-dimensional electron gas can be studied directly. The metallic state in one dimension is inherently unstable (11), and the beautiful experiments of Zeller and co-workers (10) have shown that the platinum chains do undergo a Peierls distortion (11) with decreasing temperature. The phase transition is, however, incomplete (13), probably owing to the intrinsic structural disorder (1) of the systems. The static potential fluctuations associated with random occupantion of the counterion and water sites tend to suppress the transition, reducing the mean-field transition temperature and placing weakly localized electronic states in the Peierls energy gap (12,14). Indeed, we have argued elsewhere (8,12,15) that certain features of the electron transport and low-frequency dielectric response cannot be understood in terms of the Peierls distortion alone, but arise from a finite density of localized states at the Fermi level.

For the present discussion of chemical stability, however, the disorder is of scant importance. The potential fluctuations required to explain the observed transport phenomena (8,12,15) are in this case but a small fraction of the unperturbed bandwidth (14), and the configurational entropy associated with the partially occupied interchain sites make a contribution of but order kT to the total free energy per molecule (16). The electronic entropy of the partially filled conduction band is smaller still (16). We conclude that the stability of Pt 2.3+ resides in the internal energy of the crystal, and that this is negligibly different from that of a hypothetical analogous ordered system.

We write this energy schematically as:

$$U = U_C + U_B + U_R + U_M \qquad (1)$$

where U_C is the total covalent binding energy of the complex anion, U_B the band-structure stabilization of the chains, U_R the repulsive energy between adjacent planar complexes on a chain, and U_M the remainder of the Madelung energy of the crystal. Admittedly, these distinctions are to some extent arbitrary--the first and second terms, for example, are never fully separable--but we shall find them to be of conceptual value.

Now, we can immediately dismiss on empirical grounds the possibility that the number of conduction electrons per molecule, Z, and intermetallic spacing, R, in the Pt 2.3+ Krogmann salts are determined primarily by U_M. Among the dozens of compounds in the class, structural and chemical differences which presumably lead to substantial differences in Madelung energy produce only minor differences in Z and R (1). From those of a typical example such as $K_2Pt(CN)_4Br_{0.3}\cdot 3H_2O$, these parameters change but little when Cl is substituted for Br; when halogen is excluded altogether in favor of a deficiency of cations; when oxalate is substituted for cyanide; when any alkali or alkaline earth metal is substituted for K; when the water content is changed; or when the gross crystal structure itself changes from tetragonal to triclinic, monoclinic, or orthohombic (1). It is as though the interchain species had been left to adjust themselves to dimensions and chemical composition which are essentially fixed by strong intrachain forces.

Toward an elucidation of these forces, Krogmann (1) has suggested in effect that the stability of the oxidation state Pt 2.3+ (Z=1.7) arises from an optimization of the band-structure term U_B. Such an effect would represent a rough one-dimensional analog of the well-known Hume-Rothery rules (17) for the composition of alloys.

We shall cast these arguments in a simple mathematical form. As a preamble, we observe that in calculating U_B we are dealing with a single conduction band, not degerate with any other band at the Fermi level. This we affirm by using the Peierls distortion as a probe of band occupancy. The distortion reflects the underlying divergence of the one-dimensional electronic response functions at wavenumber $q=2k_F$, where k_F is the Fermi wavenumber (18). Where the conduction electrons are shared by two one-dimensional bands which overlap in energy at the Fermi level, the Fermi "surface" consists in general of two distinct values of k_F; only when such degeneracy is absent does k_F assume the single value $\pi Z/2R$. Now, it is found (13) experimentally that with the small variations in Z which do occur among the different Krogmann salts (inferred from their stoichiometries), the period of the distortion also varies, and always corresponds precisely to a reduced wavenumber of $\pi Z/R$. We conclude that only one hand in the system is partially occupied.

III. Band-Structure Energy U_B in the Tight-Binding Model

For simplicity of illustration, we treat this band initially in the one-electron tight-binding limit (19), and defer to the next section a discussion of the adequacy of this approximation and the roles of hybridization and the Peierls distortion. Let E° be the energy of the relevant molecular orbital $|i\rangle$, centered on site i of the chain, in the presence of the Madelung field but the absence of any interaction between adjacent molecules on the chain. When the state $|i\rangle$ is perturbed by the lattice potential of the chain, $H_1=\sum_{j\neq i} V_j$, the energy of the one-electron conduction-band state of wavenumber k is:

$$E_k = E^\circ - \alpha - 2\beta \cos kR \tag{2}$$

Here $-\alpha$ is the band shift $\langle i|H_1|i\rangle$, and $-\beta$ the charge-transfer integral $\langle i|H_1|i\pm1\rangle$. As usual, overlap integrals S of the form $\langle i|i\pm1\rangle$ have been ignored.

Clearly, E° contains local contributions to U_C and U_M. To evaluate U_B, we integrate the quantity $E_k - E^\circ$ over all occupied states $|k\rangle$ of the band:

$$U_B = - Z\alpha - \frac{4\beta}{\pi} \sin \left(\frac{\pi Z}{2}\right) \tag{3}$$

We note that both α and β are functions of R and Z, and that the principal dependence of β on these parameters is surely exponential. Taking the radial screening constant for the wavefunction $|i\rangle$ to have roughly the Slater form (20), we write this dependence as

$$\beta \sim \beta_o \exp [-(\mu - \eta Z) R] \tag{4}$$

where β_o, μ, and $\eta \ll \mu$ are constants.

In contrast, the variation of the small (19) energy α with each of the parameters R and Z is at most a power-law dependence, comparable with those of typical contributions to the Madelung energy U_M. This observation has special significance for the filled-band case Z=2, where U_B according to Equation 3 is simply -2α. Under these circumstances, we expect R to be determined largely by U_M rather than U_B. We thereby account for the wide variation in R among the divalent parent compounds (1), in partial answer to question (b).

With partial oxidation of these materials to form Krogmann salts, the strongly R- and Z- dependent second term in Equation 3 enters U_B, and the situation is altered drastically. On the basis of our discussion in the preceding section, it is not unreasonable to regard the interchain channels as a "charge sink", capable of accommodating as many counterions as are necessary to balance any Z in the range of interest. To the extent that this is true, the chains constitute a chemically unique system in which the number of bonding electrons may be regarded as a variational parameter. If, as suggested by Krogmann (1), Z is determined by U_B, then for given R the observed Z must correspond to a minimum in U_B:

$$\left.\frac{\partial U_B}{\partial Z}\right|_R = 0 \approx - \left(\alpha + 2\beta \cos \frac{\pi Z}{2} + \frac{4}{\pi} \left.\frac{\partial \beta}{\partial Z}\right|_R \sin \frac{\pi Z}{2}\right) \tag{5}$$

When Equations 4 and 5 are combined, a transcendental equation is obtained relating R and Z. A typical solution is shown in Figure 1. Here we have simply determined μ and η according to Slater's rules (20), and adjusted the ratio β_o/α so as to reproduce the experimental value Z=1.7 for R=2.88Å. More elaborate fitting procedures are certainly available, but we have found no physically reasonable choice of parameters which substantially alters Figure 1 or our final conclusions.

Under the condition imposed by Equation 5, we are now in a position to calculate U_B from Equations 3 and 4. The result is

$$U_B/\alpha = - Z + \frac{\sin(\pi Z/2)}{(\pi/2)\cos(\pi Z/2) + (\partial \ln\beta/\partial Z)\sin(\pi Z/2)} \tag{6}$$

Equation 6 is plotted in Figure 2, which represents the variation of U_B with R under the constraint that each R, Z is self-consistently to be readjusted so as to minimize U_B.

Qualitatively, the calculation is consistent with the

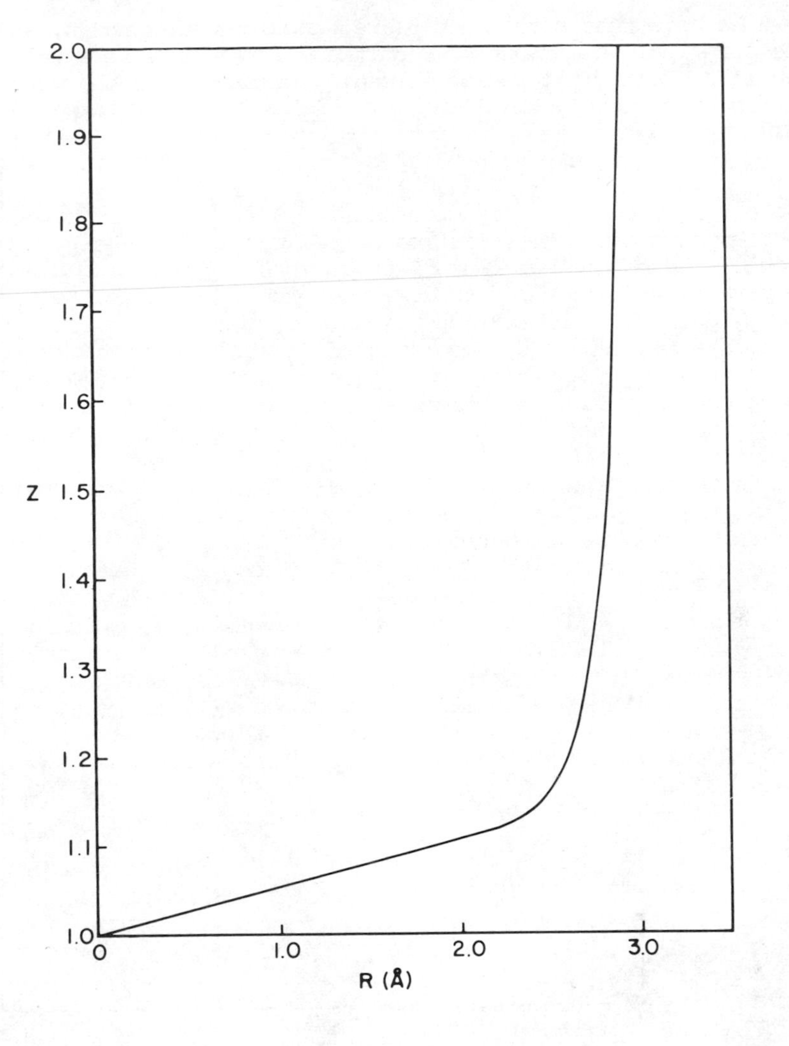

Figure 1. Variation of conduction-band occupation, Z, *with lattice constant,* R, *for a one-dimensional tight-binding system under the constraint of Equation 5*

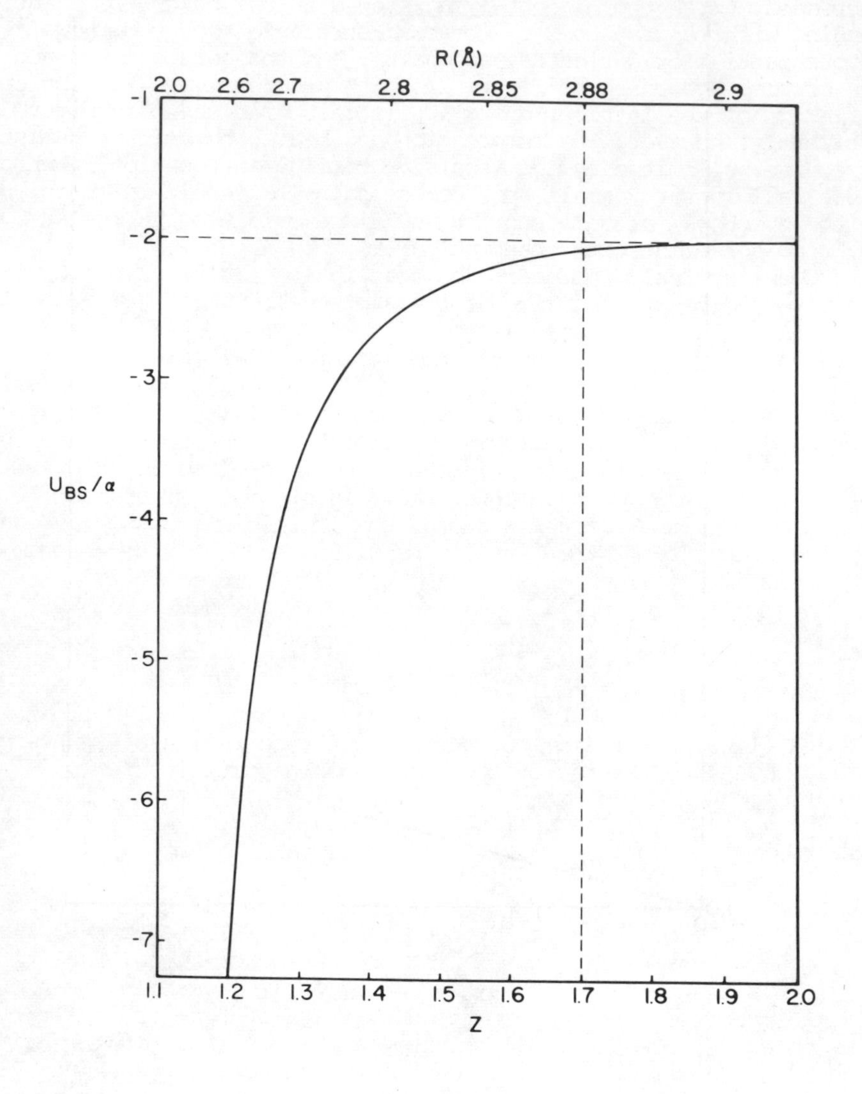

Figure 2. Variation of band-structure stabilization U_B *with* R *and* Z *for a one-dimensional tight-binding system under the constraint of Equation 5. At the experimental values* Z = *1.7 and* R = *2.88* A, $U_B(Z)–U_B(2)$ *is much too small to account for the stability of Pt 2.3+.*

description offered by Krogmann (1). For large R, we have Z=2 and U_B/α independent of R. With decreasing R, the bandwidth 4β is increased, until the highest-lying states in the band become anti-bonding with respect to E°. The band-structure energy U_B now favors partial oxidation of the chains, and the parameters Z and R are coupled according to Figure 1. With progressive oxidation and shrinkage of the lattice, energy U_B is gained as shown in Figure 2. Ultimately, however, the compression and hence the oxidation must of course be limited by the repulsive term U_R in Equation 1, which rises rapidly for R small and decreasing. The result is a minimum in the sum U_B+U_R at some short R and fixed oxidation state $1<Z<2$, in broad agreement with observation.

Quantitatively, however, the description fails. Since U_R is positive and monotonic in R, the depth of the minimum in U_B+U_R must be smaller than $U_B(2)-U_B(Z)$. According to Figure 2, this can be a large energy for $Z<1.3$, but for the observed Z of 1.7 it is only about 0.05α. Now, α is a small energy (19): it is characteristically no more than a few tenths of an electron volt, and in the present case surely no larger than the important contributions to U_M. We conclude that within the tight-binding model, the depth of the hypothetical minimum in U_B+U_R at Z=1.7 is of order kT at room temperature. Obviously, such a broad, shallow minimum cannot account for the stability of the five-sixths filled band and the absence of intermediate oxidation states between Pt 2.0+ and Pt 2.3+.

Essentially, the difficulty arises from the square-root singularity in the density of states at the top of a one-dimensional conduction band. When a symmetric, one-dimensional, tight-binding band is half-emptied, a substantial gain in kinetic energy (of order β) results. But when only one-sixth of the electrons are removed, most of the emptied states lie in the region of the singularity, close to the top of the band, and the gain in energy is slight.

IV. Beyond the Tight-Binding Approximation

How heavily model-dependent are the results of the preceding section? Let us examine some physically plausible departures from the simple tight-binding approximation considered there.

Extra stability is certainly conferred, for example, by the incipient (9,21) Peierls gap in the density of states; the contribution to U_B+U_R, however, will only be of the order of the mean-field transition temperature (18,21) of a few hundred degrees Kelvin. Further, this contribution appears at all fillings $Z<2$, and is largest for Z=1 (11). In effect, then, the Peierls contribution simply aggravates the trend of Figure 2.

Nor is distortion of the tight-binding bands through hybridization of the basis molecular orbitals likely to play an important role. It is easy to show that the s-d hybridization suggested by Zeller (10) has no effect upon the center of the conduction band, but lowers the energy of the remaining states, with the largest effect coming at the top and bottom extremes of the band. Compared with the symmetric band of the previous section, the result is a compression of the top half of the band on the energy scale and an elongation of the bottom half. This amounts to a further destabilization of Pt 2.3+ relative to Pt 2+. The p-d mixing suggested by Krogmann (1) has the opposite effect: it is symmetry-forbidden at the center and edges of the Brillouin zone and strongest at the center of the band. Thus it reduces the density of states in the top half of the band and does tend to help stabilize Pt 2.3+. This effect is partially cancelled by the s-d hybridization, however, and we have satisfied ourselves that the $5dz^2$-6s and $5dz^2$-6p energy differences deduced from spectroscopic (22) and theoretical (23) studies of the free complex are in any case too large for such mixing to play a significant role in the structure of the conduction band.

Within the standard tight-binding approximation, then, there appears little room for a contribution to U_B which accounts for Pt 2.3+. At this point it is natural to ask to what extent this approximation itself is valid for the Krogmann salts. There is strong experimental evidence that it is not.

Consider, for example, the optical reflectivity (5,7,10), which shows a well-defined plasma edge near 2 eV, polarized along the chain axis and responsible for the characteristic coppery luster of the Krogmann salts. In a quasi-one-dimensional conductor, the plasma frequency is (25):

$$\omega_p^2 = 8Ne^2 v_F/\hbar\varepsilon_\infty \tag{7}$$

where N is the number of chains per unit cross-sectional area, v_F the Fermi velocity, and ε_∞ the background dielectric constant. For the Krogmann salts, with $\varepsilon_\infty \sim 2.2$, the Fermi velocity deduced in this manner (24) would correspond in tight-binding theory to an unreasonably large bandwidth of ca. 8.5 eV. Indeed v_F is very close to its free-electron value (24).

This circumstance has led Zeller (10) to suggest that due to a fortuitous amount of s-d mixing, the entire conduction band in the Krogmann salts has nearly the free-electron form, with an effective mass m* of unity. In our view there is little foundation for this proposal, either in theory or in experiment. One does not achieve a spatially uniform charge distribution through superposition of the rapidly oscillating 5d and 6s wavefunctions, nor does one escape the tight-binding limit simply by adjusting the geometries of the basis orbitals. To put the matter another way, the nearly-free-electron representation is fundamentally a statement about the effectiveness of the screening of the atomic pseudopotential, and not about the degree of configuration interaction between atomic orbitals.

In light of these objections it is not surprising to find the nearly-free-electron model inconsistent with experiment. For example, the thermoelectric power is small and positive (6,26-28), indicating that in the vicinity of the Fermi level the band is hole-like rather than electron-like. In other words, although the first derivative v_F of the electronic dispersion curve has approximately the free-electron value at the Fermi level, the second derivative has the wrong sign.

Further, in a nearly-free-electron system with a five-sixths filled conduction band and $m^*/m=1$, it is readily verified that the threshold for the first interband optical transition should occur at or above 23,000 cm^{-1}. No such threshold occurs in the observed optical spectrum (7,24,29): indeed, the only distinct feature found in the entire visible and near-ultraviolet region above the plasma edge is the weak plasmon absorption (24,25,29) centered at 15,800 cm^{-1}. This absorption has been studied in detail by one of us and collaborators (24,25). We found that for a quasi-one-dimensional structure it does not occur in the free-electron model, and is appreciable only in the case of a relatively narrow band for which the bandwidth W is comparable with, or less than, $\hbar\omega_p$. We conclude that the nearly-free-electron model is inconsistent with the observed optical properties of the material.

The failure of both the free-electron and tight-binding models suggests the need for an intermediate description within which the observed plasma frequency and thermoelectric power can be reconciled with the upper limit on W inferred from the plasmon absorption. Foward such a description, we note that for a band as narrow as W~2 eV, the departures from the simple tight-binding model are unlikely to be severe. This encourages us to represent the corrections simply by retaining the basic tight-binding formalism, but including the overlap integrals $S=\langle i|i\pm1\rangle$ which were neglected in the standard Equation 2. In light of the recent molecular orbital calculations of Interrante and Messmer (23), such neglect appears unjustified. This extension of the tight-binding approximation is the precise analog of the Wheland extension of the simple Hückel approximation in one-electron π-molecular orbital theory (30).

With inclusion of overlap, the one-electron energies of Equation 2 are replaced by

$$E_k = E^\circ - \frac{\alpha+2\beta \cos kR}{1+2S \cos kR} \tag{8}$$

with group velocity

$$\hbar v_k = \frac{(\beta-\alpha S)(2R \sin kR)}{(1+2S \cos kR)^2} \tag{9}$$

Now, in a quasi-one-dimensional metal, the sign of the thermoelectric power is determined by the logarithmic derivative of the density of the states $\rho(E)$ with respect to energy, evaluated at the Fermi level:

$$\left.\frac{\partial \ln\rho}{\partial E}\right|_{E_F} = \left.\frac{1}{\rho}\frac{\partial\rho}{\partial k}\frac{\partial k}{\partial E}\right|_{E_F} = \left.\frac{1}{\hbar}\frac{\partial}{\partial k}\left(\frac{1}{v_k}\right)\right|_{k_F} \tag{10}$$

Substituting (9) into (10) and assuming $\beta >> \alpha S$, we find that for Z=1.7 the thermoelectric power is positive for small S, and crosses zero near S=0.37. From the very small positive values observed, we estimate S~0.35. Then using Equations 7 and 9 and the observed plasma frequency (7,24,29) of 15,800 cm^{-1}, we have $\beta - \alpha S \sim 0.32$ eV. The bandwidth

$$W = \frac{4(\beta - \alpha S)}{1-4S^2} \sim 2.5 \text{ eV} \tag{11}$$

is then comparable with $\hbar\omega_p$. The observation of a well-defined plasmon absorption (24,29) suggests that the true value of W is somewhat smaller still, but the overall agreement with experiment is good enough to demonstrate that a minimal improvement upon the simplest tight-binding approximation, such as Equation 8, is a reasonably accurate representation of the conduction band in the Krogmann salts.

In accounting for the stability of Pt 2.3+, however, the improved description fares no better than the old. Indeed, the band-structure term U_B now tends to drive Pt 2.3+ less stable than Pt 2+. Integrating Equation 8 (31), recalling that S scales roughly as β, and proceeding as in the previous section, we find that with the same values of μ and η, U_B for Z=1.7 is positive with respect to Z=2 by about 0.1 eV. This result is not substantially altered for any physically reasonable choices of μ and η. Improving the accuracy of our representation of the conduction band has simply re-affirmed our conclusion that the stability of Pt 2.3+ is not associated with the kinetic energy of the conduction electrons.

V. The Covalent Energy U_C

In the preceding sections we have argued that the variations in $U_M+U_B+U_R$ in Equation 1 are too small, and possibly of the wrong sign, to be responsible for the peculiar stability of Pt 2.3+. The remaining term is the covalent binding energy U_C of the complex cyanoplatinate ion itself, and we now suggest that this term accounts for the observed effect.

It is well known that in square-planar complexes of d^8 transition metals the occupied molecular orbitals uppermost in energy are weakly antibonding. The complex is nevertheless stable because the repulsive contribution from these orbitals to U_C is more than compensated by the attractive contribution from other orbitals, principally the strongly σ-bonding a_{1g}, b_{1g}, and e_u (23,32). Let us consider the oxidation of such a complex under conditions where, as in the Krogmann salt structure, the (time-averaged) occupation Z of a high-lying orbital may be regarded as a continuous rather than a discrete variable.

Since the removal of an infinitesimal amount of charge from the d^8 complex lowers the occupation of an antibonding orbital, it certainly tends to stabilize the complex. With progressive oxidation, however, the many-electron wavefunction is readjusted and the effective one-electron energy-level scheme renormalized so as to accommodate the change in electron-electron interaction and screening of the core potential. In particular, the high-lying orbitals must drop in energy, reflecting the increasing difficulty of further oxidation; eventually the highest occupied orbital becomes bonding, and the oxidized species displays substantial electron affinity.

In the case of square-planar platinum complexes, this has apparently happened by the time one full electron is removed. Hence Pt 3+ is unstable, and four-coordinated Pt 4+ does not occur in the square-planar geometry. Viewed from this perspective, the usual Pt 2+ in the free complex does not represent a true minimum in U_C as a function of Z, as the presence of filled antibonding orbitals attests. Rather, Pt 2+ is the best compromise which can be attained under the constraint that Z be an integer. When this constraint is lifted, the real minimum occurs at that fractional oxidation state for which the energy of the partially occupied orbital passes through zero. Clearly, this occurs somewhere between Pt 2+ and Pt 3+, and we suggest that in the cyanoplatinate ion it occurs close to Pt 2.3+.

It is hardly necessary to emphasize that to test this conjecture requires more detailed claculations than have been presented here. Nevertheless, we are impressed at this preliminary stage by its plausibility and by the lack of a reasonable alternative. Unlike the weak (or non-existent) minimum in U_B+U_R, the minimum in U_C occurs at a sharply defined value of Z, and presumably represents a substantial gain in the binding energy of the complex. The existence of such a minimum at a fractional valence is not of course unique to cyanoplatinate, but probably occurs for any molecule whose ground state includes filled antibonding orbitals. The realization of the minimum in practice, however, requires a unique structure, such as that of the Krogmann salts, in which covalent, metallic, and ionic bonding coexist.

VI. Electrochemical Oxidation of Crystalline $K_2Pt(CN)_4$ to a Solid-State Galvanic Cell

Before concluding, we digress to remark that the existence of two stable oxidation states in the same general crystal structure has interesting practical ramifications. In particular, the materials are known to exhibit ionic conductivity. It has been shown by Gomm and Underhill (33), and independently by others (28,34), that sufficiently strong electric fields reduce the Krogmann salt $K_2Pt(CN)_4Br_{0.3}\cdot 2.3H_2O$ to a compound of divalent platinum.

We have found that, contrary to an assertion by Gomm and Underhill (33), the reverse reaction can also be induced: hydrated single crystals of $K_2Pt(CN)_4$ are easily and reversibly oxidized by application of modest electric fields along the needle axis, using mercury or silver-paste electrodes. After the field is raised above a threshold of ca. 100 V/cm, the resistance falls abruptly by two orders of magnitude, and a dark, copper-colored area forms at the anode and spreads toward the cathode, where vigorous evolution of hydrogen gas is observed. When the polarity is reversed the reaction is reversed also, and the sharp boundary between the two phases recedes until the original white crystal is recovered.

At constant temperature, humidity, and applied voltage, the current does not diminish as the reaction proceeds to completion. The (hydrated) product crystal displays the coppery lusture and dichroism characteristic of Krogmann salts (1), and has shrunk to about 90% of its former length; this corresponds to the typical difference in the lattice spacing R between Krogmann salts and their divalent parent compounds (1).

By monitoring the infrared transmission spectrum of the crystal as the reaction proceeds, we confirm that the product is a Krogmann salt, and that Pt 2+ and Pt 2.3+ are the only oxidation states present in measurable quantity. We have accumulated evidence that the reaction proceeds via a proton-transfer-and-reduction mechanism similar to that suggested by Lecrone and Perlstein (35) for the electrochemical reduction of mixed-valence oxaloplatinate systems, and we assign to the product the previously unreported formula $K_2Pt(CN)_4(OH)_x \cdot nH_2O$, where $x \sim 0.3$.

If the applied field is removed before the reaction is complete, a potential difference exists between the oxidized and unoxidized sections, and the partially converted crystal acts as a galvanic cell. The Krogmann-salt region is the anode. Because of the high internal resistance associated with the $K_2Pt(CN)_4$ section, this miniature solid-state battery functions as a current rather than a voltage source. As in the initial oxidation, the current is quite sensitive to water vapor in the surrounding atmosphere, and is in fact a measure of relative humidity. Near 100% humidity and 22°C, we find typical current densities of ca. 5 ma/cm^2 and an open-circuit potential of 1.35 V.

VII. Summary and Conclusions

We summarize our conclusions as follows. The partially oxidized Krogmann salts are characterized by a single five-sixths-filled conduction band which is adequately described neither by the nearly-free-electron model nor by conventional tight-binding theory. Instead, a reasonable intermediate representation, consistent with experiment, is obtained by extending the tight-binding formalism to include the effects of orbital overlap between neighboring molecules.

The electronic kinetic energy of such a band cannot account for the stability of the five-sixths filled band (Pt 2.3+). Rather, we suggest that the appearance of this phase is a molecular effect, representing the minimization of the covalent binding energy of the complex anion with respect to the band occupation Z. Such an effect can only occur, of course, in an unusual crystal structure in which the oxidation state of the complex is continuously adjustable.

Once Z is fixed in this manner, the intermetallic spacing R is closely controlled by the band energy U_B as suggested by Krogmann (1) and depicted in Figure 1. In contrast, R in the Pt 2+ parent compounds is determined by the Madelung energy U_M, and hence varies widely from compound to compound.

Thus we answer questions (a) and (b) posed in the Introduction. In response to (c), we suppose that only in the oxalo- and cyanoplatinate complexes is the balance between σ-bonding and π-"backbonding" such that the $5a_{1g}(dz^2)$ orbital lies high enough in energy to form a band which can be oxidized to Z=1.7. For example, this band must not intersect any of the flat bands derived from the other high-lying occupied orbitals (23) above the Z=1.7 Fermi level.

If these conclusions are correct, it appears likely that the partially oxidized Krogmann salts are chemically unique systems in which, by accident, a set of rather exacting energetic and structural criteria are simultaneously satisfied. If this is so, then the synthetic search for analogs is likely to be frustrating.

Finally, we remark that these considerations probably do not apply to most of the organic TCNQ salts (12,27,36), whose structural and physical properties are in some respects similar to those of the Krogmann salts. Here packing considerations seem to preclude any adjustment of the counterion population so as to minimize U_C, and the small value of W severely limits the influence of U_B. A partial exception is the newly developed class of organic semimetals such as TTF-TCNQ (36), in which the conduction electrons are distributed between two different species of conducting chains. In these, the best organic conductors known, there is real promise of adjusting Z, and hence the electrical properties, through chemical control of U_C (36).

Literature Cited

1. Krogmann, K., Angew. Chem. Internat. Ed., (1969), 8, 35, and references therein.
2. Evstaf'eva, O. N., Russ. Jour. Inorg. Chem., (1966), 11, 711.
3. Rousseau, D. L., Butler, M. A., Guggenheim, H. J., Weisman, R. B. and Bloch, A. N., Phys. Rev. B, 10 (in press).
4. Butler, M. A., Wertheim, K., Rousseau, D. L., and Hüfner, S., Chem. Phys. Lett., (1972), 13, 473.
5. Zeller, H. R., Adv. Sol. St. Phys. XIII, (1973), and references therein.
6. Minot, M. J. and Perlstein, J. H., Phys. Rev. Lett., (1971), 26, 371.
7. Bernasconi, J., Brüesch, P., Kuse, D., and Zeller, H. R., Brown-Boveri Res. Report KLR-73-05 (Brown-Boveri Res. Center, Baden, Switzerland, 1973).
8. Bloch, A. N., Weisman, R. B., and Varma, C. M., Phys. Rev. Lett., (1972), 28, 753.

9. Lee, P. A., Rice, T. M., and Anderson, P. W., Phys. Rev. Lett., (1973), 31, 462.
10. Zeller, H. R., this volume, and references therein.
11. Peierls, R. E., "Quantum Theory of Solids", pp. 108 ff, Clarendon Press, Oxford, 1955.
12. Bloch, A. N., in Masuda, K. and Silver, M., ed., "Energy and Charge Transfer in Organic Semiconductors", p. 159, Plenum Press, New York, 1974, and references therein.
13. Zeller, H. R., private communication.
14. Sen, P. and Varma, C. M., Bull. Am. Phys. Soc., (1974), 19, 49.
15. Bloch, A. N. and Varma, C. M., J. Physics C., (1973), 6, 1849.
16. Weisman, R. B., unpublished work.
17. See, for example, Heine, V. and Weaire, D., Sol. St. Phys., (1970) 24, 250.
18. See, for example, Rice, M. J. and Strassler, S., Sol. St. Comm., (1973), 13, 697.
19. See, for example, Friedel, J. in Ziman, J. M., ed., "The Physics of Metals I: Electrons", pp. 340 ff, Cambridge University Press, Cambridge, 1969.
20. See, for example, Eyring, H., Walter, J., Kimball, G. E., "Quantum Chemistry", pp. 162-3, Wiley, New York, 1944.
21. Lee, P. A., Rice, T. M., and Anderson, P. W., Sol. St. Comm., (1974), 14, 703.
22. Piepho, S. B., Schatz, P. N., and McCaffery, A. J., J. Am. Chem. Soc., (1969), 91, 5994.
23. Interrante, L. V. and Messmer, R. P., this volume.
24. Williams, P. F., Butler, M. A., Rousseau, D. L., and Bloch, A. N., Phys. Rev. B, (1974), 10, 1109.
25. Williams, P. F. and Bloch, A. N., Phys. Rev. B, (1974), 10, 1097.
26. McKenzie, J. W., Wu, C., and Bube, R. H., Appl. Phys. Lett., (1972), 21, 1.
27. Schegolev, I. F., Physica Stat. Solidi, (1972), A12, 9, and references therein.
28. Minot, M. J., (Ph.D. Thesis, The Johns Hopkins University, 1973). The experiments of References 26 and 28 show that the small negative thermopower reported by Kuse and Zeller [Sol. St. Comm., (1972), 11, 355] is probably the result of using partially dehydrated crystals.
29. Wagner, H., Geserich, H. P., Baltz, R. V., and Krogmann, K., Sol. St. Comm., (1973), 13, 659.
30. See, for example, Daudel, R., LeFebvre, R., and Moser, C., "Quantum Chemistry: Methods and Applications", pp. 67 ff, Interscience, New York, 1959.
31. Gradshteyn, I. and Ryzhik, I. M., "Tables of Integrals, Series, and Products", p. 148, Academic Press, New York, 1965.
32. See, for example, Ballhausen, C. J, "Ligand Field Theory", McGraw-Hill, New York, 1962.
33. Gomm, P. S. and Underhill, A. E., Chem. Comm., 1971, 511; J. Chem. (Dalton), 1972, 34.
34. Würfel, P., Hausen, H. D., Krogmann, K., and Stampfl, R.,

Phys. Stat. Sol., A10, (1972), 537.
35. Lecrone, F. and Perlstein, Chem. Comm., 1972, 75.
36. Bloch, A. N., Cowan, D. O., and Poehler, T. O., in Masuda, K. and Silver, M., Ed., "Energy and Charge Transfer in Organic Semiconductors", p. 167, Plenum Press, New York, 1974, and references therein.

Acknowledgments

We are grateful to the Advanced Research Projects Agency of the Department of Defense and to the National Science Foundation for partial support of this research.

25

The Peierls Transition in One-Dimensional Solids

H. R. ZELLER and P. BRÜESCH

Brown Boveri Research Center, CH-5401 Baden, Switzerland

In his book "Quantum Theory of Solids" (1) Peierls shows that a one-dimensional metal is inherently unstable and will undergo a phase transition into a semiconducting state. Almost simultaneously and independently the theory of this phase transition was worked out by Fröhlich (2). Fröhlich showed that the phase transition may result in a low temperature state which is not semiconducting but superconducting with transition temperatures not restricted to the cryogenic range.

Recently experimental systems have been studied which undergo a Peierls transition and which are potential candidates for the Fröhlich mechanism of superconductivity. In particular the transition was shown to occur in K_2 $[Pt(CN)_4]$ $Br_{0.30}\cdot 3(H_2O)$ (KCP) and related salts (3). There is evidence that a Peierls transition also takes place in TTF TCNQ (4).

In the following we will restrict the discussion to the best understood system, i.e., K_2 $[Pt(CN)_4]$ $Br_{0.30}\cdot 3(H_2O)$. For a general introduction the reader is referred to reference (3).

Before we turn to the discussion of the Peierls instability, there is one point which should be made in connection with extended metal-metal interactions. At first sight it would seem that due to the small overlap a tight binding model for the band structure should be a very good approximation, i.e., the carriers would be holes in a $d_z{}^2$ band. It came as a big surprise when we discovered that the conduction band is definitely not a sinusoidal tight binding band but rather a parabolic free electron band with effective mass $m^* \approx me$. Although this sounds highly implausible at first sight, it can be explained as typical dimensionality effect. Nearly free electron behaviour means nearly constant electron density. As can be seen from Fig. 1 a nearly constant electron density along the strand axis can be achieved at a relatively modest overlap of the wave functions (in this case $d_z{}^2$-s orbitals). This is not possible in two or three dimensions. The essential features of the argument can also be visualised as follows: In an array of spheres it is always possible to have them touch in one line but there is always empty space between the spheres in a plane or in 3-d space. Thus as a function of overlap, free electron behaviour is reached much earlier in 1-d systems than in 2-d and 3-d ones. Of course free electron behaviour is restricted to the strand axis (5).

In the following we will discuss the Peierls-Fröhlich transition and the possibility of high temperature superconductivity based on the Fröhlich mechanism. Below the phase transition a sinusoidal distortion of the strands takes place such that an energy gap at the Fermi energy is created (6). For instance in a quarter filled band this distortion will have the period of four lattice spacings, splitting the conduction band into a filled and three empty bands. Above the transition temperature a precursor shows up in the form of a soft lattice vibration which corresponds to the low temperature static distortion. Due to the one-dimensional nature of the system fluctuations are extremely important. There is in general no sharp transition but a very gradual transformation from a metallic into a semiconducting state.

Associated with the sinusoidal distortion is a sinusoidal charge density wave (CDW). Fröhlich had realized that within a continuum model the free energy of the system does not depend on the phase of the CDW. This implies that the CDW can be shifted freely up and down the strands without any activation energy. As in conventional superconductivity the presence of an energy gap effectively inhibits scattering processes. Thus the system should behave as a superconductor with the electrons surfing on the propagating lattice distortion (Fig. 2).

The existence of a Peierls distortion in KCP was clearly demonstrated by diffuse x-ray scattering (7) and inelastic neutron scattering (8) experiments. Also the Peierls gap shows up in the optical spectra at low temperatures at about 0.2 eV (9).

Next we turn to the central question whether the Peierls-Fröhlich transition really leads to superconductivity. For a conventional BCS superconductor the conductivity $\sigma(\omega)$ is represented by a δ-function at $\omega = o$ and a peak at energies corresponding to the breaking of a Cooper pair. In an analogous fashion the ideal Fröhlich superconductor should exhibit a δ-function at $\omega = o$ and a peak at $\omega \approx Eg$, where Eg is the Peierls gap. Figure 3 shows the experimental result on KCP at 40°K obtained from Kramers Kronig analysis of reflectivity data. The peak at $\omega \approx 1600\ cm^{-1} \approx 0.2$ eV corresponds to excitations across the Peierls gap. But instead of a δ-function at $\omega = o$ Fröhlich collective mode produces a peak with finite width centered at about 2 - 4 meV dependent on sample perfection. At higher temperatures the peak gets broader and disappears around 200°K.

In their fundamental paper Lee, Rice and Anderson (10) have discussed why no true superconductivity based on the Fröhlich mode is expected. Due to commensurability with the lattice parameter, to random potentials provided by impurities or disorder and to 3-d coupling the translational invariance is broken and the CDW is pinned. This corresponds to a spring constant and

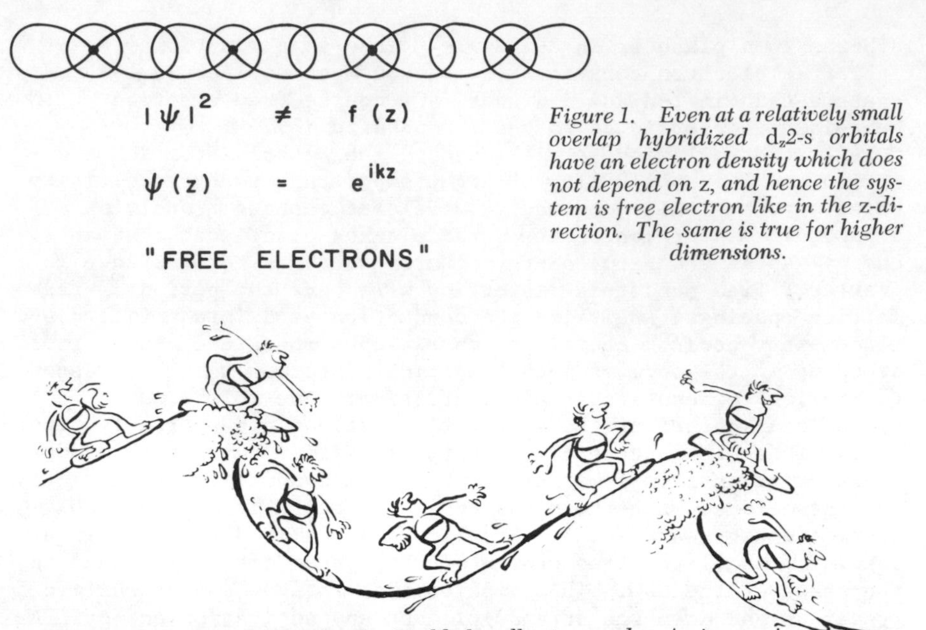

Figure 1. Even at a relatively small overlap hybridized d_z2-s orbitals have an electron density which does not depend on z, and hence the system is free electron like in the z-direction. The same is true for higher dimensions.

Figure 2. Electron transport by the Fröhlich collective mode. As in any insulator the electrons are bound to a periodic potential. In the Peierls Fröhlich state the periodic potential is not fixed in space but propagating and able to carry an electric current along the strands.

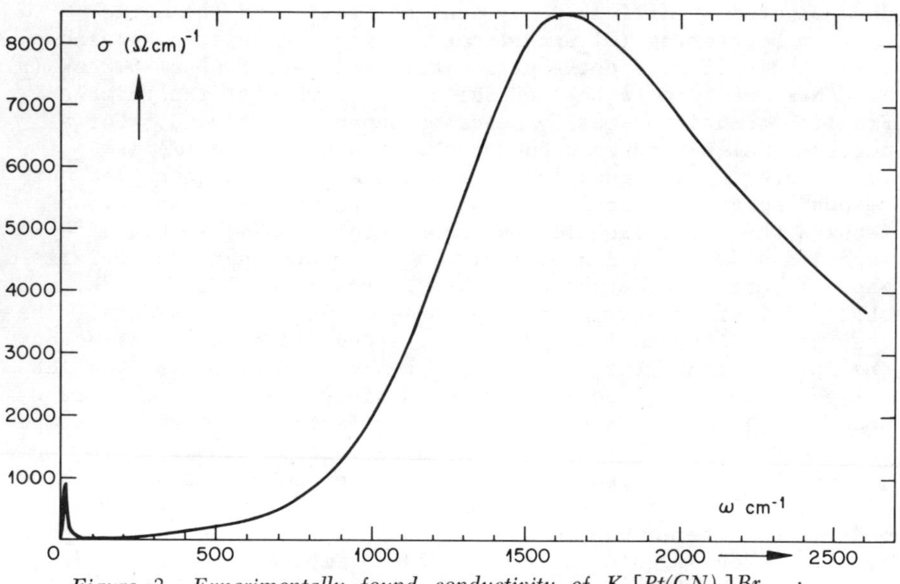

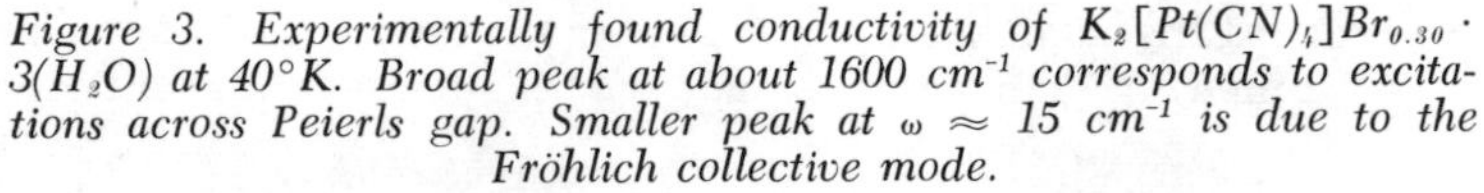

Figure 3. Experimentally found conductivity of $K_2[Pt(CN)_4]Br_{0.30} \cdot 3(H_2O)$ at 40°K. Broad peak at about 1600 cm^{-1} corresponds to excitations across Peierls gap. Smaller peak at $\omega \approx 15$ cm^{-1} is due to the Fröhlich collective mode.

hence the conductivity peak will be centered not at ω = o but at a small (compared with Eg) but finite frequency. Lifetime effects which are particularly important at higher temperature cause a finite width of the peak instead of a δ-function. As a consequence in the region of the transition temperature the Fröhlich mode may contribute to or even dominate the dc conductivity. Whether this is the case or not can most easily and directly be determined from measurements of R (ω) in the microwave and far infrared region. Due to the large effective mass of the Fröhlich mode its oscillator strength is small and it can at most form a narrow peak in σ (ω) superimposed on the very broad single particle conductivity.

From what we have learned on the model system KCP it seems feasible to synthesize systems which exhibit a sufficiently small pinning force such that high dc conductivities based on the Fröhlich mode can be achieved. A large part of the work on KCP described in this paper was carried out in collaboration with D. Kuse, M.J. Rice and S. Strässler. We also wish to acknowledge stimulating discussions with P. Fulde and T.M. Rice. Fig. 2 is due to L. Niemeyer.

References

1. Peierls R.E., "Quantum Theory of Solids" (Oxford University Press, London (1955).
2. Fröhlich, H., Proc. Roy. Soc., London (1954) A 223, 296.
3. Zeller, H.R., Advances in Solid State Physics, (1973) 13, 31.
4. Coleman, L.B., Cohen, M.J., Sandman, D.J., Yamagishi, F.G., Garito, A.F. and Heeger, A.J., Solid State Comm. (1973) 12, 1125.
5. The above arguments are due to P. Fulde and S. Strässler.
6. Rice, M.J. and Strässler, S., Solid State Comm. (1973) 13, 125.
7. Comès, R., Lambert, M., Launois H. and Zeller, H.R., Phys. Rev. (1973) B8, 571.
8. Renker, B., Rietschel, H., Pintschovius, L., Gläser, W., Brüesch, P., Kuse, D. and Rice, M.J., Phys. Rev. Letters (1973) 30, 1144.
9. Brüesch, P. and Zeller, H.R., to be published.
10. Lee, P., Rice, T.M. and Anderson, P.W., preprint.

26

The Preparation of and the Anisotropic Dielectric Properties of $K_2Pt(CN)_4Br_{0.30} \cdot 3H_2O$

R. B. SAILLANT and R. C. JAKLEVIC

Ford Motor Co., P.O. Box 2053, Dearborn, Mich. 48121

The purpose of this report is to describe the variations in the composition of $K_2Pt(CN)_4Br_{0.30} \cdot 3H_2O$, the chemical methods used to minimize these variations, and a physical method that reflects the intrinsic physico-chemical properties but is sensitive to small material variations.

Early attempts to grow crystals of the mixed-valence platinum salt resulted in crystals which were pitted, exhibited transverse striations and tended to grow very rapidly in the needle direction. Cavities and pits had been reported by other workers (1). Neutron activation analysis of typical crystals revealed chloride contamination of the bromide salt. Results of this study have appeared (2). The crystal growth affords a partitioning between chloride and bromide which strongly favors the chloride complex in the solid state.

The solution behavior of $K_2Pt(CN)_4Br_{0.30} \cdot 3H_2O$ was examined to determine if exchange reactions might occur since Pt(II) is known to catalyze Pt(IV) substitution reactions (3). The electronic spectrum from 10,000 to 40,000 cm^{-1} of $K_2Pt(CN)_4$ was taken and compared to that of $K_2Pt(CN)_4Br_2$, fig. 1. One maximum was observed for each complex in this region. $K_2Pt(CN)_4$ exhibited a transition at 35,900 cm^{-1}, $\varepsilon = 1528$, previously identified as a metal to ligand charge transfer band (4). $K_2Pt(CN)_4Br_2$ exhibits a maximum at 29,200 cm^{-1}, $\varepsilon = 1280$, which is located in an $\varepsilon = 0$ region of the $K_2Pt(CN)_4$ spectrum. The energy separation of these two maxima and knowledge of the extinction coefficients enable one to obtain an exact ratio of Pt(II)/Pt(IV) and, therefore, fix the stoichiometry of the mixed-valence complex. When, however, crystals of the mixed-valence complex are dissolved in aqueous solution, fig. 2, the spectrum is not that expected from the proportional mixing of the starting compounds. The crystals must be dissolved in 0.5N KBr in order to suppress interference due to aquation and hydrolysis products. When the spectrum is obtained in 0.5N KBr solutions, the ratio of Pt(II) to Pt(IV) is determined to be 3.333 ± .004. This corresponds to 0.30 bromide atoms per complex if the platinum (IV) is present as $K_2Pt(CN)_4Br_2$. Other

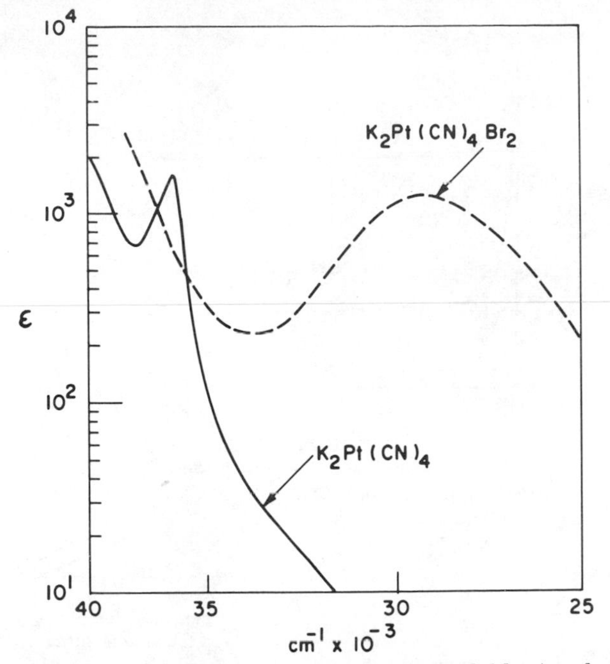

Figure 1. Solution spectra of $K_2Pt(CN)_4$ (solid line) and $K_2Pt(CN)_4Br_2$ (broken line)

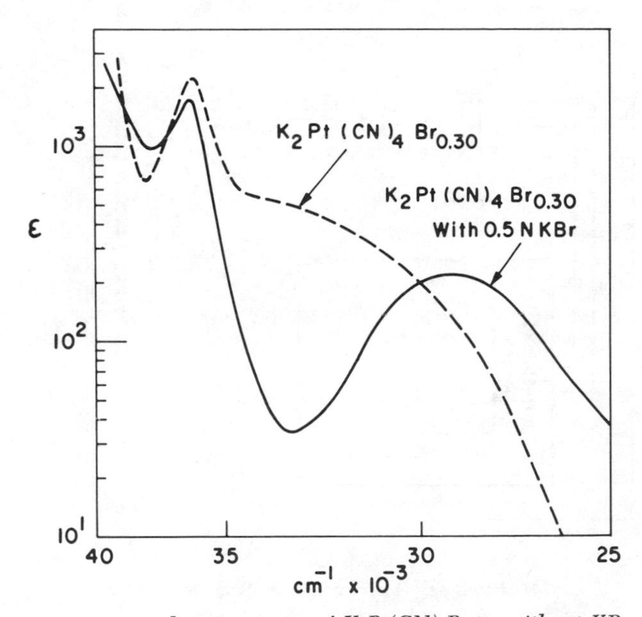

Figure 2. Solution spectra of $K_2Pt(CN)_4Br_{0.30}$ without KBr (broken line) and with 0.5N KBr (solid line)

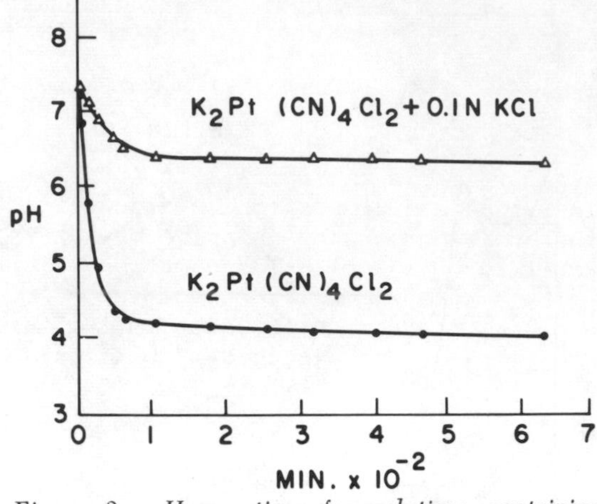

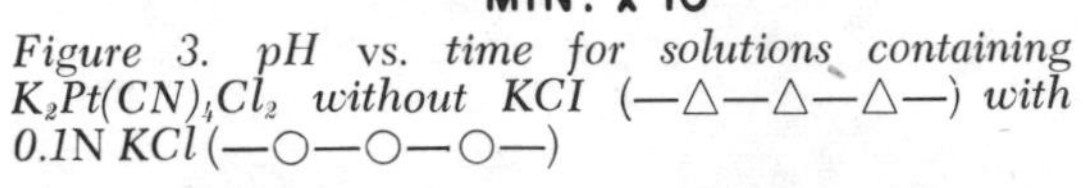
Figure 3. pH vs. *time for solutions containing* $K_2Pt(CN)_4Cl_2$ *without KCl* (—△—△—△—) *with 0.1*N *KCl* (—○—○—○—)

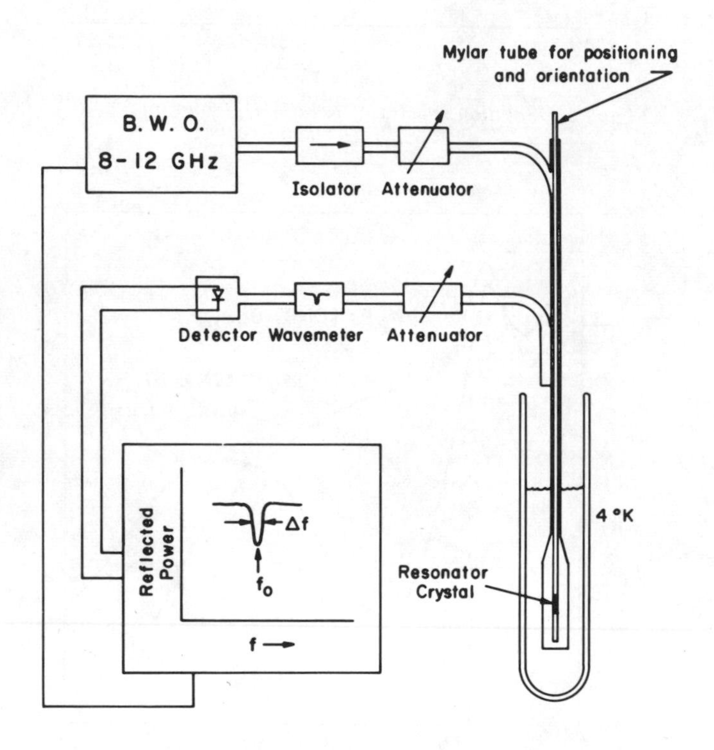

Microwave Reflectance Spectrometer

Figure 4. Experimental setup, B.W.O. = backward wave oscillator

methods including X-ray fluorescence suggest the ratio of K:Pt:Br as 2:1:0.30 (5) which generally supports the bromide concentration.

The hydrolysis of the platinum (IV) complex may be followed by observing the change in pH with time (6). The rate and extent of hydrolysis of the chloro-complex is much more rapid than the bromide complex and is shown in fig. 3. It is possible to suppress hydrolysis by the addition of the halide salt. The contamination of the bromo-complex by chloride may be the result of the increased lability of the chloride compound. This increased lability might play a role in the growth of the solid phase, which contains square planar platinum, from the solution phase which contains both square planar and octahedral platinum complexes. In the following

$$X{-}Pt{-}X^{2-} + Pt^{2-} \rightleftharpoons 2X{-}Pt^{2-} \qquad (1)$$

$$(Pt)_n + X{-}Pt^{2-} \rightleftharpoons (Pt)_n{-}Pt{-}X^{2-} \qquad (2)$$

$$(Pt)_n{-}Pt{-}X^{2-} + Pt^{2-} \rightleftharpoons (Pt)^{1-}_{n+2} + X{-}Pt^{3-} \qquad (3)$$

reaction sequence the cyanide ligands have been omitted for clarity. The more labile the halide is in the left hand species in reaction 3, the more rapidly the solid phase would grow. Thus, the competition of chloride and bromide depends on their respective labilities at this step (2).

Special care must be taken to grow $K_2Pt(CN)_4Br_{0.30} \cdot 3H_2O$ from only the purest materials and the salt is prepared from high purity platinum metal. The best crystals are grown from solutions which are tenth normal in KBr and 1 molar in urea. These crystals show no pits, no transverse etch marks and no channels. In addition, the urea acts to retard the growth along the needle axis and the result is large 0.8 x 0.8 x 2 cm crystals suitable for various types of physical measurements (2).

The physical measurement used in this study involves the unusually high dielectric constant reported for $K_2Pt(CN)_4Br_{0.30} \cdot 3H_2O$. The condition for resonance of a dielectric is that the size must be $\sim (\lambda o/\sqrt{\varepsilon})$ where λo is the free space wavelength and ε is the dielectric constant. Since ε has been reported to be about 10^3 (1), millimeter type dimensions are to be expected at 10^{10}Hz frequencies.

The dielectric resonator is analogous to a resonant metal cavity in that the highly reflecting dielectric walls serve to trap waves in the dielectric. However, the size of the resonant dielectric is reduced by the factor $\varepsilon^{-1/2}$ and therefore, for large dielectric constant materials, the crystal can be placed directly inside a standard microwave waveguide and the resonance observed by finding the wavelength at which the crystal absorbs power.

For an anisotropic material, with values $\varepsilon_{\parallel}$ and $\varepsilon_{\perp}$ parallel to and perpendicular to the c-axis respectively, a rectangular parallelepiped of dimensions a, b and c will have resonant frequencies f_o given by

$$\left(\frac{2f_o}{v_c}\right)^2 = \frac{1}{\varepsilon_{\parallel}}\left[\left(\frac{\ell}{a}\right)^2 + \left(\frac{m}{b}\right)^2 \right] + \frac{1}{\varepsilon_{\perp}}\left(\frac{n}{c}\right)^2 \quad (4)$$

where the integer ℓ or $m > 0$ and the integer $n > 0$ and v_c = 3×10^{10} cm/sec. These modes are TM(H) modes in which the magnetic field is everywhere transverse to the c-axis and the external fields are magnetic multipole in character. The lines of electric field tend to be circular while magnetic lines are perpendicular to the boundaries. The TE modes also exist but do not have low frequency resonant modes since only $\varepsilon_{\perp}$ is involved in the frequency equation and $\varepsilon_{\perp} \simeq 4.0$. To derive Eq. 4, the so-called open-circuit boundary conditions are used in which the parallel and perpendicular components of E and H respectively are a maximum at the boundary, while their respective perpendicular and parallel components are zero. These are the exact opposites of the conditions imposed at a metal boundary, and are appropriate for a nearly infinite dielectric constant material. Hence Eq. 4 is only approximate and experience has shown that it can be off by as much as 25% for low order modes even when $\varepsilon \gtrsim 100$ (7). Theoretical arguments show that the direction of the error is such that observed resonant frequencies are higher than predicted by Eq. 4; hence one would underestimate $\varepsilon_{\parallel}$ or $\varepsilon_{\perp}$. Higher order modes are expected to obey Eq. 4 more accurately. An accurate theory for anisotropic dielectric resonators does not exist at present.

The microwave measurements were done in a simple reflectance spectrometer between 4 and 20°K. The experimental detail has been presented elsewhere (8). The results provide values for $\varepsilon_{\parallel}$ and $\varepsilon_{\perp}$ of 3000 and 4.0 respectively which must be considered lower limits. This method is very sensitive to impurities, crystal imperfection and dehydration, and is non-destructive.

Literature Cited

1. Berenblyum, A.S., Buravov, L.I., Khidekel, M.D., Shchegolev, I.F. and Yakimov, E.B., JETP (1971), 13, 440.
2. Saillant, R.B., Jaklevic, R.C. and Bedford, C.D., Mat. Res. Bull. (1974), 9, 289.
3. Mason, W.R., Coord. Chem. Rev. (1972), 7, 241.
4. Moreau-Colin, M.L., Struct. Bond. (1972), 10, 167.
5. Saillant, R.B., Jaklevic, R.C. and Jaklevic, J., to be published.
6. Chernyaev, I.I., Babkov, A.V., and Zheligoyskaya, N.N., J. Inorg. Chem. (U.S.S.R.) (1963), 8, 1279.
7. Yee, H.Y., IEEE Trans. on Microwave Theory and Techniques MTT-12 (1965), 256 and references cited therein; Gastine, M., Courtois, L. and Dormann, J.L., IEEE Trans. on Microwave

Theory and Technique MTT-15 (1967), 694.
8. Jaklevic, R.C. and Saillant, R.B., to be published in Solid State Communications (1974).

27

A SCF-Xα-SW Investigation of Solid State Interactions in $Pt(CN)_4^{n-}$ Complexes

L. V. INTERRANTE and R. P. MESSMER

General Electric Corporate Research and Development, P. O. Box 8, Schenectady, N.Y. 12301

Introduction

Recent work on the compounds, $K_2Pt(CN)_4X_{0.3}\cdot nH_2O$, (X = Cl, Br, hereafter referred to as KCP) and some related non-stoichiometric, mixed-valence platinum and iridium complexes, has evidenced highly directional electrical and optical properties in the solid state, consistent with largely one-dimensional interactions among the constituent planar complex units (1). X-ray diffraction studies suggest that the primary interatomic interactions are within the linear, parallel chains of closely spaced metal atoms which result from the columnar stacking of these units (2).

The nature of these interatomic interactions and their relationship to the unusual solid state behavior of these systems is not very well understood at present. It has been suggested that the interactions of primary importance are those involving the "d_{z^2}" orbitals on the metal atoms which overlap in the solid to produce a one-dimensional band of states of appreciable width (1,2). For the $Pt(CN)_4^{n-}$ complexes, an energy level scheme deduced from spectral and magnetic circular dichroism data which places the "d_{z^2}" orbital as the highest occupied level in the free $Pt(CN)_4^{2-}$ ion (3) is generally assumed as the appropriate starting point for the discussion of the electronic structure of the solid complexes. In the case of the KCP derivatives, this leads directly to the conclusion that the Fermi level position must lie within a d_{z^2}-like band (Figure 1) (4).

Recent ESR results lend support to this conclusion regarding the Fermi level position in KCP (5); however, the assumption that the "d_{z^2}" orbital is also the highest occupied level in the free $Pt(CN)_4^{2-}$ ion is still open to question. Indeed, the order of the d-like orbitals in this ion have been the subject of much controversy in the past and at the present time the experimental evidence does not permit any definite conclusions in this regard (6).

As part of a detailed theoretical study of the electronic structure of the KCP system we have carried out some molecular orbital calculations using the self-consistent-field-Xα-scattered wave (SCF-Xα-SW) method on the $Pt(CN)_4^{2-}$ ion and a dimer unit, $[Pt(CN)_4^{2-}]_2$, arranged in the configuration found in KCP solid (2). The intent of these studies is to investigate the intermolecular orbital interactions and, in general, the transformation in electronic structure that occurs on formation of a molecular solid such as KCP from its constituent planar complex units.

Preliminary results of the $Pt(CN)_4^{2-}$ calculation have been reported elsewhere (7). This paper describes further details regarding this calculation and the results of the study of the dimer unit.

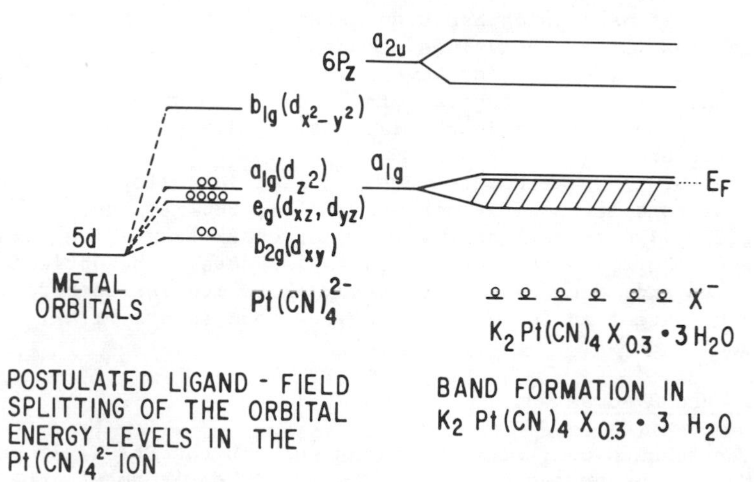

Figure 1. Previously postulated energy level scheme for $Pt(CN)_4^{2-}$ and the $K_2Pt(CN)_4X_{0.3} \cdot 3H_2O$ complexes (cf. *Ref.* 1 *and* 4)

Method of Calculation

The SCF-Xα-scattered wave method (8,9) used in these calculations employs the Xα theory of Slater (10) and the multiple scattering formalism of Johnson (11) to solve the Xα equations. The Xα theory is a one-electron method and is thus similar in this respect to the more traditional Hartree-Fock theory. However, unlike the Hartree-Fock theory which has a non-local exchange potential, the Xα has a local exchange potential proportional to the cube root of the electronic charge density.

The latter fact makes it possible to set up a multiple scattering formalism for a molecule or cluster of atoms which is similar in many respects to the Korringa-Kohn-Rostoker method (12) in the theory of energy bands in solids.

In the scattered wave method, the space occupied by a molecule is divided into three regions. Region I consists of spherical volumes located about the centers of the respective atoms, so constructed that the spheres are tangent to one another [or more generally, are somewhat overlapping (13,14)]. Region II is the volume outside the atomic spheres and inside an "outer sphere" which surrounds the molecule and is centered at the molecular center; this outer sphere is constructed to be tangent to the atomic spheres at the extremities of the molecule. Region III is the space outside the outer sphere. In the case of ionic species, such as the ones considered here, the electrostatic potential provided by the surrounding medium in solid or solution is approximated by surrounding the ion by a sphere [the so-called Watson sphere (15)] with a charge opposite to that on the ion.

In each of the atomic spheres a value of α, the exchange parameter, is used which has been determined from atomic calculations (16). The values of α and the atomic sphere radii used for the $Pt(CN)_4^{2-}$ case have previously been given (7); these same values are employed here for the dimer. The outer sphere and Watson sphere radii in the case of the dimer were both set at 7.828 Bohr.

The SCF-Xα-SW method has been applied to analogous square-planar systems (17,18,19) such as $PtCl_4^{2-}$, $PdCl_4^{2-}$ and $[PtCl_3(C_2H_4)]^-$, as well as a variety of other transition metal complexes (20,21) with a great deal of success. The optical properties and, when available, x-ray photoemission spectra in each case have been found to be in very good accord with experiment.

Results and Discussion

The results of the calculations on both the $Pt(CN)_4^{2-}$ and $[Pt(CN)_4^{2-}]_2$ units are given in Figure 2 in the form of one-electron orbital energy level diagrams.

For the $Pt(CN)_4^{2-}$ ion there are several features of the calculation which are particularly noteworthy. First of all, contrary to previous assumptions, we find extensive admixture of metal "d" orbitals with ligand σ and π orbitals in many of the molecular orbitals of the complex suggesting an appreciably covalent metal-ligand bonding interaction. This is illustrated in Table I which lists the proportion of orbital charge in metal and ligand atomic spheres for several of the highest occupied molecular orbitals. Appreciable metal d-orbital character is evident in the relatively deep $1e_g$, $1b_{2g}$ and $4a_{1g}$ bonding orbitals as well as their antibonding counterparts, $2e_g$, $2b_{2g}$ and $5a_{1g}$ at higher energy. The extensive metal-ligand overlap in the $1b_{2g}$ orbital, for example, is illustrated in Figure 3.

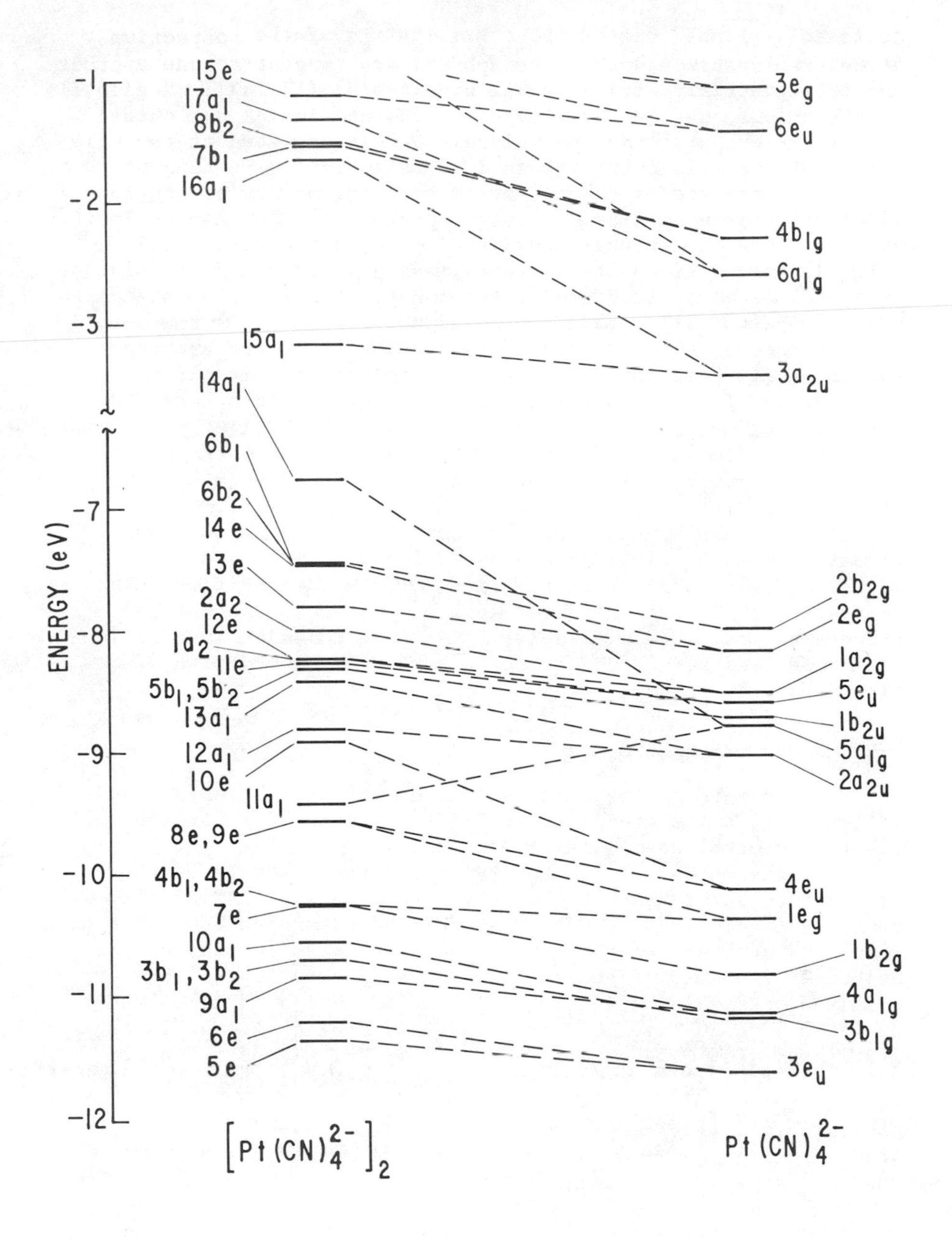

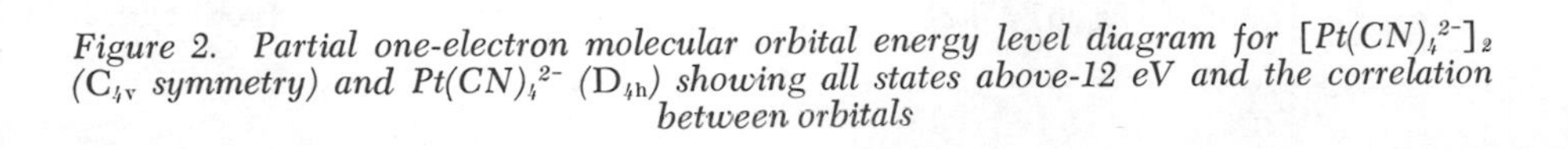
Figure 2. Partial one-electron molecular orbital energy level diagram for $[Pt(CN)_4^{2-}]_2$ *(C_{4v} symmetry) and* $Pt(CN)_4^{2-}$ *(D_{4h}) showing all states above-12 eV and the correlation between orbitals*

Table I

Molecular Orbital Charge Distribution For $Pt(CN)_4^{2-}$

Orbital	% of Orbital Charge				
	Pt	C	N	Inter-sphere	Outer Sphere
$2b_{2g}$	34.0	10.8	28.4	24.7	2.0
$2e_g$	25.9	14.4	31.1	26.9	1.8
$1a_{2g}$	0.0	23.9	39.3	35.0	1.8
$5e_u$	1.3	27.7	34.5	34.8	1.7
$1b_{2u}$	0.0	24.3	36.9	37.0	1.7
$5a_{1g}$	72.7	4.2	1.4	21.3	0.4
$2a_{2u}$	1.3	24.4	32.1	40.4	1.8
$4e_u$	13.0	37.8	19.7	26.6	2.9
$1e_g$	65.8	9.5	6.4	17.9	0.3
$1b_{2g}$	57.1	13.1	7.6	21.8	0.4
$4a_{1g}$	4.3	8.8	58.3	17.6	11.0
$3b_{1g}$	2.2	8.4	60.0	17.7	11.8
$3e_u$	3.5	17.5	46.2	25.5	7.3

Of these d-like molecular orbitals the $5a_{1g}$ is rather unique in its concentration of charge on the metal. As is shown in Figure 4, this orbital is substantially antibonding with respect to the metal-ligand interaction and has a high component of metal d_{z^2} character with a large proportion of the electronic charge extending out perpendicular to the $Pt(CN)_4{}^{2-}$ plane along the z axis. It also appears that there is significant s character in this orbital, as is suggested by the Pt-C overlap and the general shape of the orbital. Thus the likelihood of extensive interaction among such orbitals in a solid such as KCP is quite evident.

Along with the general strong metal-ligand mixing, the uppermost "d-like" orbitals are not energetically separated from the

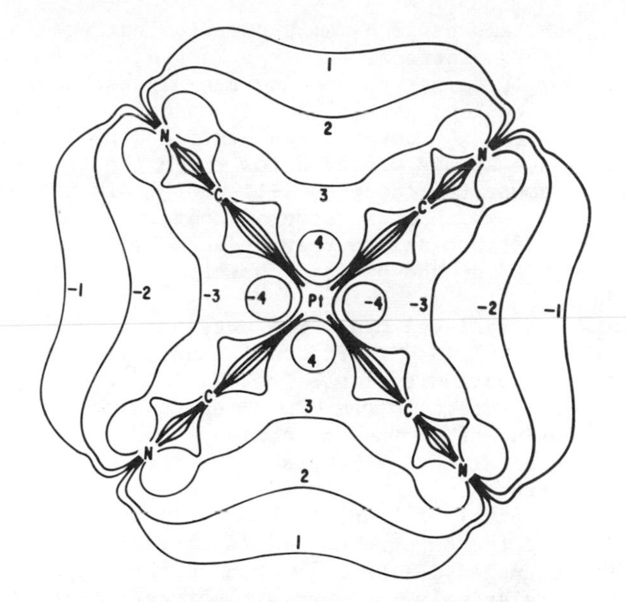

Figure 3. Contour plot of $1b_{2g}$ molecular orbital wave function. Amplitude of wave function increases by a factor of 5 with each increase in absolute value of contour label. Sign of labels gives sign of orbital lobes. Interior nodes at various atoms are not shown for clarity.

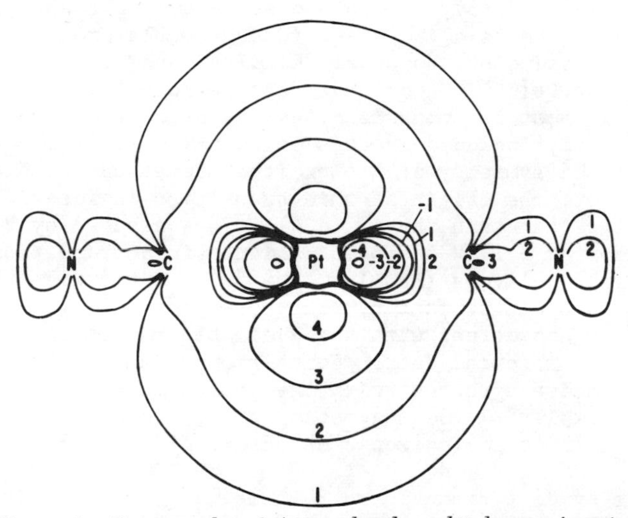

Figure 4. Contour plot of $5a_{1g}$ molecular orbital wave function, labeled as in Figure 3 except that each integral increase in value of contour label is a factor of 3 increase in orbital amplitude

ligand-based orbitals as is usually supposed but are found in the same energy region. In fact, the important d_{z^2} (+ s) molecular orbital (the $5a_{1g}$), rather than being the highest occupied orbital in the free ion, as has been assumed in most of the solid state studies of the KCP derivatives, is found to lie below three CN^- based orbitals and is 0.8 eV below the highest occupied energy level. Among the occupied d-like orbitals the energy order is $d_{xy}>d_{xz,yz}>d_{z^2}$, which is the same as that previously proposed by Mason and Gray (22) but unlike that deduced by Piepho, Shatz and McCaffery (PSM) (3) on the basis of spectral and magnetic circular dichroism data.

The same energy level order was obtained in our earlier studies of the MCl_4^{2-} (M = Pd,Pt) complexes (18) where again the highest occupied energy level was the b_{2g} (d_{xy}). The major differences with respect to the MCl_4^{2-} energy level diagrams are the relative positions of the unoccupied 4 b_{1g} ($d_{x^2-y^2}$) and $3a_{2u}$ (M p_z+L) orbital energy levels which are reversed in the two types of complexes.

This $3a_{2u}$ orbital is found to be mainly localized on the nitrogen atoms of the CN^- groups and is considerably closer in energy to the occupied d-like MO's than in the MCl_4^{2-} case, which should lead to relatively low energy electronic transitions with appreciable M → L charge transfer character.

In a manner analogous to that previously reported for the MCl_4^{2-} complexes (18), transition state calculations were carried out on the $Pt(CN)_4^{2-}$ ion leading to calculated electronic transition energies which are in good agreement with both spectral and magnetic circular dichroism experiments as well as photoelectron data. The comparison with the spectral data is shown in Table II.

Following PSM (3), the lowest energy visible absorption band has been assigned to a E_u' state (using double group notation) produced by mixing of the usual "singlet" and "triplet" excited states of the $Pt(CN)_4^{2-}$ ion under the influence of spin orbit coupling (It must be emphasized that spin orbit coupling has not been explicitly included in this calculation, only its effect on the number and symmetry of the excited states has been considered). Here, however, the origin of this transition is identified as the $2b_{2g}$ level rather than the $a_{1g}(d_{z^2})$ as was assumed by PSM. Also the shoulder at 5.1 eV has been assigned to a transition from one of the ligand π-levels ($1a_{2g}$) to the $3a_{2u}$. Otherwise the assignments are in basic agreement with those proposed previously and are entirely consistent with the available magnetic circular dichroism and spectral data. Furthermore, the results are also in good qualitative agreement with the photoelectron data for the $Pt(CN)_4^{2-}$ complex (7,23) suggesting that, indeed, a satisfactory description for the electronic structure of this ion has been obtained.

As is evident from Figure 2, combining two such $Pt(CN)_4^{2-}$ units at 2.89 Å separation in the staggered configuration found

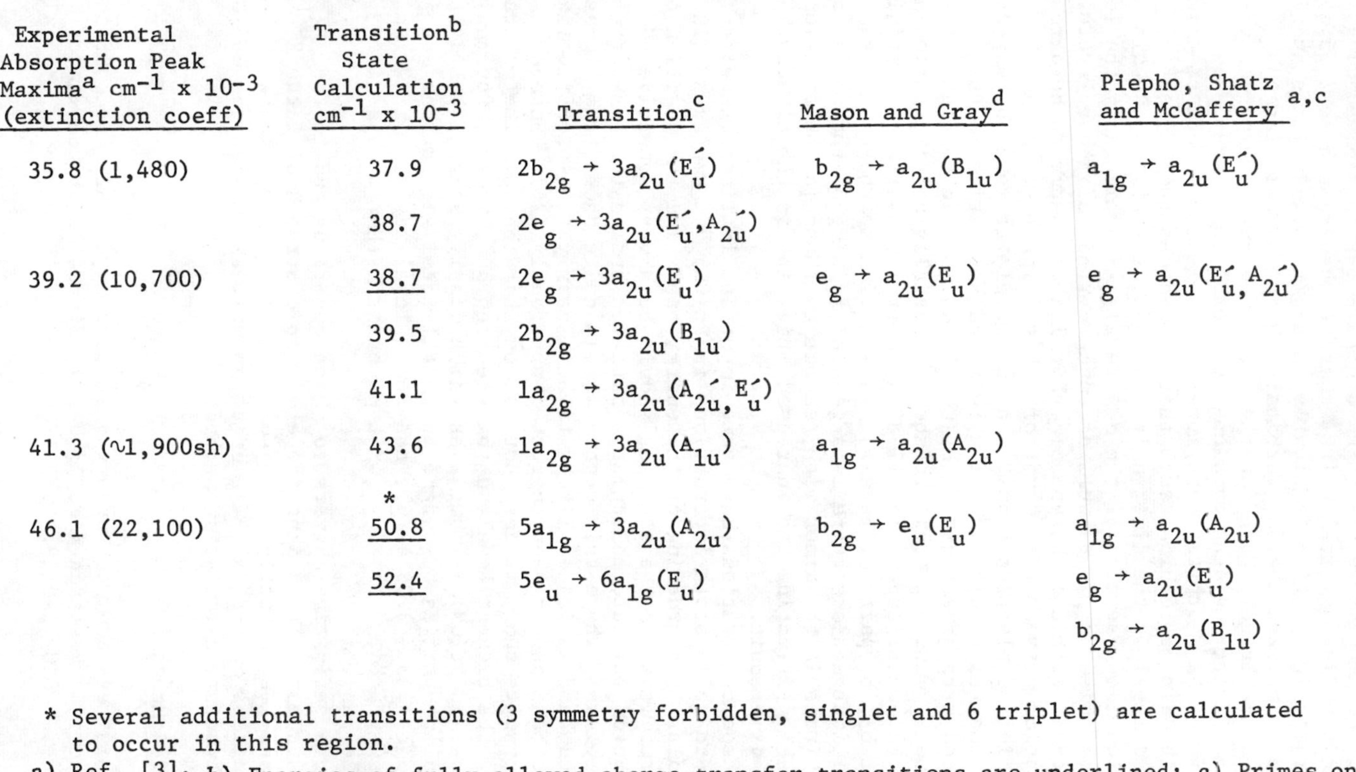

Table II. Experimental and Theoretical Transition Energies for Transitions in $Pt(CN)_4^{2-}$

Experimental Absorption Peak Maxima[a] $cm^{-1} \times 10^{-3}$ (extinction coeff)	Transition[b] State Calculation $cm^{-1} \times 10^{-3}$	Transition[c]	Mason and Gray[d]	Piepho, Shatz and McCaffery[a,c]
35.8 (1,480)	37.9	$2b_{2g} \rightarrow 3a_{2u}(E_u')$	$b_{2g} \rightarrow a_{2u}(B_{1u})$	$a_{1g} \rightarrow a_{2u}(E_u')$
	38.7	$2e_g \rightarrow 3a_{2u}(E_u', A_{2u}')$		
39.2 (10,700)	38.7	$2e_g \rightarrow 3a_{2u}(E_u)$	$e_g \rightarrow a_{2u}(E_u)$	$e_g \rightarrow a_{2u}(E_u', A_{2u}')$
	39.5	$2b_{2g} \rightarrow 3a_{2u}(B_{1u})$		
	41.1	$1a_{2g} \rightarrow 3a_{2u}(A_{2u}', E_u')$		
41.3 (~1,900sh)	43.6	$1a_{2g} \rightarrow 3a_{2u}(A_{1u})$	$a_{1g} \rightarrow a_{2u}(A_{2u})$	
	*			
46.1 (22,100)	50.8	$5a_{1g} \rightarrow 3a_{2u}(A_{2u})$	$b_{2g} \rightarrow e_u(E_u)$	$a_{1g} \rightarrow a_{2u}(A_{2u})$
	52.4	$5e_u \rightarrow 6a_{1g}(E_u)$		$e_g \rightarrow a_{2u}(E_u)$
				$b_{2g} \rightarrow a_{2u}(B_{1u})$

* Several additional transitions (3 symmetry forbidden, singlet and 6 triplet) are calculated to occur in this region.

a) Ref. [3]; b) Energies of fully allowed charge transfer transitions are underlined; c) Primes on symmetry labels denote a double group state derived from spin orbit splitting of the appropriate triplet excited state. Only the resultant symmetry allowed transition states are noted; d) Ref. [22]

in the unit cell of the KCP complexes results in substantial interaction among several of the orbitals leading to "bonding" and "antibonding" orbital pairs separated by as much as 2.65 eV. The largest interaction is clearly among the a_{1g} orbitals and in particular between the $5a_{1g}$ orbitals which our calculations indicate are largely based on the metal and have a high concentration of electron density along the metal chain axis. In the case of the $5a_{1g}$ orbital this interaction is sufficient to raise an antibonding a_1 orbital such that it becomes the highest occupied level in the dimer unit. Large interactions are also indicated among several of the a_{2u} levels - particularly the lowest unoccupied $3a_{2u}$ - as well as among certain of the e levels. In general, within each type of orbital there seems to be a rough correlation between the amount of orbital interaction and the proportion of electron density on the metal, as would be expected due to the relatively large inter-ligand separation in the staggered arrangement of the $Pt(CN)_4{}^{2-}$ units. In contrast, the in-plane b_{1g} and b_{2g} orbitals show essentially no interaction at this metal-metal separation (2.89 Å) carrying over as essentially unsplit b_1, b_2 pairs of levels. In addition to these direct orbital interactions there is a general raising of the energy of the orbitals in the dimer with respect to those in the free ion, presumably reflecting the increased shielding in the closely spaced dimer unit.

These observations have some important implications with regard to solid state interactions in $Pt(CN)_4{}^{n-}$ complexes. In particular, the ordering of the energy bands and the position of the Fermi level in these solids should be a very sensitive function of the intermolecular separation and in the case of the KCP complexes, where this separation is quite short, the highest occupied band is indeed likely to be a nearly free-electron like d_{z^2} + s band as has been suggested previously (24). However, as these results show, it does not follow that the d_{z^2}-like orbital is also the highest level in the free ion and, indeed, for many of the solid $Pt(CN)_4{}^{2-}$ complexes which display unusual solid state spectral properties but where the metal-metal separations are in the range 3.63 - 3.09 A (6), the Fermi level may well lie within quite narrow "d_{xy}" or "$d_{xz,yz}$" type bands rather than a wide d_{z^2}-like band.

More definitive answers to these questions and to the central question regarding the electronic structure of the KCP complexes are currently being sought in further SCF-Xα-SW calculations on $[Pt(CN)_4{}^{n-}]_m$ units as well as through the use of band structure calculation methods.

Acknowledgments

This research was supported in part by the Air Force Office of Scientific Research (AFSC), United States Air Force, under Contract F44620-71-0129.

Literature Cited

1. Zeller, H. R., Advances in Solid State Physics (1973), 13, 31.
2. Krogmann, K., Angew. Chem. Internat. Edit., (1969), 8, 35.
3. Piepho, S. B., Shatz, P. N., and McCaffery, A. J., J. Am. Chem. Soc., (1969), 91, 5994.
4. Minot, M. J. and Perlstein, J. H., Phys. Rev. Letters (1971), 26, 371.
5. Mehran, F. and Scott, B. A., Phys. Rev. Letters (1973), 31, 1347.
6. Moreau-Colin, M. L., Struct. Bond. (1972), 10, 167 and references therein.
7. Interrante, L. V. and Messmer, R. P., Chem. Phys. Letters (1974), 26, 225.
8. Slater, J. C. and Johnson, K. H., Phys. Rev. (1972), 5B, 844.
9. Johnson, K. H. and Smith, F. C., Jr., Phys. Rev. (1972), 5B, 831.
10. Slater, J. C., "Advances in Quantum Chemistry", Vol. 6, P.O. Löwdin, ed., p. 1, Academic Press, (1973).
11. Johnson, K. H., "Advances in Quantum Chemistry", Vol. 7, P.O. Löwdin, ed., p. 143, Academic Press, (1973).
12. Kohn, W. and Rostoker, N., Phys. Rev. (1954), 94, 1111.
13. Rösch, N., Klemperer, W. G. and Johnson, K. H., Chem. Phys. Letters (1973), 23, 149.
14. Herman, F., Williams, A. R. and Johnson, K. H., to be published.
15. Watson, R. E., Phys. Rev. (1958), 111, 1108.
16. Schwarz, K., Phys. Rev. (1972), B5, 2466; Schwarz, K., to be published.
17. Messmer, R. P., Intern. J. Quantum Chem. (1973), 7S, 371.
18. Messmer, R. P., Interrante, L. V. and Johnson, K. H., J. Am. Chem. Soc. (1974), 96, 3847.
19. Rösch, N., Messmer, R. P. and Johnson, K. H., J. Am. Chem. Soc. (1974), 96, 3855.
20. Norman, J. G., Jr., Kolari, H. J., J. C. S. Chem. Comm. (1974) 303.
21. Rösch, N. and Johnson, K. H., Chem. Phys. Letters (1974), 24, 179.
22. Mason, W. R. and Gray, H. B., J. Am. Chem. Soc. (1968), 90, 5721.
23. Bastas, R., Cahen, D., Lester, J. E., and Rajaram, J., Chem. Phys. Letters (1973), 22, 489.
24. Zeller, H. R., this volume

28

One-Dimensional and Pseudo-One-Dimensional Molecular Crystals

ULRICH T. MUELLER-WESTERHOFF and FRIEDRICH HEINRICH

IBM Research Laboratory, San Jose, Calif. 95193

In this concluding talk of the symposium we want to discuss new systems which we are looking at in order to obtain some clarification concerning the Peierls instability and how to circumvent it. Since it is the last talk, I also think that a brief outlook to promising future work is in order. First of all, we want to distinguish the systems that we are working on and which we want to call Pseudo-one-dimensional Molecular Crystals, from the one-dimensional square planar complex systems like Krogmann's Salt and the not so strictly one-dimensional molecular crystals of the TTF-TCNQ type. Our systems are composed of either mixed valence cations of square planar ligand-bridged bis-transition metal chelates with acceptor anions such as $TCNQ^-$ as the counterion or of donor cations with ligand bridged bifunctional transition metal complex counterions. In both cases we will deal with radical ion salts of charge transfer systems. The particularly interesting aspect of these compounds lies in the coexistence of intramolecular as well as intermolecular exchange and electron transfer interactions between the transition metals involved. These systems are designed to maintain the anisotropic properties of one-dimensional molecular crystals and have incorporated in them the first possibility of avoiding the transition to an insulating Peierls state. Eliminating the metal-insulator transition of highly conducting compounds is of course equivalent to creating organic and organometallic metals in the actual sense. However, the molecular design does not eliminate the possibility that we may create a true insulator, although from all energetic considerations this would not be very likely.

One-dimensional Systems

In order to better explain the reasons for our suggesting bifunctional ligand bridged systems for this investigation, we like to first take a look at the known inorganic and organic one-dimensional molecular crystals.

Inorganic Linear Systems

Several talks in this symposium discussed properties of a one-dimensional inorganic salt, which currently is very much en vogue: the so-called Krogmann's Salt ("KCP", $K_2Pt(CN)_4 0.3Br3H_2O$) in which during the transformation from the neutral Pt(II) complex by bromine oxidation to the mixed valence compound there occurs a substantial decrease in the intermetallic (in this case equal to the interplanar) distances from 3.2 to 2.88 Å, by means of which the overlap between the platinum $5d_{z}^{2}$ atomic orbitals is increased. This gives rise to metallic behaviour in one dimension along the platinum backbone of the essentially one-dimensional molecular crystal. For similar systems like the mixed valence oxalates and others, one-dimensional metallic behaviour and high anisotropy ratios of optical and other related properties have been established. At low temperatures, KCP behaves like a semiconductor (1).

Organic Linear Systems

The organic counterpart to KCP is the class of TCNQ radical anion salts (2), which is currently being studied in a number of laboratories. The unique feature of TCNQ salts is that, due in part to the polarized charge distribution in the radical anion, Coulomb interactions favor a linear arrangement within the molecular crystals. Exceptions are known, e.g. TMPD-TCNQ, in which the separation of the positive charge centers in the cation is identical to the separation of the negative charge centers in TCNQ anion; this could well be the main reason for a D-A-D-A-D-A stacking being energetically favored by coulomb attraction. While in charge transfer systems the overlap of the π-systems will generally favor a linear arrangement, the symmetry of the interacting molecular orbitals is rarely as suited for the formation of segregated stacks as in TCNQ radical anion salt. Due to the rather delicate balance between Coulomb and packing forces, the molecular symmetry is directly related to the anisotropy of molecular crystals. Two main classes of TCNQ salts have to be distinguished:

Class C: TCNQ radical anion salts of Closed shell cations; within this class, there are such diverse systems as LiTCNQ, $Cs_2 (TCNQ)_3$, NMP-TCNQ, NEtP-TCNQ, Q-TCNQ, Et_3NH-$TCNQ^2$ etc.

Class O: TCNQ radical anion salts of Open shell cations, comprising the most interesting systems like BFD-$(TCNQ)_2$, TTF-TCNQ, TMPD-TCNQ and others.

The cations in class C compounds do not directly participate in the electronic conduction process, although in some of them the qualities of polarizability and permanent dipole moments have bearing on the Coulomb potential along the conductive $TCNQ^-$ chains.

Class O compounds are true charge transfer salts of closed shell donor and acceptor molecules. It should be kept in mind, however, that a typical class C compound can also be a CT salt: Li^+TCNQ^- is a true charge transfer salt but of an open shell donor and a closed shell acceptor, leading to a closed shell donor cation and an open shell acceptor anion. In Class O salts, the coexistence of radical cations and anions is the key to high conductivities, since in addition to the properties available also in class C compounds there is the possibility in donor and acceptor ions for both electron and hole conduction.

In the TCNQ complex of bisfulvalene-diiron BFD-$(TCNQ)_2$, which represents the first mixed valence complex salt of TCNQ, the intramolecular electron transfer within the Fe(II)-Fe(III) systems seems to play an important role. However, here the limit of intermolecular interactions is the same as in TTF-TCNQ: it is the overlap between π-electron systems. (3)

This kind of limitation has prompted us to consider mixed valence systems with direct intermetallic bonds as the most promising approach to stable highly anisotropic organometallic metals.

Transition Metal Complexes and TCNQ Salts

It is a tempting approach, to try to combine the two known linear systems: linear mixed valence square planar transition metal complexes (for which in the further discussion we will only consider platinum as the classic example) and organic or organometallic acceptors (for which we will take TCNQ as the example of choice - although in its acceptor properties it is no longer particularly outstanding).

Consider the two approximate spacing parameters of the following illustration describing a TCNQ stack and a linear Pt chain: In TCNQ the interplanar spacing is about 3.2 Å, the overlap is roughly between the center of the molecule to the center of the malonitrile group of its neighbors - giving us a packing distance of 4.2 Å as a average value. In KCP the Pt-Pt distance is 2.88 Å. Since we are dealing in this case with square planar units with a total negative charge of 2.3 units, it is reasonable to expect that due to the lesser coulomb repulsion in a neutral complex a Pt-Pt distance of 2.7 to 2.8 Å can be obtained, with this number being an important factor for determining the stoichiometry of square planar mixed valence systems containing stacks of $TCNQ^-$ as the counterions. As shown in the following diagram there is only the one allowed ratio of 2.8/4.2 = 0.67 TCNQ's per Pt in uniformly stacked solids.

Although several attempts have been made in other laboratories to produce conductive organometallic mixed valence systems of the above kind, such approaches cannot be successful since there is a discrepancy between the Pt-Pt distance of 2.8 Å for optimal interactions and the minimal distance (3.2 Å) for

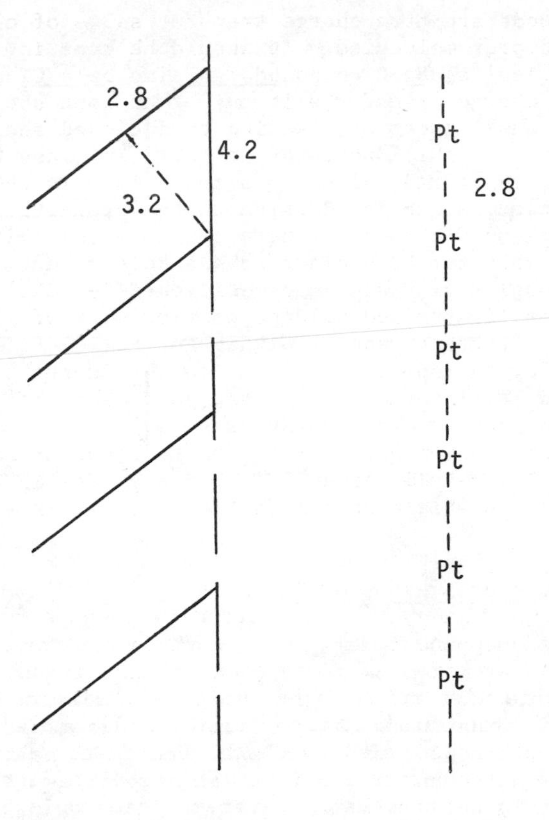

Pt/TCNQ = 4.2/2.8 = 1.5 (i.e.: TCNQ/Pt = 0.67)

organic ligand molecules imposed by repulsion forces. The only way by which in such compounds significant metal-metal interactions can be obtained is to distort the molecules from their planar ground state and to create dimeric species. Such structures are reminiscent of the Peierls state of one-dimensional systems.

A thorough crystallographic study (4) of the unsubstituted dithiene-platinum complex showed such dimers. Schematically, the distortion is shown in the following diagram.

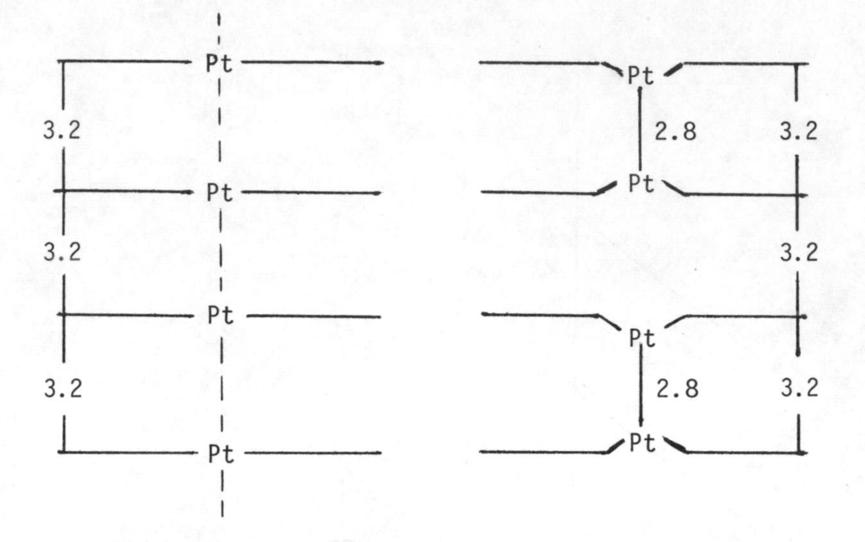

Uniform (no significant Pt-Pt Interactions) Distorted (strong pairwise Pt-Pt Interactions)

We consider this distortion to be one of the main reasons for the absence of organometallic transition metal complexes of more than semiconductor qualities (with the BFD system above being the only exception so far). It is reasonably safe to predict that there can be no metallic systems based on square planar linearly overlapping organometallic transition metal complexes. One alternative structure that avoids the steric problems is based on the use of polymeric associates of the $(C_5H_5Tl)_x$ and $(C_8H_8Eu)_x$ type. For CpTl, a linear alternating Cp and Tl arrangement in the solid is known and partial oxidation of such a stack by TCNQ could lead to Tl(I)-Tl(III) mixed valence cations segregated from stacks of TCNQ anions. Attempts in our laboratory to prepare such systems have been unsuccessful.

Our approach to avoid the ligand repulsion problems and to still utilize the advantages of direct metal-metal bonding through d_{z^2} orbitals as in KCP lies in the construction of bifunctional organometallic species, which offer several surprising advantages.

Pseudo-One-Dimensional Systems

The principle of our approach lies in the synthesis of two classes of π-ligand bridged bimetallic planar transition metal complexes. The significant point in pursuing the synthesis of these complexes lies in their unique possibility of avoiding the Peierls State. As a matter of fact, these compounds have to undergo a distortion in order to become metallic!! Herein lies their particular promise. The displacement out of the compound plane should have a negligble influence on the intramolecular interactions.

We are currently looking at two main classes of complexes. In the first class, transition metals are complexed by rigid bifunctional ligand systems. The synthesis of such ligands poses severe difficulties, which are understandable to anyone

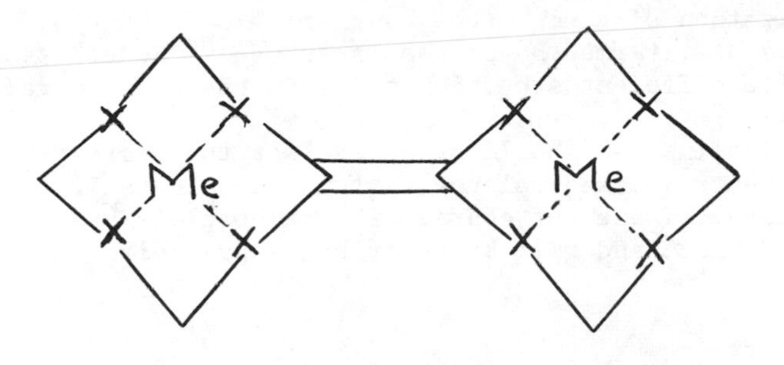

familiar with the preparation of macrocyclic compounds. However, we have recently succeeded in preparing such compounds and complex systems - although these experiments do not as yet allow the preparation of ligands in useful quantities.

The second approach is to use ligand exchange reactions to construct complexes that exhibit strong intramolecular interactions. In this approach, we sacrifice molecular rigidity for synthetic simplicity. Useful bridging ligands which are incorporated in this synthesis are of the 2,2'-bipyrimidyl type to give us complexes of the following general type.

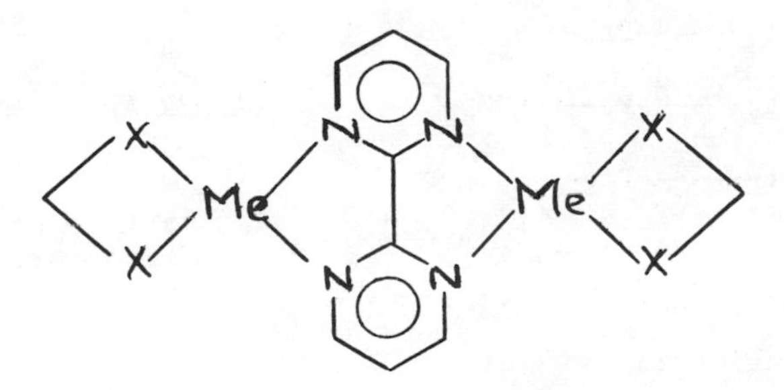

Both methods produce planar systems containing two metal atoms. For the following discussion, we wish to represent these systems by the notation

----------- Pt -------- Pt ----------

which schematically equals a side view of these systems. The entire complex carries either a positive or negative partial charge, depending on the counterion involved.

The Two Stacking Arrangements

There are two principal ways in which complexes of this type together like a linear stack of counterions can be arranged in regular stacks in molecular crystals and still fully utilize the stabilization offered by the overlap of d_{z^2} orbitals. These two arrangements, termed by us the "step" and "ladder" stackings require quite different stoechiometries in their mixed valence state.

As a sideline, it should be noted that there are presently also a few neutral mixed valence system under investigation in our laboratory. These, of course, form a completely different class of compounds and will be described separately.

```
---Pt-------Pt---

         ---Pt-------Pt---
                                                STEP Arrangement
                  ---Pt-------Pt---

                           ---Pt-------Pt---


         ---Pt-------Pt---

         ---Pt-------Pt---

         ---Pt-------Pt---                     LADDER Arrangement

         ---Pt-------Pt---

         ---Pt-------Pt---
```

The STEP Arrangement

Considering spacing and stoechiometry for a molecular crystal made up from stacks of TCNQ radical anions and bifunctional mixed valence cations in the step arrangement, we have only one free parameter that determines the relation between stoechiometry and intramolecular Pt-Pt distances. In both stacks the interplanar spacing is 3.2 Å, the TCNQ chain spacing is 4.2 Å, which has to be matched as illustrated in the following diagram. For a 1:1 stoechiometry (one Pt per TCNQ), the intramolecular Pt-Pt distance has to be 7.8 Å. For non-integral stoechiometries, the values in the following table apply.

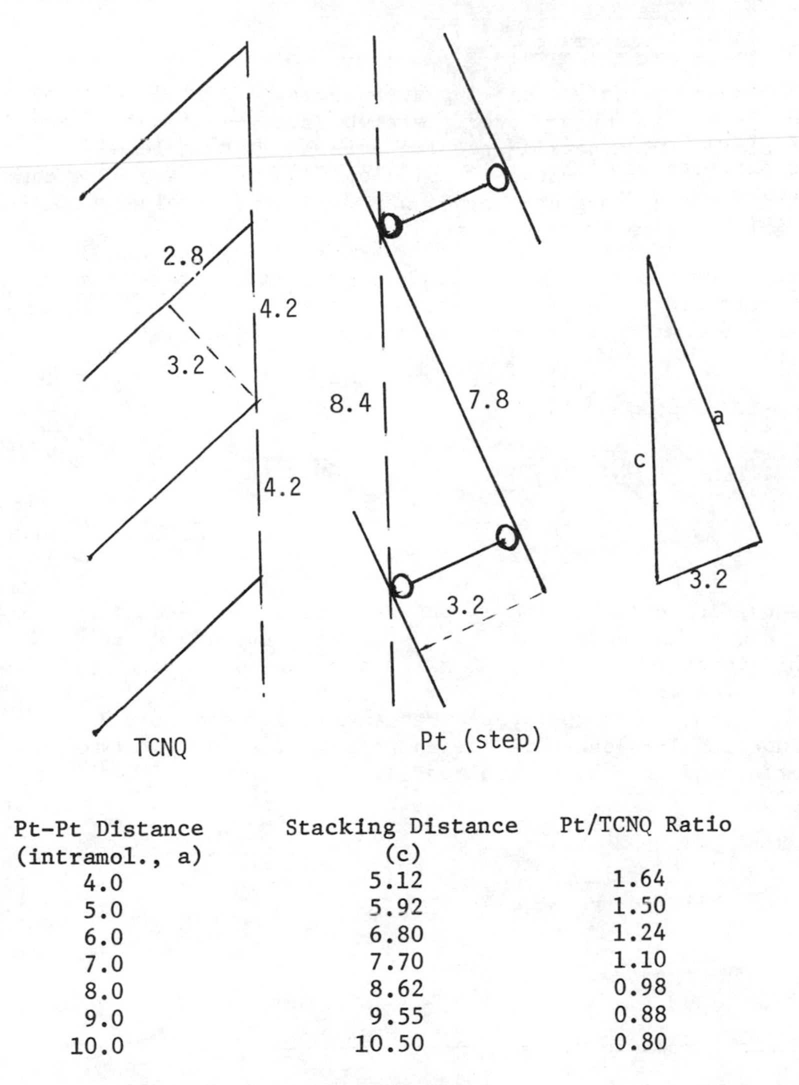

Pt-Pt Distance (intramol., a)	Stacking Distance (c)	Pt/TCNQ Ratio
4.0	5.12	1.64
5.0	5.92	1.50
6.0	6.80	1.24
7.0	7.70	1.10
8.0	8.62	0.98
9.0	9.55	0.88
10.0	10.50	0.80

In order to achieve significant intermolecular metal-metal overlap, the molecules have to distort in the same way as it was found for the dithiene-Pt complex dimer. However, since the dimers in this case of bifunctional systems are connected by a conductive π-electron ligand bridge, they are in essence stabilizing the metallic state through this distortion. As is evident from the above sketch, we now have systems with intramolecular interactions as well as direct intermolecular metal-metal coupling. The step arrangement appears to be stable and not subject to further distortions.

The LADDER Arrangement

The alternative stacking arrangement depicted above offers significant advantages over the step arrangement, since the relation between packing requirements and intramolecular Pt-Pt distance has been removed. Bridging ligands of any size can be accomodated as long as they do not deviate from planarity. As is evident from the following diagram,

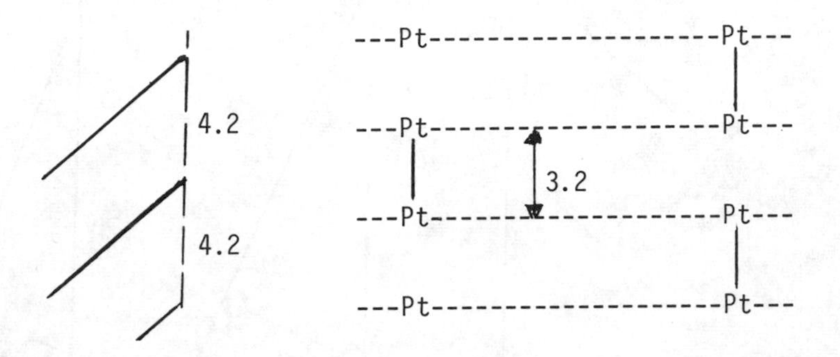

the confining limitation in the spacing of complex units relative to TCNQ units leads to a fixed stoechiometry of (Pt-Pt)/TCNQ = 1.31, which for simple TCNQ salts corresponds to a Pt oxidation state of 2.66.

Again, as in the step arrangement, a distortion of the bifunctional molecule is needed to maximize the metal-metal overlap and to achieve the metallic state:

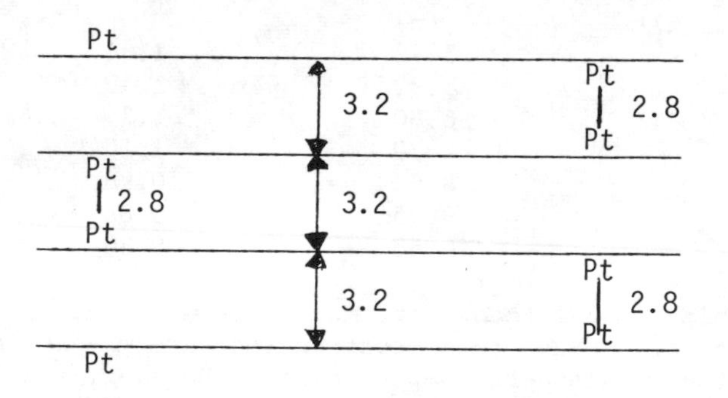

Although the advantages of this arrangement are obvious, there is one possibility which might lead to a pairwise "in phase" distortion as illustrated below, which would lead to a much less conductive state. In this case, the properties will be similar to the mixed valence BFD-$(TCNQ)_2$ system mentioned earlier. A priori, there are no criteria to safely predict which of the two distortions will occur - the stabilized metallic state or the real Peierls state. Since extended interactions will stabilize the total system, we would expect the "out of phase" metallic state to be energetically favored. It is important

to realize that the two states should not interconvert, since to do so, we would have to break a metal-metal bond, move the metal through the molecular plane and reform a metal-metal bond on the other side. Thus, even if metallic conduction is not achieved, we are eliminating the metal-insulator transition and are beginning to understand which influence the deviations from strict one-dimensionality have on the preservation of regularly spaced transition metal complex systems.

Literature Cited

1. See proceeding papers by K. Krogmann, A. N. Bloch and H. R. Zeller for details and reference on KCP.
2. Coleman, L. B., Cohen, M. J., Sandman, D. J., Yamagishi, F. G., Garito, A. F. and Heeger, A. J., Solid State Commun. (1973) 12, 1125 and Ferraris, J., Cowan, D. O., Walatka, J. and Perlstein, J. H., J. Amer. Chem. Soc. (1973) 95, 948 describe TTF-TCNQ and give reference to earlier TCNQ work.
3. Mueller-Westerhoff, U. T. and Eilbracht, P., J. Amer. Chem. Soc. (1972) 94, 9272. Other BFD salts were also prepared by Cowan, D. O. and LeVanda, C., ibid. (1972) 94, 9271.
4. Browall, K. W., Bursh, T., Interrante, L. V., and Kasper, J. S., Inorg. Chem. (1972) 11, 1800.

INDEX

A

Absorption ... 343
 solid state ... 277
 techniques, single crystal ... 301
Absorptivity, molar ... 309
Acetonitrile ... 70
Acid activity, Lewis ... 328
Addition reactions, oxidative ... 325
Adduct structures, solvent ... 38
Amine–halide complexes ... 255
Amine–TaS_2 complexes ... 29
2-Aminoanthracene ... 27
Anion bridged compounds ... 239, 248
Anion bridges ... 236
Anionic chlorines ... 196
Anisotropy ... 199, 346, 375
 Ising-like ... 202
 magnetic ... 210
Antiferromagnetic coupling ... 59, 202
Antiferromagnetic interactions ... 241
Antiferromagnetic salts ... 185
Arsenic ... 14

B

Band, parabolic free electron ... 372
Band-structure energy ... 359
Biscyclopentadienyl titanium(III), linear chain complexes from ... 142
Bonding contacts, hydrogen- ... 85
Bridges, anion ... 236
Bridging
 angle ... 101
 hydroxo group ... 121
 ligand ... 76
  interactions through a ... 66
 units, symmetry of the ... 81
Bronzes, perovskite ... 2

C

Cations, shielding between dimeric ... 77
Cesium metal triiodides ... 182
Cesium nickel trihalides ... 188
Chain compounds
 mixed valence metal ... 246
 partially oxidized ... 333
 single and mixed valency ... 234
 single valence metal ... 239
 transition metal ... 236
Chain metal complexes ... 314
Chains, extended metal ... 276
Chains, linear ... 194
 optical properties of ... 164
Chalcogenides ... 23
 extended interactions in ... 1
 transition metal ... 12
Charge densities, nuclear quadrupole resonance derived ... 29
Charge transfer ... 304
Chlorines, anionic ... 196
Chloro-bridged copper(II) dimers ... 131
Chromium(III) dimers ... 94
Chromium doped V_2O_3 ... 16
Chromium(III) ions ... 164
Cobalt ... 12, 194
Cobalt(II) ... 328
 dimers ... 76, 90
Collective states ... 235
Complex compounds, square planar ... 350
Complexes
 directed synthesis of linear chain metal ... 314
 square planar d^8 ... 301
 RMX_3 ... 154
 of tantalum sulfide, organic intercalation ... 23
Compounds
 metallic ... 3
 semiconducting ... 3
 TCNQ ... 350
 transitional ... 3
Computation of field-swept epr spectra ... 40
Conduction band ... 372
Conductivity, electrical ... 341
Cooperative magnetic order ... 34
Cooperative transitions ... 36
Copper(II) ... 94
 complexes ... 109, 124
  superexchange interactions in ... 108
 dimers ... 76, 84
  chloro-bridged ... 131
Correlation of structural and magnetic properties ... 118
Corundum ... 17
Counterion tetraphenylborate ... 76
Coupling
 antiferromagnetic ... 59, 202
 exchange ... 143
 ferromagnetic ... 167
 hyperfine ... 55
 spin orbit ... 224
Covalency effects ... 159
Covalent energy ... 366
$Cr(urea)_6^{3+}$–$Cr(urea)_6^{3+}$ pairs ... 59
Crystal
 chemistry of V_2O_3, high temperature ... 16
 polarized reflectance spectra, single ... 301
 structure of $V(urea)_6I_3$ ... 54
 studies, mixed- ... 291
Crystalline $K_2Pt(CN)_4$, electrochemical oxidation of ... 367
Crystals
 magnetically dilute molecular ... 34

Crystals *(Continued)*
pure stoichiometric single 2
rhodium 306
with square complexes, electronic spectra of 254
$CsMI_3$ transition metal iodides 182
Cubic tungsten bronzes 2
$[Cu(diamine)OH]_2^{2+}$ 112
$[Cu(pyrazine)(NO_3)_2]_n$ 132
Cyanides 298
Cyanoplatinates 347

D

Davydov components of transitions 246
Decylamine–TaS_2 complex 24
Delocalized excited state 293
Delocalized molecular orbitals 66
Destabilization of Pt 2.3+ 364
Dibenzylamine 24
Di-*n*-butyl formamide 29
Diethylamine 24
Diformylhydrazine 29
Diimine ligands, ferrous chloride polymers of 205
Dilute molecular crystals, magnetically 34
Dimeric cations, shielding between 77
Dimers, chloro-bridged copper(II) 131
Dimers, chromium(III) 94
Dioxalatoplatinates 350
Directed synthesis of linear chain metal complexes 314
Directly interacting metal ions 242, 250

E

Eigenfield equation 41
properties of 42
Electrical conductivity 341
Electrical resistivities of $CsNiI_3$ 192
Electrochemical measurements 72
Electrochemical oxidation of $K_2Pt(CN)_4$ 367
Electron
-accepting ligands 323
band, parabolic free 372
density enhancement at the Fermi level 27
gas, one dimensional 358
Electroneutrality 350
Electronic
and magnetic properties 142
spectra 189
of crystals with square complexes 254
structure of Krogmann salts 356
structure of planar d^8 complexes 302
transition 318
Energy, covalent 366
Entropy change 196
EPR spectra for coupled $Cr(urea)_6^{3+}$–$Cr(urea)_6^{3+}$ pairs 59
EPR spectra for systems with large interelectronic interactions 40
Equation
properties of eigenfield 42
Van Vleck 95, 111
ESR transitions, singlet–triplet 87
Exchange coupling 143
Exchange paths 36
Excited state, delocalized 293
Exciton–exciton transitions 171
Exciton+magnon transition 169
Exciton model 310
Extended interactions in transition metal oxides and chalcogenides 1
Extended metal chains 276

F

$Fe(bipyridine)Cl_2$ 205
$Fe(morphylyldtc)_2I$ 38
$Fe(phenanthroline)Cl_2$ 205
Fermi level 390
electron density enhancement at the 27
Ferrocene polymers 72
Ferromagnetic coupling 167
Ferromagnetic interaction 205
Ferromagnetism 218
Ferrous chloride polymers of diimine ligands 205
Field dependent susceptibility 224
Field-swept epr spectra for systems with large interelectronic interactions 40
Fields, perturbation formulation to predict transition 41
Formic acid 351
Free electron band, parabolic 372
Free electron model 365
Frequency-shift perturbation 45
Frohlich transition, Peierls- 373

G

Gas, one dimensional electron 358
Glycinato complex 95
Guaninium complex 130

H

Halide-bridged outer-sphere dimers 89
Halide ions 346, 350
Halide lattices, multisite magnetic interactions in 51
Halides, hydrated ternary transition metal 238
Halogen bonding, metal– 183
Heat capacity 196
Heisenberg interaction, isotropic 36
Hemin, imidazolate bridged polymer of an iron(III) 221
Hexaflurophosphate salts 67
Hexaurea–metal(III) halide lattices 51
High temperature crystal chemistry of V_2O_3 16
Hydrated ternary transition metal halides 238
Hydrazines, substituted 27
Hydrogen atoms, hydroxo-bridge 119
Hydrogen-bonding contacts 85
Hydroxo group, bridging 121
Hyperfine coupling 55

I

Imidazolate bridged polymer of an iron(III) hemin 221
Increased interlayer spacings 31
Infinite RMX_3 linear chain complexes 142
Inorganic linear systems 393
Instability, Peierls 392

Insulator transition, metal– 16
Interacting metal ions 242
Interaction, ferromagnetic 205
Interaction, isotropic Heisenberg 36
Interactions
antiferromagnetic 241
through a bridging ligand 66
in copper(II) complexes, superexchange 108
in $Cr(urea)_6X_3$ 54
in hexaurea–metal(III) halide lattices, multisite magnetic 51
interatomic 381
interdimer 128
intermolecular 321
metal–metal 66
in molecular crystals, intermolecular magnetic exchange 34
Ru–Ru 69
systems with large interelectronic 40
in transition metal oxides and chalcogenides, extended 1
weak metal–metal 68
Interatomic interactions 381
Intercalation complexes of tantalum sulfide, organic 23
Intercalation of *N*-methylformamide 29
Interdimer interactions 128
Interelectronic interactions 40
Interligand repulsion 350
Intermolecular interactions 321
Intermolecular magnetic exchange interactions in molecular crystals 34
Intermolecular separation 390
Intervalence transfer bands 73
Iodides, $CsMI_3$ transition metal 182
Ionic states, transitions to 266
Ions, d^8 323
Ions, directly interacting metal 242, 250
Ions, halide 350
Iridium complexes, platinum and 381
Iron(III) bis(dithiocarbamates) 34
Iron(III) hemin, imidazolate bridged polymer of an 221
Ising-like anisotropy 202
Ising model, one-dimensional 199
Isomer shifts, Mössbauer 215
Isomorphism 294
Isotropic Heisenberg interaction 36

K

K_2PdCl_4 289
K_2PtBr_4 257
K_2PtCl_4 257, 289
Kramers doublet 37
Kramers–Kronig analysis 280
Krogmann salts, electronic structure of 356

L

Ladder arrangement 400
Lattices, hexaurea–metal(III) halide 51
Lewis acid activity 328
Ligand
bridged compounds, polymeric 71
-bridged ruthenium complexes 68
interaction through a bridging 66
Ligand *(Continued)*
mixing, metal 388
parameters, variation of 319
Ligands, electron accepting 323
Ligands, organic 205
Linear chain complexes from biscyclopentadienyl titanium(III) 142
Linear chain metal complexes, directed synthesis of 314
Linear chain series RMX_3 194
Linear chains, optical properties of 164
Linear systems, inorganic 393
Linear systems, organic 393
Localized and collective states 235
Lowest-energy allowed electronic transition 276

M

Magnetic
anisotropy 210
exchange interactions 76
in hexaurea–metal(III) halide lattices, multisite 51
in molecular crystals, intermolecular 34
moments 207
order 37
cooperative 34
properties of chromium(III) dimers 94
properties, correlation of structural and 118
properties of $CsMI_3$ transition metal iodides 182
susceptibility 152, 185
and thermal properties of $[(CH_3)_3NH]MX_3 \cdot 2H_2O$ 194
Magnetically dilute molecular crystals 34
Magnetization studies 207
Magnon assisted transitions 173
Magnus green salt 261, 285, 319
solid solutions 331
Magnus salts 276
Manganese(II) dimers 90
Metal
chain compounds, mixed valence 246
chain compounds, single valence 239
chain compounds, transition 236
chains, extended 276
complexes, transition 394
compounds, polymeric, mixed-valence transition 66
-halogen bonding 183
-insulator transition 16
iodides, $CsMI_3$ transition 182
ions, directly interacting 242, 250
-ligand mixing 388
-metal interactions 66
weak 68
oxides and chalcogenides, extended interactions in transition 1
plane, tetra coordinated 350
-sulfur bonds 12
systems, triatomic 154
Metallic compounds 3
Metallic Pt salts, 1-D 346
Metalloporphyrins 221
Metals, 1-D 331
N-Methylformamide, intercalation of 29

N-Methylformanilide 29
1-Methyl-1-phenylhydrazine 31
Microspecular reflectance attachment 302
Mixed-crystal studies 291
Mixed metal Ti(III) compounds 143
Mixed valence metal chain compounds 234, 246, 331
Mixed valence solids 324
Mn(II) in tetrahydrofuran 150
Model, molecular orbital 120
Molar absorptivity 309
Molar susceptibility 185
Molecular crystals 392
intermolecular magnetic exchange interactions in 34
magnetically dilute 34
Molecular orbital model 120, 312
Molecular orbitals, delocalized 66
Molecular structure studies 213
Mössbauer isomer shifts 215
Mössbauer spectroscopy 318
Multisite magnetic interactions in hexa-urea–metal(III) halide lattices 51

N

Nickel 14
compounds 173
dimers 77
dimethylglyoxime 282
dimethylglyoximate 244
trihalides, cesium 188
NMR spectroscopy 318
Nuclear quadrupole resonance derived charge densities 29

O

One-dimensional
behavior 164
electron gas 358
Ising model 199
solids 372
systems 392
Optical properties of linear chains 164
Orbital overlap 368
Order, magnetic 37
Organic intercalation complexes of tantalum sulfide 23
Organic ligands 205
Organic linear systems 393
Outer-sphere dimers, halide-bridged 89
Overlap, orbital 368
Oxalato complex 99
Oxidation of crystalline $K_2Pt(CN)_4$, electrochemical 367
Oxidative addition reactions 325
Oxyfluorides 7
Oxidized chain compounds, partially 333

P

Packing 37, 77
Palladium(II) 254
Parabolic free electron band 372
Paramagnet 196
Paramagnetic species 63
Partially oxidized chain compounds 333
Peierls–Frohlich transition 373
Peierls transition in one-dimensional solids 372
Perovskite bronzes 2
Perovskites 236
Perturbation formulation to predict transition fields 41
Perturbation, frequency-shift 45
Phenanthroline complexes 97
π-exchange 84
Piperidine 24
Planar complex compounds, square 350
Planar d^8 complexes, electronic structure of 302
Planar d^8 systems 277
Platinum 254
chains, mixed-valence 331
complexes 381
salts 336
1,1′-Polyferrocene compounds 71
Polymer of an iron(III) hemin, imidazolate bridged 221
Polymeric, ligand-bridged compounds 71
Polymeric, mixed-valence transition metal compounds 66
Polymers of diimine ligands 205
Polymers, ferrocene 72
Polymers, ruthenium 72
Polymorphism 38
Properties of eigenfield equation 42
Pseudo-one-dimensional systems 396
$Pt(en)Br_2$ 266
$Pt(en)Cl_2$ 266
$Pt(CN)_4^{2-}$ 388
Pt salts, 1-D metallic 346
Pure stoichiometric single crystals 2
Pyridine–TaS_2 complex 24

Q

Quadrupole resonance derived charge densities, nuclear 29

R

$RbMnF_3$ 171
Reactions, oxidative addition 325
Reflectance attachment, microspecular 302
Reflectance spectra, single crystal polarized 301
Reflectance spectrometer 379
Reflection spectroscopy, specular 278
Refluxing, formic acid 351
Repulsion, interligand 350
Resistivity 9
Resonance derived charge densities, nuclear quadrupole 29
$Rh(CO)_2$ acac 301
Rhodium crystals 306
Rhodium derivatives 328
RMX_3 systems 154, 164
Ru–Ru interactions 69

Ruthenium complexes, ligand-bridged 68
Ruthenium polymers 72

S

Salt, Magnus green 285
  solid solutions 331
Salts
  antiferromagnetic 185
  1-D metallic Pt 346
  electronic structure of Krogmann 356
  hexaflurophosphate 67
  platinum 336
  TCNQ 394
SCF-Xα-SW scattered wave method 382
Selenocyanate dimers 81
Semiconducting compounds 3
Shielding between dimeric cations 77
Shift perturbation, frequency- 45
σ-exchange 84
Single crystal absorption techniques 301
Single crystals, pure stoichiometric 2
Single and mixed valency chain compounds 234
Single phase solid solutions 337
Single valence metal chain compounds 239
Singlet–triplet ESR transitions 87
Singlet–triplet splitting 101
Solid-state absorption 277
Solid state galvanic cell 367
Solid state interactions in $Pt(CN)_4^{n-}$ complexes 381
Solids, mixed valence 324
Solids, Peierls transition in one-dimensional 372
Solvent adduct structures 38
Spacings, increased interlayer 31
Spectra
  for coupled $Cr(urea)_6^{3+}$–$Cr(urea)_6^{3+}$ pairs, EPR 59
  of crystals with square complexes, electronic 254
  electronic 189
  isolated-molecule 277
  $Pt_2Br_6^{2-}$ 264
  single crystal polarized reflectance 301
Spectrometer, reflectance 379
Spectroscopic and magnetic properties of $CsMI_3$ transition metal iodides 182
Spectroscopic methods 318
Spectroscopy, specular reflection 278
Specular reflection spectroscopy 278
Spin density 38
Spin Hamiltonian parameters 160
Spin orbit coupling 224
Spin wave 167
Splitting, singlet–triplet 101
Square complexes, spectra of crystals with 254
Square planar complex compounds 301, 350
Stacking arrangement 398
Step arrangement 398
Structural and magnetic properties, correlation of 118
Substituted hydrazines 27
Sulfur bonds, metal– 12
Superconductivity 24
Superexchange 101, 164
  interactions in copper(II) complexes 108
Susceptibility
  field dependent 224
  and magnetization studies 207
  molar 185
  temperature variation of the inverse 126
Symmetry of bridging units 81
Synthesis of linear chain metal complexes, ... 314

T

Tantalum sulfide, organic intercalation complexes of 23
TaS_2 complexes 24, 29
TCMP (tetra coordinated metal plane) 350
TCNQ compounds 350
TCNQ salts 394
Temperature crystal chemistry of V_2O_3, high 16
Temperature, transition 38
Temperature variation of the inverse susceptibility 126
Ternary transition metal halides, hydrated 238
Tetrachloroplatinate(II) 288
Tetrahydrofuran, Mn(II) in 150
Tetrakis(phenylisonitrile)cobalt(II) 328
Tetraphenylborate counterion 76
Thermal properties of $[(CH_3)_3NH]MX_3 \cdot 2H_2O$ 194
Thiocyanate dimers 81
Tight-binding model 359
Titanium(III) 142
Titanium–zinc complexes 145
$Ti(urea)_6^{3+}$ 63
Transfer bands, intervalence 73
Transition
  electronic 318
  exciton+magnon 169
  fields, perturbation formulation to predict 41
  lowest-energy allowed electronic 276
  metal chalcogenides 12
  metal complexes 394
  metal compounds 66, 236
  metal–insulator 16
  metal oxides and chalcogenides 1
  in one-dimensional solids, Peierls 372
  temperature 38
Transitions
  cooperative 36
  exciton–exciton 171
  to ionic states 266
  magnon assisted 173
  singlet–triplet ESR 87
Triatomic metal systems 154
Triplet ESR transitions, singlet– 87
Triplet splitting, singlet– 101
Tungsten bronzes, cubic 2
Tungsten(VI) oxide 2
Two-phase complexes 29

V

Vanadium(IV) oxide ... 5
Van der Waals forces ... 23
Van Vleck equation ... 95, 111
Variation of ligand parameters ... 319
V_2O_3, high temperature crystal chemistry of ... 16
$V(urea)_6I_3$, crystal structure of ... 54

W

Wave, spin ... 167
Weak metal–metal interactions ... 68

Z

Zinc complex ... 143
Zinc complexes, titanium– ... 145